OCCUPATIONAL HEALTH AND SAFETY CONCEPTS

Chemical and Processing Hazards

GORDON R. C. ATHERLEY

M.D., M.F.C.M., D.I.H.

Professor of Safety and Hygiene,
Department of Safety and Hygiene,
University of Aston in Birmingham, UK

APPLIED SCIENCE PUBLISHERS LTD
LONDON

APPLIED SCIENCE PUBLISHERS LTD
RIPPLE ROAD, BARKING, ESSEX, ENGLAND

ISBN: 0 85334 742 5

WITH 24 TABLES AND 63 ILLUSTRATIONS

© APPLIED SCIENCE PUBLISHERS LTD 1978

Printed in Great Britain by Galliard (Printers) Ltd, Great Yarmouth, Norfolk

For AUDREY, KATIE and BELLA

Human existence being an hallucination containing in itself the secondary hallucinations of day and night (the latter an insanitary condition of the atmosphere due to accretions of black air) it ill becomes any man of sense to be concerned at the illusory approach of the supreme hallucination known as death.

D. E. SELBY

From: Flann O'Brien, *The Third Policeman.*

Acknowledgements

I have received much valuable assistance in the preparation of this study. Mr Ciaran O'Gallagher provided me with material and edited the draft, Dr Charlie Clutterbuck also provided me with material. The tedious job of criticizing drafts was undertaken by my colleagues Dr Richard Booth, Mr Mike Breckin, Ms Rowena Clayton, Dr Dennis Else, Dr Ian Glendon, Mr Andrew Hale, Dr Irwin Lavery, Dr Len Levy, Mr Mike Phillips, Mr John Riley and Ms Claire Whittington. Dr Morris Cooke and Professor Bryan Harvey let me take hours of their time in discussion.

My Department's Information Coordinator, Ms Vicki Kendal, researched and sought out reference and bibliographic material for me. My Administration Assistant Ms Carol Taylor and my Secretary Ms Mary Gibbons gave me unfailing support which allowed me to devote time to the book. Ms Pat Barrett transcribed my spoken word into typed draft and the much-modified typed draft into final typescript, all of which she did with great speed and precision. Members of my family, Anthony Ranells and Paul Atherley, researched for me and Simon Atherley drew the figures. For awakening my interest in and concern about my subject I am indebted to my teachers: Professors Ronald Lane, Tommy Scott, Tim Lee and Peter Lord. Dr Bill Noble, Dr Alan Martin, Dr Stan Gibbons, numerous friends and colleagues in the academic world in Britain and elsewhere, in the Health and Safety Executive, in occupational medicine, occupational hygiene, industrial safety and several generations of students helped develop my ideas. To all, named or unnamed, I should like to express my appreciation and record my thanks. Finally, I gratefully acknowledge the forebearance of my wife for her patience during the long hours I spent in my study.

Preface

This book is aimed specifically at readers with little background in biology who nevertheless wish to know something of the biological aspects and a little of the social aspects of occupational health and safety. The subject matter is selective, reflecting that of the course taught in the Department of Safety and Hygiene at the University of Aston in Birmingham, where the subject of Occupational Health and Safety is divided into four headings: (a) human safety, (b) safety engineering, (c) law, and (d) individual and organizational behaviour. This division is convenient though far from rigid, all four subject divisions having a social as well as a technical framework.

This book is principally concerned with human safety and not with the other divisions, though there is some intentional overlap. Its specific aim is to provide a basis (in the sense of substratum) for knowledge of human safety. My purposes are to convey and explore the concepts supporting three ideas:

(1) that the biological nature of the harm occasioned by danger from work can be perceived in terms of a flow-chart or, to give it a more pretentious title, an input–output model;

(2) that the selection and implementation of action against danger is ultimately determined by social and economic pressures as well as scientific knowledge; and

(3) that strategies intended to preclude or minimize danger can be classified on a simple hierarchy.

Contents

Acknowledgements vii
Preface ix

SECTION 1: ORIENTATION

1 Introduction to Human Safety 3
2 An Input–Output Model for Human Safety 7

SECTION 2: DEFENCE PROCESSES

3 Inhalation, Respiratory Defensive Processes and the
    Measurement of Aerosols 25
4 Defensive Cells 56
5 Inflammatory Response and Immune Response . . . 68
6 Stress, Stress Responses and Resistance, and Homeostasis 76
7 Thermoregulation 85
8 Metabolic Transformation and Conjugation; Toxicity and
    Pathogenicity 96

SECTION 3: NORMAL BIOLOGICAL PROCESSES

9 Growth 109
10 Genetics of the Cell 118
11 Transport Systems: Input, Distribution and Storage, and
    Bio-dumping 130

SECTION 4: MODES OF ACTION, PATHOLOGICAL PROCESSES AND DISEASE FOLLOWING HARMFUL INPUTS

12 Dose, Effects, Quantitative Relations and Target Organs 147
13 Inflammation as a Harmful Process 162
14 Altered Sensitivity 173
15 Disorders of Repair 187
16 Disorders of Growth 200

SECTION 5: CASE HISTORIES AND EXAMPLES

17 Radiating Energies 231
18 Aromatic Amines and Occupational Cancer of the Renal
    Tract 259

19 Metals in the Disservice of Man   274
20 Asbestos   290
21 Case Study: Respiratory Disease in the Coal Industry   .   303
22 Gassing Accidents   328
23 Case History of Vinyl Chloride   344

SECTION 6: STRATEGIES

24 Strategies in Occupational Health and Safety   371

Index   399

Section 1: ORIENTATION

This section shows in broad terms how the book fits into the field of occupational health and safety and presents the input–output model as a framework for the remainder of the book.

Chapter 1

Introduction to Human Safety

OBJECTIVES OF THE CHAPTER

This chapter:
  (1) outlines the scope of human safety, and shows that it involves knowledge of human biology applied in a social setting,
  (2) specifies broad objectives for the study of occupational health and safety of which human safety forms part,
  (3) indicates the professional context of human safety, and
  (4) introduces the three principal themes of the book.

INTRODUCTION

Danger arising from work is an unwanted and unintended by-product of the processes of work. All interested parties—the public, government, the agencies of government, employers, managers, employees, representative organizations, technologists, scientists, and occupational health care personnel—are concerned with the minimization or elimination of danger associated with work. This concern is growing, and justifiably so, because the decline in mortality and morbidity from infectious diseases caused by micropredators is bringing man-made diseases into greater prominence. Moreover, 'environmental' factors are increasingly recognized as causative agents in serious disease such as cancer. Of environmental factors, a substantial proportion, though by no means all, has a human instrumentality. Included in the category of human instrumentality are occupation-linked diseases, injuries, and deaths. No longer is there unquestioned acceptance of the belief that death, life-shortening and disablement are an 'acceptable risk' that must necessarily be run by working people.

Against this background it is hardly surprising that occupational health and safety is being carefully studied, whether from motives of humanitarianism or academic interest, or because there are excellent career opportunities for those with a degree of mastery of the subject. All practitioners and students soon recognize, however, the breadth of knowledge needed to achieve awareness let alone mastery of the subject. Comprehensive knowledge is needed of a variety of scientific subjects; biological subjects as diverse as the intimacies of genetics of the cell, carcinogenesis, and epidemiology, as well as a wide

3

diversity of topics in the physical sciences. Herein lies the problem encountered by students of occupational health and safety. Almost no one enters the field of study of occupational health and safety with the necessary prerequisite knowledge of all the subject's constituent topics.

Differing views exist about policy towards health and safety problems. It is generally agreed that social policy should attend to occupational health and safety, but the specific aims of that policy and the means by which these should be met are topics on which agreement is difficult. Very often those concerned in the discussion have inadequate knowledge of the interplay of social policy and the scientific basis of occupational health and safety.

Taken together, the topics relevant to occupational health and safety constitute a massive subject for study. The task is Herculean but not impossible, because a start is being made, world-wide, in delineating areas of knowledge which are central to occupational health and safety, and in specifying the depth to which that knowledge is required. The broad objective, in academic terms, is to specify the scope and depth of knowledge required by students in order that they should:

have broad comprehension of the subject's principles,
be able to read across the subject from one topic to another, and
be able to develop a perspective for their studies.

Academic institutions throughout the world are currently working towards this objective. Although approaches and emphases inevitably vary, there is general agreement that the subject is not purely a specialism within medicine and that its study should not be restricted to graduates in medicine.

It seems likely that in Britain there are now about equal numbers of medical and non-medical graduates with specialist qualifications relative to occupational health and safety. If persons with professional qualifications in the health and safety field are included, the 'medicals' are heavily outnumbered. The trend, moreover, is towards a larger proportion of non-medical personnel. Of these (if the University of Aston's experience is at all representative) many have a limited background in the biological and/or social areas, because these are topics not normally embraced by the physical sciences, graduates of which constitute a substantial proportion of new recruits.

THE INPUT–OUTPUT MODEL: PRELIMINARY

'Input' is used to describe all harmful or undesirable factors relevant to occupational health and safety. 'Input' suffices for this purpose provided we recognize that it includes factors which may be harmful by being deficient or absent (oxygen, for example). This conceptual difficulty can be rationalized by our supposing that such inputs, in arithmetic terms, are negative quantities. 'Modes of input' is used to describe the various routes and means whereby inputs gain access to the body.

A category of output, termed 'bio-dumping', involves substances and energies being removed from the body in the course of regulatory biological activities. Another category of output, 'harm' in its broadest sense, can be used to include a continuum of effects of chemical, physical, microbiological, or psychological inputs of which effects the most serious are injury, disease, disablement, and death.

Inputs are likely to provoke reactive changes in biological processes. In this book, for descriptive purposes, three broad groupings are implied—defensive processes, corrective processes, and normal biological processes. It should be recognized, however, that these are not mutually exclusive and that particular processes may be classifiable under different headings according to circumstances.

The manifestations of disease are the changes in the biological processes. However, not all such changes result in disease being manifested; defensive and corrective processes may successfully resist the input and bring about restoration of the biological state of affairs prior to the input's presence. 'Disorder' is used here to label those changes in biological processes which result in disease being manifested. It serves present purposes well enough, although I recognize that some authorities might demur on the grounds that some of the changes so labelled are over-reactions (*hyper*-reactions) or under-reactions (*hypo*-reactions) rather than disorders. Nevertheless, these types of change can be labelled 'disorder' provided that the term is clearly used in its broadest sense.

'Mode of action' is used as a general heading for descriptions of the way in which inputs provoke changes in biological processes.

SELECTION AND IMPLEMENTATION OF ACTION IN OCCUPATIONAL HEALTH AND SAFETY

Sooner or later, the recognition of danger in a particular context is followed by talk of action against the danger. In selecting and implementing action, legislators, administrators, and enforcing authorities usually have a number of strategies to choose from. The case histories and examples, given in Section 5, show that choice of strategy is influenced by social and economic considerations as well as scientific considerations.

HIERARCHICAL CLASSIFICATION OF STRATEGIES

From almost any comparative study of actions against dangers it soon becomes apparent that there is a common pattern for the actions considered and selected, almost irrespective of the danger. Time and time again, certain strategies seem especially favoured. Close scrutiny reveals that the strategies most commonly adopted can be classified in a way that exhibits their relative effectiveness, and which also indicates why their selection and implementation are influenced by socio–economic as well as scientific factors.

Chapter 2

An Input–Output Model for Human Safety

OBJECTIVES OF THE CHAPTER

This chapter:
  (1) Describes an input–output model which provides a suitable framework on which to build an understanding of human safety, and
  (2) defines certain terms and concepts central to human safety.

THE MODEL

The input–output model shown in Fig. 2.1 is a convenient framework for important sectors of human safety and should provide readers with a usable model for understanding the subject's essential concepts and terms. The model is used as the basis for the subject matter in this book. The principal function of the model is instructional, and readers should not hesitate to adapt it for their own purposes. Because the field of human safety is rapidly advancing, the input–output model should be seen as being flexible and capable of extension to encompass new knowledge.

Human safety represents a branch of knowledge instrumental in the prevention of death, injury, and disease associated with work. In the context of this study, death, injury, and disease are undesired outputs resultant upon a variety of inputs which have activated one or more processes within the living body. Intervention with a view to prevention can be made at various points in the input–output sequence. Measurements, too, can be made at various points in this sequence; these can be used to indicate the extent to which the input is, or is likely to be, the cause of an undesired output; or measurements can be made in order to evaluate the effectiveness of intervention.

The link between a defined input and an observed output may be clear and unambiguous. On the other hand, the link may be no more than a matter of scientific speculation. This uncertainty can be reduced only by the application of rigorous scientific methods. In order to establish beyond reasonable scientific doubt that a particular output is related to a particular input, that is to say a causal relation exists, information of two basic kinds is required (Harré, 1967):
  (a) *Supporting statistics* A causal relation can be assumed to

7

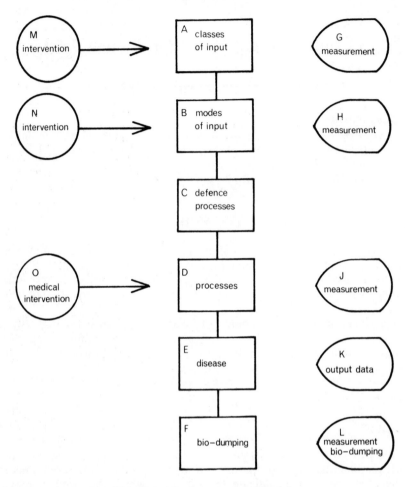

FIG. 2.1. The input–output model.

exist if the output is observed when the input is present and not observed when the input is absent. A similar assumption can be made if the output increases in incidence or magnitude as the input increases and/or the output decreases with a decrease in input. If the statistics show a complete relation which satisfies all these criteria, and certain others, it may be possible to conclude unequivocally that a link—but not necessarily a causal one—exists between the output and the input. In practice, however, all the necessary criteria are seldom satisfied, though in some instances a causal relation can still exist. For example, cancer may be triggered by a carcinogen, the removal of which does not remove or stop the cancer. In other instances the output is non-specific and can be causally related to more than one type of input. But, despite all the difficulties, the search for causal relations is a matter of practical as well as theoretical importance.

(b) *Generative mechanism* There must be a scientifically-plausible generative mechanism to explain the link between the output and the input. Sometimes the generative mechanism will be straightforward and scientifically secure, but in other instances the generative mechanism may lie at or beyond frontiers of existing knowledge. Normally, in order to scientifically establish that a link exists and that the output is not just coincidental with the input, it is necessary not only to demonstrate favourable supporting statistics but also put forward a plausible generative mechanism. Thus in the understanding of human safety, 'processes' and 'modes of action,' as understood for the purposes of this book, can be regarded as the generative mechanisms. The necessary statistics are provided by the epidemiological or other evidence relating exposure to death, disease, or injury. Thus knowledge of the processes and modes of action—the generative mechanisms—is necessary not only for the practical business of prevention but also for the scientific business of establishing that an observed output can reasonably be attributed to a particular input. There are numerous instances in occupational health and safety where statistics are uncertain and links are tenuous. Even though on humanitarian grounds action may be called for in the face of evidence which is scientifically incomplete—as, for example, in the case of carcinogens—this does not excuse us from observing the discipline of scientific methods.

ELEMENTS OF THE MODEL

The model, depicted in Fig. 2.1, comprises a number of elements which are described and subdivided in the following paragraphs. It should be noted that the flow-chart format considerably simplifies the processes involved; in particular the model does not convey any impression of the *interactions* which are possible and which, on occasion, play such an important part in the processes of occupational disease and injury.

In general it should be assumed that a member of one of the classes of input is likely to have more than one mode of input, to be involved in more than one biological process, to evoke more than one defence process and to result in more than one type of output (see Fig. 2.1).

There are a number of examples specified in later chapters where one member of a class of input may interact with another of the same class or with one or more members of other classes of input. Moreover, inputs acting in conjunction with each other behave differently to inputs acting singly. For some inputs the nature of the defence process depends on the level of the input, and this presents a further complication. 'Loopback' mechanisms are additional factors which must be considered. Any model which purports to display all possible relations between input and output is, thus, likely to be inpenetrably complex.

The model shown in Fig. 2.1 is not intended to display the inter-relations. It is merely intended to provide a serviceable framework for the remainder of the book.

(A)* Classes of Inputs

Inputs, for the purposes of the model, can conveniently be described under four separate headings: (1) chemical substances, (2) energies, (3) climates, and (4) micropredators.

(1) Chemical Substances
These include the whole range of substances and compounds embraced by organic and inorganic chemistry. The range can be subdivided into: (a) nutrients, and (b) non-nutrients.

(a) Nutrients are substances or compounds which are used by the body for the production of energy or the synthesis of tissue. Nutrients, generally, range from complex proteins to common metals.

(b) Non-nutrients are substances which do not normally take part in either energy production or biosynthesis; they are thus not required by the body for normal healthy growth and function. However, it is possible for non-nutrients to be present in the body without necessarily causing harm.

Examples of non-nutrients which are harmful and which are discussed in later chapters include the aromatic amines, asbestos, and certain inorganic substances. Also mentioned are certain metals which are essential in trace amounts but definitely harmful when excessive amounts are present in the body. Manifestly, the word 'toxic' cannot properly be used, without qualification, to describe essential micronutrients. Here we see a point which will be made again in later chapters, namely that the property of being toxic should not be considered a fixed attribute of a chemical substance. Whether a substance is or is not harmful depends on (i) the substance, (ii) the

*The capital letters correspond with those shown on the input–output model.

amount taken into the body, (iii) the state of the body when exposure takes place, and (iv) the future state of the body, during the remainder of life in the case of substances which enter the storage 'compartments' of the body.

(2) *Energies*
As will be discussed in Chapter 17 it is convenient to discuss the energies under five general headings: mechanical energy, thermal energy, electrical energy, radiant energy, and ionizing radiations. In the living body, however, energies are readily transformed from one form to another; such *transformations* are often involved in production of injury to living tissues by energies.

(3) *Climates*

General climates   These refer to the external environment in which the human is called upon to function. The French physiologist Claude Bernard (1813–1878) used the term *milieu interieur* to describe the normal state of balance, in the process and counterprocess, in the living organism's physiology. Body temperature provides an example from everyday experience. The temperature at the core of the human body remains remarkably constant through a wide variation in climate and bodily activity. The constancy of the internal milieu generally is a remarkable physiological phenomenon. It is maintained against wide variations in the external milieu. However, there are extremes— climatic extremes, for example—which put the physiological maintenance mechanisms under stress. Under such climatic excess these mechanisms cannot cope, and harm may result. It should be noted that a number of these climatic extremes, such as humidity, are not primarily or wholly energy-related.

Psycho–social climate   This term is intended to describe psychological and sociological influences which operate on an individual while he/she is at work. Such influences can be regarded as inputs for the purposes of the input–output model. Manifestly, not all psychological and sociological influences are relevant to occupational health and safety. However, the heading embraces a number of such influences which are of great importance. Some examples are mentioned in later chapters.

(4) *Micropredators*
These are the parasites of man, ranging from the viruses to insects.*

*The term and concept of micropredators comes from Carr (1972).

Laboratory accidents have caused occupational exposure to viruses, exposure which has resulted in disease in those work people in contact with the contamination. In some contexts the infestation of the body by living insects—for example scabies—is a relevant consideration in occupational safety and hygiene. Bacteria are probably responsible for the largest number of diseases of occupational significance, wool-sorter's disease (anthrax) being one of several classical occupational diseases caused by bacteria.

Further information about micropredators is given in Appendix 2.1.

(B) Modes of Input

The four headings for inputs embrace a very large number of individual agents. In order for any one of these to cause harm to the human body it must first gain access to the body. Most of the individual agents embraced by the four headings of inputs enter the body by one or more of the following modes, that is to say routes, of input:

Inhalation
A whole range of substances and microorganisms can be carried into the body on the breath. The lungs behave as a highly effective filter which is self-cleaning. However, there is a critical size-range within which airborne particles are capable of penetrating the entire length of the airway, and there are substances and micropredators in these ranges which are able to cause harm.

Ingestion
Nutrients, non-nutrients and microorganisms are carried into the gut, from which some will pass into the body by absorption. Like the lung, the gut behaves as a selective filter which keeps out many but not all harmful agents presented to it. But micropredators, like certain chemical substances, can do harm within the gut without absorption.

Pervasion
The skin, if intact, is proof against most but not all inputs. There are certain substances and microorganisms which are capable of passing straight through the intact skin into underlying tissues or even into the blood stream, without apparently causing change in the skin itself.

Implantation
A forceful breach of the skin—even a tiny puncture—can carry substances or microorganisms through the skin barrier.

Surface Penetration (Imbruement).
Certain substances, certain microorganisms, and certain energies pass into the epidermis but not through it. These may cause damage such as dermatitis and certain tumours, without causing any damage to deeper tissues.

Irradiation
This describes the process of exposure of the body to ionizing and non-ionizing radiation, with or without body surface penetration and with or without deep penetration.

Information
This term is used to describe the total information input from the external milieu and from certain monitoring systems in the body. Afferent information provides us with our awareness of the world about us and, to a certain extent, the world within us. Only part of the information input is at a conscious level—the part is termed sensory information. Sometimes in sensory information there is a strong element of dissonance, for example where one type of sensory input conflicts with another. Pilots, in learning to fly by instruments without reference to an external horizon, have to suppress the sensory afferent information from their organs of balance, which readily becomes dissonant with the sensory information conveyed by the aircraft's instruments.

Dissonance in information may lead to stress or errors and/or accidents; hence, 'information' may be considered a mode of input of relevance in human safety, even though it has many other functions of no special relevance to human safety.

(C) Defence Processes

Generally speaking, each mode of input is guarded, so to speak, by one or more *defence processes.* As the name implies, these processes function against any harmful inputs. It should be noted that the bodily defence processes are not, of course, activated only by harmful inputs of an occupational nature. Indeed, it is safe to assume that most of the body's defence processes are operating, to some degree at least, throughout life. As we shall see, the destruction of certain of the body's defensive processes explains one of the forms of harm occasioned by exposure to ionizing radiations. The defensive process in question normally safeguards the body against the effects of micro-predators gaining access to the body system. Normally, it deals with these quickly and effectively, but where it has been impaired by the radiation the micropredators may gain the upper hand, with fatal results.

The defence processes discussed in this book are:

Respiratory Filtration
As previously mentioned, this refers to the ability of the respiratory pathways to filter out and remove to the exterior most but not all airborne particles.

Cellular Defence
Certain cells in the body are specialized for a defensive function. These have an extremely important role in certain occupational diseases.

Inflammatory Response
A crucial and basic defence process. Like many defence processes it can itself become disordered even to a harmful degree.

Immune Response
Where immunity forms the basis of an important defence process. But there are instances of the immune response being definitely harmful, and there are other instances where it is neither defensive nor harmful. Terminology varies from authority to authority but throughout this book 'immune response' will be used to signify those examples of altered sensitivity which can be regarded as defensive.

Homeostasis
This describes the various processes by means of which the internal milieu of the body is maintained in a stabilized condition.

Stress Resistance
This describes the body's ability to resist or cope with stress, an ability linked with an important hormone called cortisol, which appears to be the most important factor in determining the body's resistance to stress.

Thermoregulation
By means of a series of complex mechanisms the core temperature of the living body is maintained at a constant level through a wide range of climatic and other changes.

Metabolic Transformation
This refers to the alteration of a harmful substance into a less harmful form. Often this is more water soluble and it is then excreted via the kidneys.

(D) Processes

Processes is the general term used to describe (1) normal biological processes taking place in the body throughout life, (2) the modes of action of inputs which have gained access to the body and which have evaded the biological defence processes, and (3) pathological processes evoked by the input.

(1) Normal Biological Processes
This term really refers to the whole of human biology—much too broad a scope for this book; fortunately attention can be concentrated on just a few of the normal biological processes: *growth* of living tissues, *cell genetics* (by which vital information is transmitted from one generation of cells to another), and *transport systems* (by which substances and energies are carried from one place to another within the body).

(2) Modes of Action
There are, obviously, many ways in which harmful inputs can act on living tissues. However, the range of modes of action is much narrower than the range of harmful inputs because many inputs evoke a common mode of action. Some of the commoner modes of action are identified below, while others are dealt with in various chapters throughout the book. Discussion of modes of action in subsequent chapters is generally combined with that on pathological processes.

(3) Pathological Processes
Action by a harmful input leads to the initiation of pathological processes which may subsequently lead to disease. The sequence of events just described appears to imply that pathological processes are the half-way houses to disease. That is sometimes true but not always.

The words *pathological* and *pathology* are used in common language to signify that which is morbid or diseased. Although it is true that pathology is the study of disease and the processes of disease, a number of processes labelled pathological are in fact crucial defence processes. There are numerous instances where physiological processes and pathological defence processes overlap. The distinction between physiological and pathological should not, therefore, be drawn too finely. Nevertheless there are pathological changes which are definitely harmful. Sometimes these may be regarded as aberrations of physiological or pathological defensive mechanisms but the common feature is that the end result is undesirable and harmful. There is an important intermediate group of changes of a pathological kind which in some contexts appears to be beneficial, and in others appears to be harmful.

The pathological processes discussed are:

Inflammation, degeneration, and cell death   These describe the tissue or cellular response to insult and damage.

Altered sensitivity   This describes phenomena widely understood as 'allergies' in which an individual displays a marked change in response towards a particular form of input. In fact, as will be shown, several important and complicated reactions are involved.

Disorder of growth   Growth normally takes place as a result of cells dividing and reproducing themselves, thereby increasing the bulk of the tissue. Cancer represents a loss of control over the negative feedback which governs the ordinary process of cell division. Mutagenesis (abnormalities passed from one generation of cells to another) and teratogenesis (abnormalities passed from one generation of offspring to another) can also be regarded as disorders of growth.

Disorder of repair   A normal repair response to tissue injury is the formation of scar tissue. Under certain circumstances the repair process gets out of hand; e.g. pneumoconiosis, an important group of diseases caused by dust. Certain of the dusts cause overproduction of persistent scar tissue in the lungs, so certain types of pneumoconiosis represent a disorder of a normal process of repair.

(E) Disease

For the purposes of the input–output model the word 'disease' is used in the broadest possible sense and is not confined to states of illness necessitating medical treatment, the common usage of the word. The word 'disease' is used here to denote any unhealthy condition or response of whatever duration in the body or by the body resultant upon one or more specific inputs.

Disease should, in general, be regarded as the quantitative alteration (increase or decrease) of existing biological processes, rather than the imposition of new or different structure and/or function.*

In connection with disease the following general headings are used:

---

*This concept of disease stems from Forbus's *Reaction to Injury* (1943) as discussed by Hill in La Via and Hill (1975).

Injury
This implies, here, physical damage to the body following, for example, the traumatic application of energy. It should be noted that 'injury' is also used to describe any damage to cells or living tissues, from whatever cause. Which meaning of 'injury' is intended is normally evident from the context. Disease results, generally, from the biological reaction to injury at the cellular level. In many instances it is appropriate to regard the cellular injury as the trigger for the disease process.

Fatigue
This is a complex phenomenon with physiological as well as psychological components.

Errors and Accidents
It is convenient to disassociate errors from accidents because the former do not necessarily lead to the latter.

Absence from Work
This form of output provides data on which 'safety performance' and 'health statistics' are often judged.

Disablement
Used in its broadest sense this implies a change in the individual's ability to perform any work or function of which he/she was previously capable. Disablement commonly follows injury or disease but the injury or disease process associated with it may be difficult to identify, especially where the disablement is of a psychological nature.

Death
The attribution of death to occupation may or may not be readily establishable.

(F) Bio-dumping

The ungainly word 'bio-dumping' is used to describe, for example, the output of waste or inactivated substances from the human body. Bio-dumping needs to be considered as part of the input–output model because its study provides important information about the way in which the body deals with harmful inputs and, moreover, the bio-dumping process can itself become the locus of disease. Modes of bio-dumping include:

(1) sweating,
(2) urine formation,
(3) formation of faecal matter,
(4) respiration,
(5) radiation—the body bio-dumps thermal energy by means of radiation where an appropriate thermal gradient exists, and
(6) exfoliation—this refers to cells lost from the body surfaces (whether internal or external) through the normal processes of growth and wear and tear of body surfaces.

(G) Measurement of Quantity of Specific Input

Often measurements are made to determine the degree of contamination of a workplace by chemical substances or energies, concentrations of which lead them to be regarded as inputs. Such measurements determine the concentration in the workplace generally; further information is normally needed in order to deduce the 'dose' of the input to which individual employees are exposed. Dose, as will be stated later, is a complicated concept.

(H) Measurement of an Input in Relation to a Specified Mode or Input

Under this heading is included measurements aimed to identify 'dose' of harmful input being received by an individual; sometimes individual dose will correspond closely to the extent of general contamination—but this is not always the case and personal dosimetry is a technique of importance for evaluation of 'personal' dose of, for example, ionizing radiations and certain dusts.

(J) Measurements of Processes' Activity by Biochemical or Functional Estimation

There are numerous indicators which can be used to monitor biological processes; the intention behind many such measurements is to gauge the extent to which harm is taking place or therapy is being effective.

(K) Measurements of Output

Data relating to injury and disease are often used as indicators of extent of danger in occupational milieux. Absences from work are often used as indicators but these data are prone to interference from factors unrelated to danger. Epidemiology is an important field of study which provides suitable techniques for making measurements and drawing inferences from a wide variety of data relating to injury and disease.

(L) Measurements of Outputs in the Form of
Bio-dumping

These may be used to provide data from which conclusions can be drawn about the extent to which inputs are being coped with by biological defence processes, and sometimes these data are used to infer the effectiveness of strategies aimed at minimizing harmful inputs.

INTERVENTION

The model displays a number of points at which intervention is possible. In connection with the inputs, intervention is aimed at eliminating or minimizing the dangerous input. There is an important dichotomy in approach to danger; the dichotomy raises issues of a socio–economic as well as of a technical nature. Thus the choice of strategy against danger can, at times, be a matter of great complexity or controversy; strategies against danger are discussed in Chapter 24.

(M) Intervention by Elimination or Minimization of Danger
from a Specified Input

Obviously, if a harmful input is removed from all possibility of human contact, harm to human beings will be avoided altogether.

(N) Intervention Designed to Prevent a Specified Input
Gaining Access by a Specified Mode of Input

Respiratory protection, for example, might be used to limit input (by inhalation) from the spraying of agricultural chemicals. But by itself that protection would not limit the input by other modes, for example pervasion.

(O) Intervention by Medical Treatment (Therapy) Aimed at Curing or
Countering Effects of a Harmful Input

Questions of medical treatment are not within this book's scope.

In the chapters which follow various elements identified in the model are described and the reader should be able to relate this information to the input–output model. Once the reader has gained familiarity with the concepts involved, the model should provide a useful framework for approaches to the whole range of problems in human safety.

APPENDIX 2.1:   MICROPREDATORS

The micropredators of man include:
  (1) microorganisms—fungi, protozoa, bacteria and related organisms, and viruses,
  (2) arthropods and helminths which become external internal parasites of man.
One large category of microorganisms can be excluded from our discussion: the saprophytes. These microorganisms live freely in the external milieu without becoming in any way parasitic on man. Our concern is with those microorganisms which have adapted to a parastic existence on or in man's body. These microorganisms can be categorized according to the success in their degree of adaptation.

Commensals

This describes the well-adjusted parasites which inhabit the skin and other parts of the healthy human body. These microorganisms derive their nourishment from the human host normally without any deleterious effects. Occasionally, however, harm of one kind or another diminishes the body's resistance to commensals, and disease results. For example, the forceful breach of skin may carry into the subcutaneous tissues microorganisms of the genus *Staphylococcus*. Occasionally these staphylococci set up infection in the wound.

Symbionts

A few microorganisms have reached the stage of adaptation where their presence is not only harmless to the host but also positively beneficial. Thus bacteria in the lower intestine synthesize a variety of vitamins such as vitamin B12. Interference with symbionts may result in vitamin deficiency disease. Disturbance of symbionts is probably a rare cause of occupational disease.

Pathogens

These are microorganisms which adapt badly in the human host. Anthrax, for example, fails to adapt to the human organisms and produces a rapidly fatal disease (unless there is successful treatment). Only a minority of pathogens appear to completely lack the ability to adapt to the human host. The majority are capable of more or less satisfactory adaptation. Partial adaptation results in sub-clinical disease. The uneasy adaptation of these microorganisms to the human host is readily disturbed by other harmful inputs. The importance of harmful inputs from work in this context is difficult to determine. But it is likely that harmful inputs in the occupational context do, from time to time, result in disturbance of the adaptation of pathogenic organisms, resulting in disease. The role of occupation in the cause of non-epidemic disease from micropredators is an area in need of research.

Parasitic Arthropods and Helminths

This refers to insects and worms capable of parasitizing the human body. The infection by the parasite may take place at work but the source of the input is more likely to be non-occupational. The parasite may adapt badly to the human body, reducing resistance to other harmful inputs. The question of parasitic infestation (this covers arthropods as well as helminths) may decrease body resistance to harmful inputs from occupation. A large proportion of the world's population suffers from infestation by badly adapted organisms; many of these people are also exposed to harmful inputs at work.

The debilitating effects of the infestation, together with the diminished resistance which may accompany the infestation, must affect the susceptibility of large numbers of people to harmful inputs at work. This is a point to be remembered in any consideration of 'universal' hygiene limits for use at work. Limits which are validly applied to well-nourished European populations may be wholly inapplicable where infestation, or malnutrition is endemic. The additional interaction of psycho–social conflict with harmful inputs from work in people already debilitated by malnutrition and infestation poses a series of challenging questions to which, at present, there are few answers.

REFERENCES

CARR, I. (1972). *Biological Defence Mechanisms*, Oxford: Blackwell.
HARRÉ, R. (1967). *An Introduction to the Logic of the Sciences*, London: Macmillan.
LA VIA, M. F. and HILL, R. B. (eds) (1975). *Principles of Pathobiology*, 2nd Edition, New York: Oxford University Press.

Section 2: DEFENCE PROCESSES

The following six chapters outline certain defence processes of the human body. Selected for Section 2 are those defence processes which appear most relevant for the study of human safety. However, it should be noted that the body possesses defence processes other than those mentioned in the section.

Whether a specified input actually causes disease depends, *inter alia*, on the adequacy of the body's defence processes. The purpose, therefore, of discussing these is to lay the foundations for the understanding of the cause (aetiology) of occupational disease. It should also be noted that if certain defence processes are called into action for an excessive time, or to an excessive extent, they can themselves be the locus of disease. Defence processes, it might be said, can have a harmful as well as a protective mode. Various harmful modes of certain defence processes are explored in Section 4.

Section 2's objectives are to give:
(a) an outline of key defence processes,
(b) an outline knowledge of certain biological terms and concepts necessary for later sections, and
(c) an indication of how knowledge of respiratory defence processes is used in the design of measuring instruments.

Chapter 3

Inhalation, Respiratory Defensive Processes and the Measurement of Aerosols

OBJECTIVES OF THE CHAPTER

This chapter is divided into two parts. Part (A) deals with inhalation and respiratory defensive processes and Part (B) deals with measurement of aerosols.

Part (A) of the chapter:

(1) indicates the scope for harm from inputs entering the body by the inhalation mode,
(2) outlines the principal defensive processes involved in the respiratory system,
(3) identifies respiratory filtration and describes its action,
(4) identifies physical factors involved in respiratory filtration,
(5) names and illustrates principal parts of the respiratory system,
(6) relates the processes of respiratory filtration to parts of the respiratory system,
(7) outlines respiratory clearance,
(8) outlines transport out of alveoli,
(9) presents a flow-chart which displays principal pathways which can be taken by a substance acting as an input in the inhalation mode, and
(10) outlines the elements of circulation and the lymphatic system relevant to inhalation and respiratory defence.

Part (B) of the chapter:

(1) outlines the rationale of size-selective aerosol measuring devices,
(2) illustrates how knowledge of respiratory filtration is applied in the design of measuring instruments for certain types of aerosols,
(3) describes briefly two types of measuring instruments, and
(4) discusses the indications for size-selective aerosol measuring devices.

PART (A):   INHALATION AND RESPIRATORY DEFENSIVE PROCESSES

It is widely understood that gas exchange takes place in the respiratory units of the lungs. These present an interface between the blood system and the air in the respiratory system. The interface is a thin living membrane with a total surface area measuring many square metres. One surface of the membrane is moist and is in contact with respired air moved to and fro, like a tide, in the respiratory pathway. The other surface of the lung membrane is covered with a network of fine capillary blood vessels, which themselves present a high surface area; the structural details are described below.

The gas exchange between the blood vessels and the respired air is facilitated by the high surface area of the lungs, the moistness of their lining, and their proximity to the blood stream. But these characteristics also facilitate the ingress of harmful inputs such as substances with chemical or mechanical harmful action, certain climatic factors, such as air of excessive temperature or dryness, and micropredators such as bacteria or viruses. Without defensive processes the human body would soon fall prey to the wide range of harmful inputs which form part of the ordinary as well as the occupational external milieu.

The respiratory defensive processes are numerous and normally very effective although there are weaknesses, as we shall see. Some of the defensive processes are readily identified: coughing results in the forceful ejection of inhaled substances and micropredators. The upper part of the respiratory tract acts as a humidifier, which is sometimes defensive in function. Certain glands in the surfaces of the respiratory system (goblet cells, see later) respond to irritation by secreting extra mucus which, forming sputum, is expectorated or swallowed. This is defensive up to a point but excessive secretion can block small airways. Here we see an example of a principle repeatedly remarked upon in subsequent chapters: that the excessive or prolonged action of a defensive process can itself become harmful.

There are other glands in the surface of the respiratory system which produce defensive agents in ways which are not so readily identified. These include antibodies against substances and micropredators, the antiviral substance interferon, and an enzyme capable of destroying bacteria: lysozyme.

In addition to those defensive processes already mentioned, and as important, is a group of processes which can conveniently be termed, collectively, respiratory filtration.

RESPIRATORY FILTRATION

The surgeon Lister* first observed the effectiveness of respiratory filtration as a defensive process safeguarding the respiratory membrane. He observed that wounds penetrating the chest frequently set up infection in the pleural space (see Figs. 3.1 and 3.2) but that air entering the pleural space by internal rupture of the lungs (spontaneous pneumothorax) seldom set up infection. He attributed the difference in response to the filtration effect of the respiratory path-

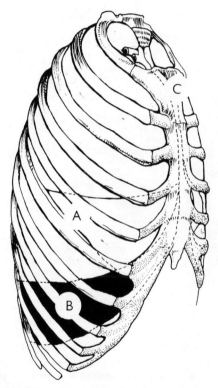

FIG. 3.1. The chest, showing: (A) the space occupied by the lungs; (B) the lower extent of the pleural linings; and (C) the sternum forming part of the rib cage. Redrawn, with acknowledgements to *Gray's Anatomy*, R. Warwick and P. L. Williams, eds, (1973) Edinburgh: Longman.

*The summary of early work is based on Chapter 1 of *Clinical Aspects of Inhaled Particles*, edited by D. C. F. Muir (1972).

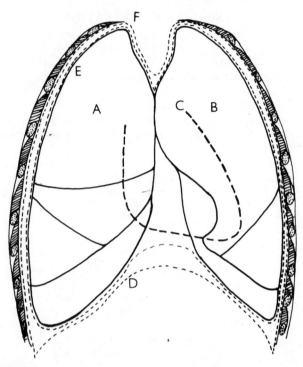

FIG. 3.2. Diagram of section through chest showing: (A) right lung divided into lobes; (B) left lung divided into lobes; (C) the outline of the heart; (D) the pleural lining on the upper surface of the diaphragm; (E) the pleura lining the outer surface of the *lung*; and (F) the pleural lining on the inner surface of the chest wall.

way. In 1870, in the course of his experiments into the scattering of light by particles, Tyndall noted that there was no scattering of light by air *exhaled* from the deep parts of the lung even though there was scattering from the air which was *inhaled*. He concluded that the floating particles in air were retained by filtration in the deep parts of the lung. These early observations demonstrated that the respiratory pathway not only filters harmful floating particles out of the inspired air but may also play some part in rendering them inactive.

In order to understand the defensive mechanism of respiratory filtration, outline knowledge is needed of:

(1) the deposition dynamics of airborne particles, and

(2) the anatomy, physiology, and pathology of the respiratory system.

DEPOSITION DYNAMICS

The term *aerosol* is used to describe all airborne particles small enough to float in the air irrespective of their chemical composition or physical nature. Aerosol thus embraces liquid droplets as well as solid particles.

For the purposes of our understanding of respiratory filtration we need to identify four headings for aerosol deposition dynamics: sedimentation, interception, impaction, and diffusion. This is the terminology used by Muir (1972) and Parkes (1974).

Sedimentation describes the settlement of particles by gravity. This is the familiar process whereby dust settles out of the air on to horizontal surfaces. The terminal velocity of the sedimenting aerosol is related to the density of the aerosol and to the square of the diameter of the aerosol. These relations hold for aerosols in their diameter range of 1 μm to 20 μm. However, many industrial aerosols are not spheres of uniform shape. They may instead be clumps (aggregates) of particles. The terminal velocity of aggregated aerosols cannot be determined by the simple relations and instead the *aerodynamic diameter* needs to be determined.

The aerodynamic diameter is defined as the diameter of a uniform sphere which has the same terminal velocity as the aggregate or other irregular particle. Where the uniform sphere has *unit density* the aerodynamic diameter is expressed as the diameter of an equivalent unit density sphere (EUDS); where the uniform sphere has the *same* density as the irregular sphere the diameter is expressed as the Stokes diameter. Both are given in micrometers. Where not specified, aerodynamic diameter is normally taken to be EUDS.

Interception describes the process by which irregular particles such as asbestos fibres or mica plates become caught on the walls of small airways. The length and the size of the aerosols in relation to the dimensions of the airway are important.

Impaction takes place when the airstream curves, and the suspended aerosols continue under momentum and collide with the airway wall. Impaction is related to the velocity of the aerosol's movement and the angular change of direction.

Diffusion describes the process in which small aerosols behave like molecules and move freely throughout an air space. Diffusion is brought about by the random bombardment of the aerosols by the molecules of the gas in which they are suspended.

RESPIRATORY ANATOMY AND PHYSIOLOGY

The respiratory tract is a collective term used to describe the following anatomical structures, identified in Figs. 3.3, 3.4 and 3.5:

> nasal cavity
> buccal cavity
> oro-pharynx
> larynx
> trachea   conducting
> main bronchi   airways
> bronchi
> bronchioles
> terminal bronchioles

> respiratory bronchioles
> terminal respiratory bronchioles   respiratory
> alveolar ducts   units
> alveoli

Those parts of the respiratory pathway down to and including the terminal bronchioles are the conducting airways. The parts beyond

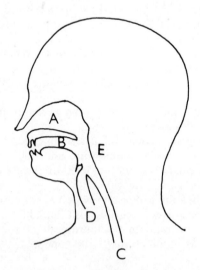

FIG. 3.3. The upper respiratory pathways, showing: (A) nasal cavity; (B) buccal cavity; (C) oesophagus; (D) larynx leading to conducting airways; and (E) pharynx.

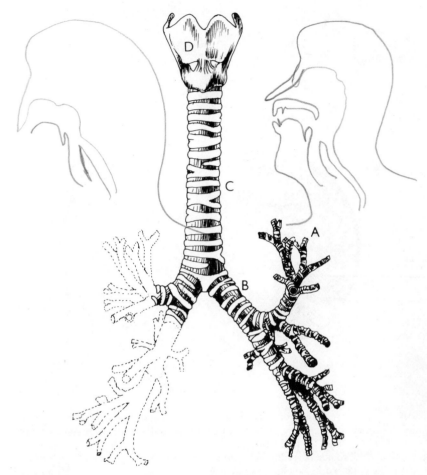

FIG. 3.4. The conducting airways, showing: (A) bronchi; (B) main bronchi; (C) trachea; and (D) thyroid cartilage. Redrawn, with acknowledgements to Lord Brock, Guy's Hospital Medical School; from *Gray's Anatomy.*

the terminal bronchioles constitute the respiratory units. Gas exchange takes place only in the respiratory unit but key elements of the filtration process take place in the conducting airways which, of course, also carry the tidal flow of air. The conducting airways are semi-flexible tubes supported by hoops of hyaline cartilage (the typical structure of which resembles reinforced hose). The internal surface of the conducting airways consists of mucous membrane. The medial surface of the mucous membrane (the medial surface is that in

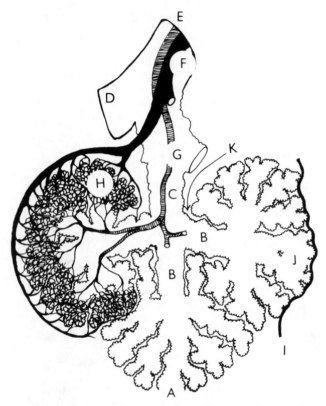

FIG. 3.5. A respiratory unit, showing: (A) alveoli; (B) alveolar ducts; (C) respiratory bronchioles; (D) terminal bronchioles; (E) bronchioles (note: bronchioles and terminal bronchioles do not form part of the respiratory unit); (F) branch of pulmonary vein; (G) branch of pulmonary artery; (H) capillary network in close contact with the alveoli; (I) pleural lining; (J) interstitial spaces enclosed by pleural lining; and (K) peribronchial interstitial space.

contact with the tidal air) is lined by special cells which have cilia—resembling flails—attached to the surface of the lining cells. The mucous membrane contains mucous glands which secrete a tacky fluid, which forms a sticky film bound up with the cilia (see Figs. 3.6 and 3.7).

The cilia, by co-ordinated movement, waft the mucous film outwards, as a slowly moving stream. Aerosols contacting the mucous surface adhere to it and are wafted outwards with the mucous stream.

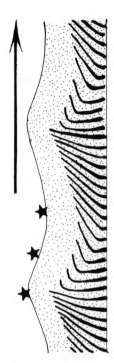

FIG. 3.6. Ciliary escalator: showing 'beat' of cilia and direction of movement of particles adhering to mucous film. Redrawn with acknowledgements to M. A. Sleigh (1962), *The Biology of Cilia and Flagella*, Oxford: Pergamon Press, and to Gray's Anatomy.

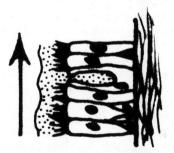

FIG. 3.7. Cells lining conducting airways showing goblet cell secreting mucous film (the goblet cell is sandwiched between the tall cells which form the mucous lining).

This part of the respiratory tract's defensive process is known as the ciliary escalator.

Aerosols which are deposited on the ciliary escalator are brought up to the pharynx within a few hours. From there they are either swallowed or expectorated. For insoluble and inactive aerosols the clearance mechanism is protective, but for soluble aerosols or for radioactive compounds the clearance mechanism may be insufficiently rapid for the purposes of preventing harm. Cigarette smoke and possibly certain other chemical aerosols are capable of inhibiting the ciliary escalator, thereby prolonging the clearance time.

DEPOSITION OF AEROSOLS IN CONDUCTING AIRWAYS

The relation between the aerodynamic diameter of equivalent unit density sphere and the deposition site or the fate of the aerosols is summarized as follows:

EUDS	Deposition site or fate
7–10 μm	ciliary escalator
0.5–7 μm	respiratory bronchioles and alveoli
≤ 0.5 μm	remain airborne and are exhaled; or diffuse, to come into contact with airway or lung membrane

The size-selection function (discussed in Part (B)) of air samplers intended to measure aerosols is designed according to the relations as summarized above. However, the summary might by itself imply an abrupt transition between the three size ranges. Examination of Fig. 3.8 shows the pattern as observed in different studies. The data are taken from Lippmann (1976) who summarized graphically the findings from eleven sources. Two observations on Fig. 3.8 need to be emphasized:

(a) there is no appearance of an abrupt transition in pattern as aerodynamic diameter increases, and

(b) there is considerable variation in the data; this is, in fact, greater than the figure suggests because no indication is given of the statistical dispersion of the data (variation, in the statistical sense is almost invariably a major factor to be considered in any biological measurement).

The deposition of particles in the nose is surprisingly effective, reaching 100% at 10 μm for particles of diameter 10 μm and above.

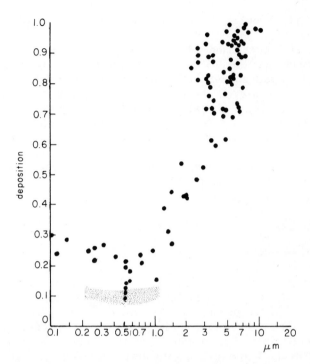

FIG. 3.8. Total deposition of aerosols in the respiratory tract during inhalation through the mouth. Ordinate: fractional deposition (1.0 equals complete deposition). Abscissa: diameter of aerosol in μm; the scale changes, it shows *aerodynamic diameter* from 1 to 20 μm and *linear diameter* from 0.1 to 1 μm. The large dots represent data from experimental observations; the shaded area represents a collation of data from one study. Redrawn, with acknowledgements, from Lippmann (1976).

Particles with a diameter in excess of about 50 μm do not normally get drawn into the respiratory system because the suction is not sufficiently powerful.

Tracheo-bronchial deposition occurs with aerosols in the range 0.5–10 μm aerodynamic diameter. Deposition in the respiratory units as well as conducting airways occurs with the smallest diameter aerosols, the likeliest explanation being diffusion.

The conclusion to be drawn from Fig. 3.8 is that aerosols of about 10 μm and less are able to successfully run the gauntlet of respiratory filtration. Below about 7 μm, deposition in the respiratory units is relatively great, though there is a minimum in the region of 0.5 μm aerodynamic diameter. These patterns are discussed further in Part (B).

The overall conclusion to be drawn so far is that aerosols of an aerodynamic diameter of $50\,\mu m$ and less have to be regarded as potentially effective inputs in the inhalation mode. But not all aerosols below $50\,\mu m$ cause harm. Some are inactive, and there are other defensive processes, as will be described. Aerosol aerodynamic diameter is obviously an important consideration in measurement of *certain* dusts for the purposes of assessment of risks to health. However, occupational exposure to dust normally involves aerosols in a wide range of aerodynamic diameters. How such problems are interpreted forms a topic in Part (B).

RESPIRATORY CLEARANCE

Soluble aerosols deposited in the alveoli dissolve in the film of moisture which lines the alveoli. Transport across the alveolar membrane then takes place bringing the soluble aerosol into the pulmonary capillaries and then into the blood stream. The moist layer of the alveoli contains surfactant—a substance which reduces surface tension. The surfactant's primary role appears to be to reduce the force required of respiratory movements in order to overcome the surface tension effects of the alveolar moisture. Surfactant is also likely to encourage the spread of aerosols coming into contact with the alveolar wall.

TRANSPORT OUT OF ALVEOLI

In Chapter 4 there is a description of macrophages—wandering, scavenging cells which occur widely throughout the body. They are found in the lungs and—of particular interest in the present discussion—in the alveoli. The alveolar macrophages scavenge the substances in the alveoli, and solid particles plastered on the alveolar walls. After scavenging, the alveolar macrophages migrate. Their migration may take them back along the respiratory pathway from the respiratory units to the conducting airways. At the terminal bronchioles they encounter the bottom-most reaches of the ciliary escalator on which they are carried, ultimately to be swallowed or expectorated. In this way, macrophages supplement the respiratory filtration.

It is not entirely clear how particles (or non-particulate substances) cross the alveolar membrane. A plausible explanation is that the alveolar macrophages having scavenged them from the alveolus, migrate through the alveolar wall to the interstitial spaces. However, this may be too simple an explanation. It is possible, for example, that cells which form part of the alveolar membrane have special properties which take part in the transport of particles across the cell membrane. Once transported these particles are picked up by macrophages on the other side of the membrane and taken to a destination such as the interstitial space.

Particles which have penetrated the alveolar membrane by whatever mechanism may, as previously stated, be taken to the interstitial spaces. From here the particles may pass into the lymphatic channels and carried to the hilar lymph glands (see below) where they may accumulate, or they may pass onwards ultimately to reach the general circulation.

Figure 3.9 identifies the pathways open to aerosols deposited in the conducting airways or respiratory units. As previously described, aerosols trapped on the ciliary escalator are either expectorated or are swallowed into the gut. The fate of ingested particles and substances will depend on a number of factors such as particle size and the chemistry of which the particles are composed. Some particles and substances will be excreted, unchanged, in the faeces. Others will be acted upon by enzymes and by other processes in the gut. Still others will be absorbed from the gut and pass within the portal vein to the liver. Here particles or droplets (because they have ceased to be airborne it is no longer appropriate to refer to them as aerosols, instead they will be termed 'substances') may be subject to the metabolic transformation and conjugation defined in Chapter 2 and described in Chapter 8. The end-products of the biochemical reactions may then be excreted in the urine. Other substances will form outputs leaving the body by the faeces. The skeleton, as a storage compartment, is discussed briefly in later chapters.

CIRCULATION AND LYMPHATIC CIRCULATION

A point has been reached in the narrative at which it is convenient to consider the circulation and, in particular, the lymphatic circulation which plays an important part in the disease processes as well as the defence of the lung. Figure 3.10 shows a diagram of the circulatory system. The system has numerous vital functions including the transport of gases to and from the lungs: oxygen is consumed by the tissues as an inseparable part of metabolism from which is formed carbon dioxide as an end product of the reaction.

Fluid, largely composed of water in which is suspended various

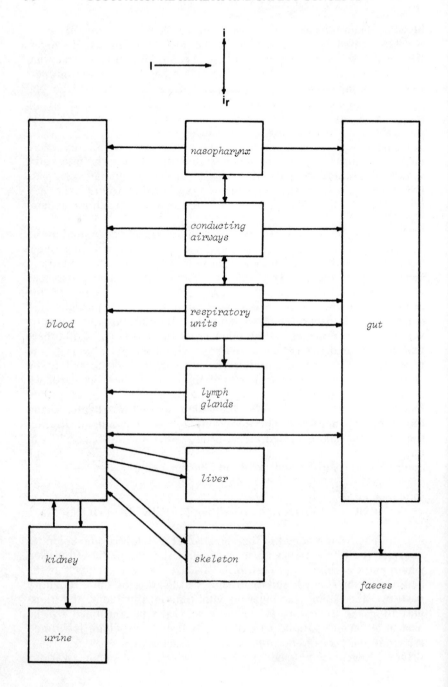

FIG. 3.9. (opposite) Pathways for respiratory deposition and clearance: (I) represents the total quantity of input entering by inhalation mode; (i) represents the portion exhaled; and (i_r) represents the portion retained by respiratory filtration. The ciliary escalator operates between the nasopharynx and the conducting airways; the respiratory units are also cleared, to some extent, by alveolar macrophages migrating to the lower end of the ciliary escalator lining the conducting airways. The liver, skeleton, and kidney have important roles in the de-activation and removal of harmful substances which have entered the body by any mode of input as well as inhalation. Blood acts as the transport medium and there is an exchange in both directions between the blood circulation and the gut (gastrointestinal tract). The faeces act as the vehicle for the bio-dumping of substances which have entered the body by means of the inhalation mode. Modified from, with acknowledgements, International Commission on Radiological Protection (Task Group on lung dynamics): Deposition and retention models for internal dosimetry of the human respiratory tract, *Health Physics* (1966) **12**, 173–207; and Hobbs, C. H. and McClellan, R. O. (1976), Radiation and Radioactive Materials, in *Toxicology*, eds L. J. Casarett and J. Doull, New York: Macmillan.

cellular and other elements, is distributed to all parts of the body by means of the circulation. Most of the fluid is retained within the vessels of the system but a proportion escapes from the capillaries in tissues. That proportion is returned to the general circulation by means of the lymphatic circulation comprising lymphatic vessels, lymph glands, and the thoracic duct (Fig. 3.10).

Three items of terminology can be dealt with now: plasma, tissue fluid, and lymph. The first of these describes the watery component of the blood; the second refers to the fluid which, having escaped from the capillaries reaches the tissues; and the third describes the fluid on its way back, within the lymphatic circulation, from the tissues to the main circulation. Although the three fluids vary in their constituents, they can be conveniently regarded as being essentially one and the same substance.

LYMPHATIC CIRCULATION IN LUNG

The lymphatic system is often thought of as a tissue drainage system. This is a useful notion in that it emphasizes the point that inputs of particles, substances, or micropredators reaching the tissues may be carried away, together with tissue fluid, in the lymphatic circulation. The inputs may be arrested at the lymph nodes (used synonymously with 'lymph gland'), shown at Fig. 3.11, where they may initiate disease. Disease-inducing arrest can happen within the lymphatic

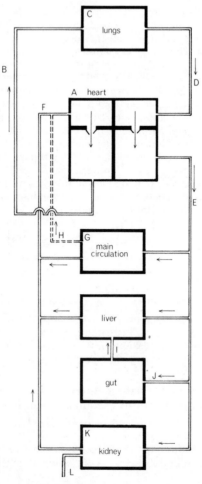

FIG. 3.10. Blood circulation, showing: (A) the heart with four chambers showing direction of flow; (B) pulmonary artery; (C) capillaries in lung; (D) pulmonary vein (carrying oxygenated blood); (E) arterial system distributing oxygenated blood throughout the body; (F) main veins returning de-oxygenated blood from tissues to heart; (G) distribution of blood to tissues; (H) lymphatic drainage from tissues via lymph glands returning to the circulation through thoracic duct; (I) the portal system, (the blood system supplying the liver); (J) blood supply to the gut; note that the veins from the gut join the portal system in the liver (thereby providing a direct pathway to liver for substances absorbed in the gut; there is, however, also a direct link between gut and the venous system generally); (K) the kidney with its blood supply and return; and (L) ureter through which urine passes from kidney to bladder.

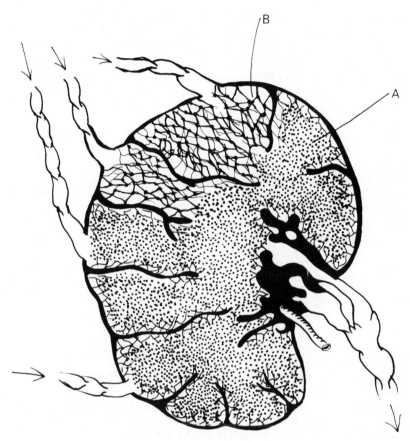

FIG. 3.11. Lymph node, diagram showing: lymphatic vessels entering lymph node, and principal lymphatic vessel leaving node. (A) Lymphocytes; and (B) fine supporting framework for lymphoid tissue. Redrawn, with acknowledgements, from Carr, I. (1972), *Biological Defence Mechanisms*, Oxford: Blackwell.

circulation in any part of the body, and it is an important factor in the development of certain lung diseases linked with occupation.

Substances which cross the alveolar membrane may reach the pulmonary capillaries; others find their way into the tissue spaces of the lung from which they are drained by means of the lymphatic circulation. Figure 3.5 shows the interstitial spaces, sometimes termed tissue spaces. The interstitial spaces of particular importance are those surrounding the respiratory bronchioles, and those lying just beneath the pleura, about to be described. Figure 3.2 shows the pleural space formed as a potential space between the two layers of pleura which line the inside of the chest wall and the lungs themselves. The pleura covering the lungs can be thought of as a 'shrink wrapping' which helps to maintain the form of the lungs and provides them with a lubricated surface for movement relative to the chest wall. Figure 3.5 indicates how the pleura seals in certain of the interstitial spaces.

Substances which reach the interstitial spaces may remain there causing the spaces to act as a storage compartment (see Chapter 11), or they may set up disease processes (see Chapter 15), or they may leave via the lungs' lymphatic circulation. The lymphatic circulation in the lungs originates in two plexuses (networks), one lying beneath the pleural lining and the other associated with the conducting airways. The two systems link up at the lymph glands situated at the root or hilum of the lung. Because, as previously mentioned, the hilar lymph glands act as filters for substances or micropredators being carried away in the lymphatic circulation, these glands are often involved in lung disease.

OUTLINE SUMMARY

Disease can occur, following inhalation of a harmful input, at several points in the respiratory pathway and also in many other organs of the body following the distribution of inhaled substances round the body in the circulation. The extent to which disease actually occurs in the respiratory pathway or elsewhere depends on the nature and quantitative aspects of the input and the effectiveness of the defensive processes to be discussed in later chapters.

PART (B):
THE MEASUREMENT—THE RATIONALE

The reader may find it convenient to return to this part after reading subsequent chapters.

AIR SAMPLING: GENERAL CONSIDERATIONS

For the majority of substances capable of acting as inputs in the inhalation mode, measurements are required of air contamination and the results are expressed either in milligrams per cubic metre or as

TABLE 3.1
SUMMARY OF PHYSICAL FACTORS NEEDING CONSIDERATION IN AEROSOL SAMPLING

(1) *Mass of particle*
The mass of a particle is, generally, related to the cube of its diameter. Hence, gravimetric methods (those which produce a sample of dust for weighing) are likely to be sensitive to sources of error which affect relatively large particles.

(2) *Fibres*
Fibres have a generally low sedimentation rate and where a dust contains the mixture of fibres and semi-spherical particles (as is often the case with asbestos) samples taken may falsely accentuate the non-fibrous particles.

(3) *Moisture effects*
Moisture can affect particles by causing them to increase in size, it can affect agglomeration, and it can also alter the weight of membranes used for collecting dust samples unless the membranes are carefully dried prior to all measurements.

(4) *Isokinetics*
Where aerosols are carried in moving air, unless the body of air is moving very slowly, errors will be introduced unless the sample of air is drawn isokinetically, that is to say at the same velocity as the moving air.

(5) *Instrument orientation and design*
Some sampling instruments are directional, that is to say the results will vary according to the orientation of the instruments; the entry of air to instruments can be disturbed by the proximity of other objects and, in the case of personal samplers (see later), the movement of an individual may affect readings taken with personal samplers. For these reasons, 'unofficial' modification to air sampling apparatus needs to be carried out with due regard to the effect on the instruments' performance.

(6) *Personal samplers*
Personal samplers are designed to be worn by individuals during the course of their work. Normally the sampling head is attached to the individual's lapel and the air pump is worn by means of a harness or belt, at the individual's side. Personal samplers are intended to provide a better estimate of dose than general atmospheric measurements made with static samplers. However, personal samplers help to overcome only some of the sources of variability in estimates of dose. In particular, personal samplers provide measurement of the individual's 'personal' atmosphere which he breathes. Where jobs must involve movement from one part of a plant to another, personal samplers are essential. However, it should be emphasized that all the limitations involved in estimating dose from air samplers apply equally to personal samplers.

parts per million. For numerous substances limiting concentrations are specified in one form or another in connection with codes, limits and standards. Atmospheric measurements are carried out in order to determine whether the general atmosphere at a place of work conforms to the specification, that is to say the levels of contamination do not exceed the limiting levels. For some purposes, for example coal dust, the limiting level is specified in legal terms which dictate the method of measurement and the instrumentation. For other purposes, such as the monitoring of effectiveness of extract ventilation, measurements may be carried out with a view to determining concentrations of substances within specified aerosol dimensions. In all the circumstances mentioned in this paragraph, and others besides, biological considerations such as mode of action, site of contact, or nature of effects in the respiratory system do not materially affect methods or instrumentation. The prime concern, as with all measurements, is accuracy and repeatability; these are likely to be upset by physical factors, examples of which are listed in Table 3.1.

In Part (A) of this chapter brief mention was made of certain quantitative aspects of the inhalation mode of input. In this part there is further consideration of those concepts of especial importance in approaches to measurement of aerosols. It will be recalled from the input–output model (Fig. 2.1) that opportunities for measurement are identified at two points on the model: at position G corresponding to classes of input, and at position H corresponding to modes of input. The former refers to measurement aimed to determine the concentration of any specified contaminant in the workplace generally (as mentioned in the previous paragraph) whereas the latter refers to estimates of the 'dose' received by individuals. Figure 3.9 depicts the routes open to substances entering the body by means of the inhalation mode. In Fig. 3.9 the following relation is depicted:

$$I = i + i_r$$

where I is the total quantity of substance entering the respiratory system by means of the inhalation mode, i is the portion exhaled, and i_r is the portion retained by respiratory filtration.

It is presumed that i leaves the respiratory pathway without having had any effect on the respiratory pathway. Consideration should therefore focus on i_r.

CONSIDERATION OF i_r

The partition between i and i_r is applicable to gases, vapours, and aerosols. In considering i_r we need, as a first step, to differentiate

between
- (a) the respiratory system as a target organ (as defined in Chapter 12), and
- (b) the respiratory system as a boundary which is crossed by substances which are then transported to a target organ remote from lung (see Chapter 11).

It should be emphasized that (a) and (b) are not necessarily mutually exclusive, though there are enough instances to justify the separation. This chapter is concerned primarily with (a) rather than (b) but this emphasis should not lead the reader to overlook the importance of the latter.

The fate and effects of the retained portion i_r depends on
- (1) the mode of action of the substance being inhaled, and
- (2) the site of contact with the respiratory pathway.

It is customary to assume that measurements of concentrations of airborne contamination yield an estimate of I (where appropriate, adjustments may be made for respiratory protection). The quantity i_r is often inferred from I by the simple expedient of ignoring i. For many purposes this gloss causes no practical difficulties but in certain instances the resultant error does. Thus, if the purpose of measurement is to relate the *dose* of a substance to its *effect* (both as defined in Chapter 12) then i_r must be estimated as accurately as possible. If i were always a constant fraction of i_r then the problem of quantitation could be readily overcome. However, the relation between the two may be complex so that, in practice, estimation of i_r may be difficult. Hence, a problem is faced when measurements are made with the intention of determining accurately the dose being received, via the inhalation mode, by individuals. As a first step it may be necessary to determine i_r directly rather than infer it from I. As a second step it may be necessary to partition i_r according to the part of the respiratory pathway at which the retention takes place. The partitioning of i_r could be as follows:

$$i_r = i_{rc} + i_{ru}$$

where i_{rc} is the portion retained in the conducting airways, and i_{ru} is the portion reaching the respiratory units.

As stated, it is being postulated for the purposes of this discussion that i_r is being measured in order to measure accurately dose for, say, the purposes of establishing a dose–effect relation (as defined in Chapter 12). Hence, preliminary consideration needs to be given to the effect of the substance being investigated. Its effects and its mode of action will, of course, be closely bound up with each other. The *locus* of the effect will be determined in part by modes of action and in part by the morphology and chemical nature (including solubility) of the inhaled substance. A further consideration is the site of contact of the substance and whether this is also the site of effect of the substance.

In order to illustrate the points just raised, examples will be used of substances which will be mentioned again in later chapters. The list of examples is by no means exhaustive but it suffices to illustrate the present discussion. Shown in Table 3.2 are ten examples of input substances, their sites of contact in the respiratory pathways (that is to say the conducting airways plus the respiratory units) and their principal effects. The question now arises, for each of these, which quantity or combination of quantities should be determined if the objective is to establish a dose–effect relation for the substance—i_r, i_{ru}, or i_{rc}? For the purposes of the following discussion the problem of measuring i will be ignored, though in a practical problem it might well have to be considered.

Scrutiny of the table shows that in all instances the conducting airways (upper or lower) are the site of contact. From information given in Part (A) it will be recalled that any aerosol below an aerodynamic diameter of less than about 50 μm is likely to reach some part, at least, of the conducting airways. Hence, the first requirement for all ten substances is a measurement of I, the concentration in air, of all aerosols of 50 μm and less. In practice, particles in excess of this size rapidly sediment out of air in which they are suspended. Therefore, no size-selection device will normally be required if air sampling is carried out so as to avoid splashes or 'injection' of particles into the sampling device. Also, from information given in Part (A), it will be recalled that particles of about 1 μm and less tend to remain permanently suspended in air and many of these are, therefore, likely to be exhaled without retention (i.e. these correspond to i). If i is being disregarded, however, there would be no need for a sampling instrument with a lower cut-off point.

Scrutiny of the effects of the ten substances shows that, with two exceptions (see opposite), the effects may arise at the site of contact between the substance and the respiratory pathway. In a case of cancer the effect may also be in other parts of the respiratory system as well as the respiratory pathway.

In the cases of the two examples of fibrogenic dust (see Chapter 15), namely silica and asbestos, the site of effect is not at the point of contact but in the peribronchial and interstitial spaces. Silica reaches those parts of the lungs by means of transport processes such as phagocytosis by macrophages. The macrophages are likely to phagocytose the silica particles within the alveoli. For silica, therefore, it would be necessary to measure the portion of airborne dust likely to reach the respiratory units; measuring instruments for this purpose

TABLE 3.2

EXAMPLES OF AEROSOL INPUTS, SITES OF CONTACT, AND PRINCIPAL EFFECTS

Example	Site of contact	Effect
(1) Soluble irritant gases	Upper conducting airways	Inflammation
(2) Low solubility irritant gases	Lower conducting airways and respiratory units	Inflammation
(3) Soluble irritant gas adsorped onto low solubility aerosol	Conducting airways and respiratory units	Inflammation
(4) Sensitizing metals	Conducting airways (dependent on aerodynamic diameter of particles)	Immunopathological reaction
(5) Tolylene di-isocyanate	Conducting airways	Immunopathological reaction
(6) Cotton dust	Conducting airways	Immunopathological reaction, possibly to bacteria in cotton dust
(7) Fibrogenic dust: (a) silica	Conducting airways and respiratory units (dependent upon aerosol dimensions)	Collagenous pneumoconiosis
(b) asbestos	Lower conducting airways and respiratory units	Collagenous pneumoconiosis
(8) Radioactive aerosols	Conducting airways and respiratory units, dependent upon morphology	Neoplasia
(9) Certain types of wood dust	Upper conducting airways leading to sinuses	Neoplasia
(10) Cigarette smoke	Conducting airways and respiratory units	Metaplasia of bronchial lining

are discussed later. In the case of asbestos it is likely that fibres penetrate the alveolar membrane as well as being picked up by macrophages. The morphology and size of fibres are important in determining the penetration (asbestos fibres are discussed in Chapter 20).

In summary, therefore, it can be stated that partitioning between i_{rc} and i_{ru} is, in practice, most useful for non-fibrous, fibrogenic dust; indeed it is for these types of dust that size selection instruments have been validated. Non-fibrogenic dusts are not normally the subject of elaborate measurement simply because they are non-fibrogenic and hence are normally regarded as harmless (though, in fact, possibility of adsorption makes this a doubtful assumption). Fibrous dust, even though fibrogenic, has aerodynamic properties in the respiratory pathway which have not, as yet, been fully reconciled with those of spherical and near spherical aerosols. Hence, as stated, partition between i_{rc} and i_{ru} is most often done for fibrogenic dusts. Those containing quartz represent an especially important group.

INSTRUMENTS FOR PARTITIONING i_{rc} AND i_{ru}

Partitioning i_{rc} and i_{ru} is by suitably designed sampling instruments. These are constructed so as to produce a sample within the respirable range as defined in Part (A). Lippmann (1976) has reviewed the central concepts involved in the design of these instruments and the following discussion is strongly influenced by his review.

It is apparent that there have been two major stimuli for *research* into aerosol measurement in the context of human safety: dust encountered in mining, and questions raised about the inhalation of radioactive aerosols. However, in *practical* application, measurements in connection with mining represent the dominant influence.

In the USA the first devices intended for size-selective measurement of dust were developed in the 1920's by the cooperative efforts of the US Bureau of Mines, the US Public Health Service, and the American Society of Heating and Ventilating Engineers. In Britain, the Medical Research Council, although its specific concern was with pneumoconiosis in the coal-mining industry, made important advances in size-selective dust measuring devices.

Lippmann summarized the work of the US Atomic Energy Commission on the retention of 'insoluble' internal emitters (radioactive aerosols) and studies by the Task Group of the International Commission for Radiological Protection. As he pointed out, there are two basic approaches to the measurement of aerosols:

(a) specification of the proportion of aerosols in terms of the complete profile of airborne dust specified by the distribution of aerosols within the entire range of aerodynamic diameters represented in the dust being measured, and

(b) by identification of the distribution of aerosols within the respirable range.

Of the two, (b) is the more practical, especially if a device is available which will preselect so that particles outside the respirable range are excluded from the sample. The utility of preselectors is thus obvious, and this recognition has led to considerable development in instrumentation over the past half-century.

Preselectors are important for several reasons. Particle-counting methods are subject to important sources of error: counting errors by the observer (which may be avoidable by 'automatic' methods such as measuring light scattering) and the possibility that in any processing of the aerosols in preparation for counting the physical form of the aerosols may be altered, for example by the breaking-up of aggregates.

The advantage of preselectors is that they restrict the range of aerosol diameters needing counting without interfering with their physical form. Preselectors, moreover, permit measurements based on weights of samples collected (gravimetric methods) as well as numbers of particles counted.

Preselectors depend upon two stages of validation: (1) that the respirable range has been defined as closely as possible on a *biological basis* with human subjects, and (2) that the preselector selects according to the specification for respirable range set biologically.

Figure 3.12, based on Lippmann's data, demonstrates the extent to which the performance of two types of preselector matches the biological data. The figure shows idealized curves for *alveolar deposition* for mouth breathing and for nose breathing, and the performance characteristics of the Medical Research Council's (MRC) preselector and those of American Conference of Governmental Industrial Hygienists (ACGIH). It can be seen that the preselector characteristics differ, although not by very much, and that the ACGIH specification lies nearer to the biological data for mouth breathing.

Lippmann attached certain qualifications to all data relating alveolar deposition studies, because in a total of 11 studies which he reviewed there was evidence of relatively large differences *between* studies. To some extent, all these differences could be minimized by attention to experimental method and standardization of certain variables. The curves relating to mouth and nose data shown in Fig. 3.8 are the medians from Lippmann's data which show, in fact, quite

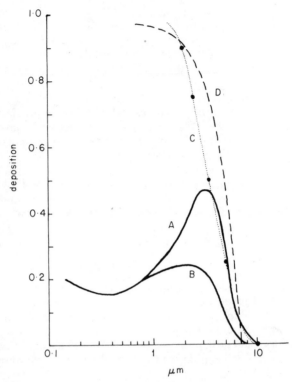

FIG. 3.12. Comparison of sampler acceptance curves: (A) alveolar deposition with mouth breathing (median from Lippmann's data, see Fig. 3.8); (B) alveolar deposition with breathing through the nose; (C) acceptance curve for ACGIH sampler; and (D) acceptance curve for the (British) Medical Research Council sampler. Ordinate: deposition fraction as in Fig. 3.8. Abscissa: aerodynamic diameter in μm, as in Fig. 3.8. Redrawn, with acknowledgements, from Lippmann (1976).

wide dispersion. Even though Lippmann's data favour the ACGIH preselector, the variability in the data is such that there is no case for the abandonment of the MRC specification in favour of the ACGIH, as he agreed (1977, personal communication).

For the purposes of this part, description is confined to two designs of preselector which appear, at present, to be in the ascendancy: elutriators and cyclone preselectors. Both preselect, in that particles outside the range of interest are separated from the stream of in-flowing air. The horizontal elutriator consists of a series of parallel metal plates lying horizontally one above the other. Air is drawn in a

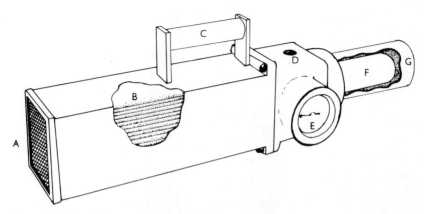

FIG. 3.13. Hexhlet-type horizontal sampler. (A) Protective gauze at sampler orifice; (B) horizontal plates forming elutriator upon which the heavy particles settle; (C) carrying handle; (D) orifice through which air is sucked by means of a suction pump; this causes air to be drawn, via a critical orifice, over the plates (B); (E) pressure gauge, for checking correct air flow; (F) porous thimble which allows air to be drawn through but retains particles; and (G) cover for thimble.

steady stream through the spaces between the plates; the 'dwell time' of the aerosols between the plates is sufficient for those of aerodynamic diameters in excess of the respirable range to settle out under the influence of gravity. Those within the aerodynamic range remain airborne in the airstream and are carried into whatever aerosol trap the instrument contains. Figure 3.13 shows one example of a horizontal elutriator and aerosol trap.

A cyclone, for vertical use as a personal sampler, is shown in Fig. 3.14. There are advantages and disadvantages for elutriators and cyclones, and for further information the reader is referred to the discussions by Roach (1973) and Lippmann (1976).

SPECIFICATION OF SIZE SELECTION

From Lippmann's review (1976) it is apparent that there are three current, major approaches to the specification of aerosol data:

(1) The Medical Research Council specifications put forward in 1952 and adopted by the Johannesburg International Conference on pneumoconiosis in 1959. These included the following specifications:

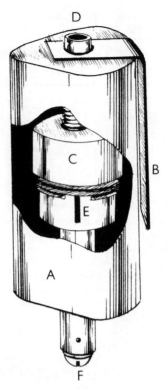

FIG. 3.14a. Drawing of a cyclone-type size selection measuring device. (A) Protective cover; (B) clip for retaining sampler to holder (the sampler pictured is designed for use as a personal sampler); (C) body of cyclone; (D) exit for air being drawn through slit (E) ((D) is attached to a tube which in turn is attached to a suction pump); (E) entry slit for dust-laden air; and (F) screw providing entry to aerosol trap. *Note*: the sampler is drawn partially dismantled, that is to say partly withdrawn from the protective cover. Normally (F) would be level with the lower edge of (A).

> ... for purposes of estimating airborne dust in its relation to pneumoconiosis, samples for compositional analysis, or assessment of concentration by bulk measurement such as that of mass or surface area, should represent only a 'respirable' fraction of the cloud.

and

> The 'respirable' sample should be separated from the cloud while the particles are airborne and in their original state of dispersion.

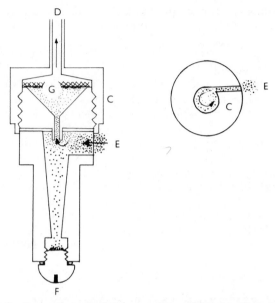

FIG. 3.14b. Cyclone-type sampler in section. (C) Body of cyclone showing slit and cylindrical volume where selection takes place; (D) tube for air exit; (E) dust-laden air entering slit; (F) size-selected particles settling into aerosol trap; and (G) holder for membrane filter upon which are collected the aerosol in the respirable range. The particles collected on the filter at (G) represent i_{ru} (see text).

and

> The 'respirable fraction' is to be defined in terms of the free falling speed of the particles, by the equation $C/C_0 = 1 - f/f_c$, where C and C_0 are the concentrations of the particles of falling speed f in the 'respirable' fraction and in the whole cloud, respectively, and f_c is a constant equal to twice the falling speed in air of a sphere of unit density 5 μm in diameter.

(2) The US Atomic Energy Commission (AEC) established in 1961 a definition of 'respirable dust' in terms of the proportion of the inhaled dust which penetrates to the non-ciliated portions (equivalent to the respiratory units) of the lung. Lippmann noted that this application of the concept of respirable dust was intended only for *'insoluble' aerosols which exhibit prolonged retention in the lung. It was not intended to embrace relatively soluble aerosols, nor those with strong biological activity—such as, presumably, irritants.* (These points emphasize those made at the start of this part.)

The specification for 'respirable dust' radioactive particles which are retained in the respiratory pathways is:

Size (EUDS in μm)	10	5	3.5	2.5	2
Respirable portion (%)	0	25	50	75	100

(3) The International Commission for Radiological Protection set up a task group concerned with retention of aerosols in the respiratory pathway. One recommendation included the observation that a single parameter, the mass median diameter (MMD), could be used to typify dust clouds. It is defined as that aerosol diameter which represents the point of division between two portions of the dust cloud where the dust cloud has been divided so that each part has the same mass of dust. Hence, half the mass is above the MMD and half below it.

There are several methods of measurement which can be used to determine MMD, but the role of MMD is not yet fully established. In particular, it is not clear whether MMD can, in fact, be used to specify aerosol data for all purposes.

SIZE SELECTION AND PERSONAL PROTECTION

Respiratory protection's effectiveness in excluding harmful agents must be a prime concern where this strategy is utilized. Size selection will not be a crucial consideration for gases, vapours and certain other aerosols, because the requirement is to remove them totally from the inspired air. For those aerosols where size selection is a relevant consideration, then size-selection capability of personal protection should also be considered. The starting points for such consideration are the protective equipment manufacturers' data and relevant national standards.

REFERENCES

LIPPMANN, M. (1976). Size-selective sampling for inhalation hazard evaluation, in *Fine Particles*, B. Y. H. Liu (ed.), New York: Academic Press.

MUIR, D. C. F. (ed.) (1972). *Clinical Aspects of Inhaled Particles*, London: Heinemann.

PARKES, W. R. (1974). *Occupational Lung Disorders*, London: Butterworth.

ROACH, S. A. (1973). Sampling air for particulates, in *The Industrial Environment—Its Evaluation and Control*, Anon, (ed.), Washington: National Institute for Occupational Safety and Health, US Department of Health, Education, and Welfare.

WARWICK, R. and WILLIAMS, P. L. (eds) (1973). *Gray's Anatomy*, 35th Edition, Edinburgh: Longman.

Further Reading

KILBURN, K. H. (1968). A hypothesis for pulmonary clearance and its implications, *American Review of Respiratory Disease*, **98**, 449–63.

NEWHOUSE, M., SANCHIS, J. and BIENENSTOCK, J. (1976). Lung defence mechanisms, Parts 1 and 2, *The New England Journal of Medicine* **295**, 990–7 and 1045–52.

Chapter 4

Defensive Cells

OBJECTIVES OF THE CHAPTER

This chapter presents an outline of:
(1) the types and function of defensive cells generally,
(2) phagocytosis,
(3) the macrophage system and the reticulo-endothelial system, and
(4) defensive cells in the blood and in the tissues, and a picture of the injury caused to defensive cells of the blood by ionizing radiation.

INTRODUCTION

Defensive cells and their functions figure in discussions about many occupational diseases. An understanding of these cells and their functions provides a helpful basis for further understanding of the processes involved in occupational diseases. Two examples will suffice to underline the usefulness of an outline knowledge of defensive cells:
(a) the defensive cell process of phagocytosis has a central role in certain occupational lung diseases which are of great importance (and which are discussed in later chapters), and
(b) blood-borne defensive cells respond quickly to injury from ionizing radiations and consequently these cells have received a lot of attention in the field of health physics.

In the previous chapter it was shown that macrophages are an important element in the respiratory defence processes. But macrophages are not confined to lung; nor are macrophages the only defensive cells. This chapter's purpose is to identify defensive cells and their functions generally. The topic is important because defensive cells by their very nature are involved from the earliest stages in the body's struggle against harmful inputs, and they may themselves be the site of developing occupational disease. Changes in defensive cells may be an early manifestation of harm from a harmful input, hence changes in defensive cells may be used as a monitor of certain harmful inputs.

Defensive cells are found in blood, and in the other fluids, and in structures and tissues of the body. Not all cells have a defensive function although all cells exhibit changes in their normal function, and sometimes structure as well as function, when they are subject to harmful inputs. Highly specialized cells—for example brain cells—are too specialized to react other than by dying or degenerating. Some cells, though specialized to a certain function, can under certain circumstances exhibit defensive reactions. Others appear to be involved in no other process besides defensive functions.

DEFENSIVE PROCESSES OF CELLS

Defensive processes of cells include or are associated with the processes of:
  (a) phagocytosis,
  (b) inflammatory response,
  (c) immune response,
  (d) repair,
  (e) secreting defensive substances (such as α-1-antitrypsin which inhibits proteolytic enzymes normally present in lung and damaging to lung tissue), and
  (f) preventing excessive blood loss from the circulatory system.

Phagocytosis is discussed in this chapter, inflammatory response and immune response are discussed in Chapter 5, and repair is discussed in Chapter 15. The way in which secretory defensive substances are produced is touched upon in Chapter 10. Coagulation of the blood, important though this process is, is not discussed.

Of the five functions at least three—inflammatory response, immune response, and repair—are subject to disorders of considerable significance in human safety; these are discussed in Chapters 13–15.

Phagocytosis

The Russian zoologist Ilya Mechnikov (1845–1916) observed that there were apparently two families of phagocytic cells, the large, and the small. He termed these macrophages and microphages. Phagocytosis is a vital element in the body's defence against harmful particles of all kinds, including micropredators. Many different types of cell are phagocytic. Phagocytosis and the closely related pinocytosis may in fact be the way in which many types of living cell take in nutrients, (discussed in a later chapter). Indeed, it is thought that phagocytosis and pinocytosis are characteristics of living cells generally. Although the following description may apply, in part, to the process of phagocytosis generally, the emphasis is placed on the defensive aspects of the process.

Phagocytosis consists of several stages which are listed below and also depicted in Fig. 4.1.

(1) Chemotaxis of the phagocytic cell by the particle or substance, (for simplicity's sake 'particle' is used to signify both particle and substance even though substances, strictly speaking, may not be in particulate form). Chemotaxis is a phenomenon widely recognized in biology. It describes processes whereby growth or movement is

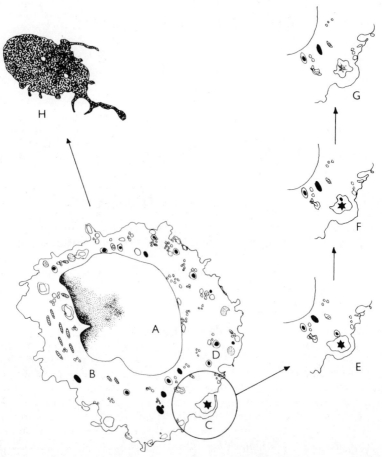

FIG. 4.1. Sequence illustrating phagocytosis by macrophage: (A) nucleus of macrophage; (B) cytoplasm of macrophage; (C) ingestion of particle; (D) lysosome; (E) phagosome about to be formed by discharge of attached lysosome's contents; (F) phagosome formed; (G) digestion of particle almost complete; and (H) macrophage killed by ingestion of cytotoxic particle.

influenced in direction by small quantities of chemical substances. If phagocytic cells can be said to eat and drink, then chemotaxis represents the whiff of the forthcoming feast.

(2) Adhesion of the particle to the phagocytic cell. The immune response, involving binding between cells and particles, is probably an important basis for the adhesion.

(3) Ingestion of the particle by ameoboid movement of the phagocyte.

(4) The intra-celluar inclusion of the particle attracts a lysosome. This is a 'bubble' of enzymes. The lysosome moves through the plagocyte's cytoplasm and becomes attached to the particle. The lysosome discharges the enzymes into the space surrounding the particle. The mixture of enzymes and the particle is known as a phagosome.

(5) The enzymes may chemically digest the particle. The phagosome may remain in the phagocyte's cytoplasm as a residual body, apparently in an inactive state. Alternatively, the phagocytic cell may deposit the residue of the phagosome in its wake.

(6) The particle may be indigestible and remain as an intracellular inclusion while the phagocytic cell migrates to other tissues.

Harmful particles, and micropredators engulfed by phagocytes may be rendered harmless. On the other hand, phagocytes may be killed by a particle or a micropredator. Phagocyte death results in the formation of pus, which should then be discharged. The space left following the discharge of the pus is made good by scar tissue and cell regeneration, although the extent to which cell regeneration takes place is variable.

Macrophages and the Macrophage System

Some macrophages are found circulating in the blood stream; others are found throughout the body, in two broad groups: *fixed macrophages*, which form part of certain tissues. Certain 'fixed' macrophages appear to be capable of some migration, which makes for difficulties in distinctions between fixed macrophages and *free macrophages*, which are found wandering throughout the tissues of the body.

Macrophages are regarded as belonging to a widespread system of highly phagocytic cells called the macrophage system. 'Reticuloendothelial system' can be used to describe the fixed part of the macrophage system.

Table 4.1 summarizes elements of the macrophage system. Some of these elements are not the subject of further discussion but are included, in parentheses, for completeness.

TABLE 4.1
SUMMARY OF MACROPHAGE SYSTEM

Site	Name	Comments
Connective tissues generally (see Chapter 16 for discussion on connective tissues)	Tissue macrophages, histiocytes	Free cells
Lung	Pulmonary macrophages	Free and fixed?
Blood	Monocytes	Free; may be tissue macrophages having migrated from blood stream
Lining blood spaces in:		
bone marrow		Fixed
spleen	Von Kuppffer cells	Fixed
liver		Fixed
(adrenal medulla)		Fixed
(pituitary gland)		Fixed
Within 'resident' populations of:		
spleen		Free
lymph glands		Free
bone marrow		Free
(tonsils)		Free
Nervous tissue:		
connective tissue	Microglial cells	Fixed
membranes lining brain	Meningocytes	Fixed

Microphages

As their name implies, the microphages are smaller than macrophages. They belong to the white cell family in the blood. Of the blood's white cells the *neutrophil leucocytes* are unquestionably phagocytic. *Eosinophils* are capable of phagocytosis of antibody antigen complexes (see also Chapter 5). The non-phagocytic member of the white cell family is the *basophil.* All these cells normally circulate in blood; they appear in tissues, that is external to the blood vessels, in response to harmful processes taking place in the tissues.

The Blood's White Cell Family

Collectively the members of the blood's white cell family are known as leucocytes. Of these the monocytes, Fig. 4.2, are the largest in size; these are macrophages.

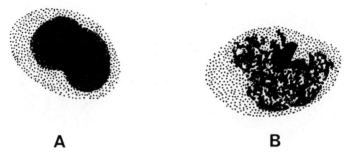

FIG. 4.2. Normal and abnormal monocytes. (A) Normal monocyte drawn from blood smear; and (B) drawing of monocyte on the tenth day following accidental radiation exposure, the lacy appearance of the nucleus is due to vacuolation. Redrawn, with acknowledgements, from Hempelman *et al.* (1952) (see text). Nucleus normally stained blue.

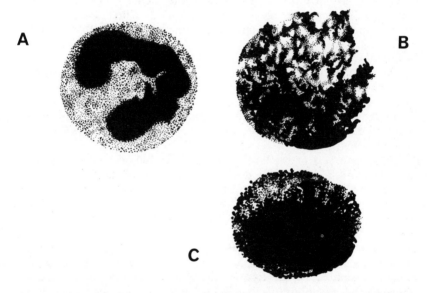

FIG. 4.3. Normal and abnormal neutrophil polymorphs. (A) Normal polymorph showing irregularly shaped nucleus (from which the cell derives its name); (B) drawing of young neutrophil in the blood smear 24th day of acute radiation syndrome; the cell contains large deeply-staining toxic granules; the nucleus (lower half) is not easily distinguished from the rest of the cell; and (C) neutrophil polymorph on the 24th day of acute radiation syndrome; the cell is tightly packed with toxic granules. Nucleus normally stained blue. Scale different from that in Fig. 4.2. (Redrawn, with acknowledgements, from Hempelman *et al.* (1952).)

FIG. 4.4. The eosinophil and basophil. (A) Normal eosinophil (nucleus stained purple-blue, and cytoplasm stained pink-red); and (B) basophil nucleus obscured by cytoplasm with dense blue stain. Nucleus normally stained blue.

Another group of cells belonging to the family share the common feature of having an irregularly shaped nucleus. The nucleus is said to be polymorphous in form, and cells with this feature are referred to collectively as polymorphs, *or* granulocytes, *or* polymorphonuclear granulocytes.

The polymorph group of cells is characterized by the granules contained in the cells' cytoplasm (Figs. 4.3 and 4.4); the terminology of the group derives from the affinity of the granules to certain dyes. Thus the cells with red staining granules are called eosinophils, those with blue staining granules basophils and those with neutral staining granules neutrophils.

Other members of the leucocyte family include the *large* and *small lymphocytes* (Fig. 4.5). The defensive functions of the leucocytes are summarized in Table 4.2.

Defensive Cells in Tissues

As has already been mentioned, certain defensive cells in the blood may have a common origin to a particular type of cell which they closely resemble, in the tissues. Monocytes resemble tissue macrophages; the basophils of the blood resemble mast cells which are found in tissues. Lymphocytes (and the closely related plasma cells) are common to blood and tissues, and that these circulate has long been recognized.

One type of cell found only in tissue and without any counterpart in blood are fibroblasts. These cells are either elongated or stellate in form (Fig. 4.6). Fibroblasts synthesize collagen, a fibrous protein normally giving mechanical support to tissue. The defensive function

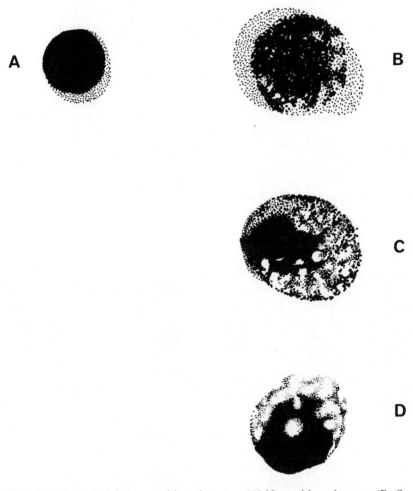

FIG. 4.5. Normal and abnormal lymphocytes. (A) Normal lymphocyte; (B, C, D) drawing of lymphocytes 24 days post accidental radiation exposure. In (B) the lymphocyte is considerably larger than normal and the nucleus is somewhat less well defined; in (C), apart from the dark section of the nucleus on the left hand side of the lymphocyte, most of the nucleus has almost completely disappeared; and (D) lymphocyte showing prominent vacuoles in the nucleus. (Abnormal lymphocytes redrawn from Hempelman et al., 1952—see text.)

TABLE 4.2
DEFENSIVE CELLS: SUMMARY

Blood	Tissues	Defensive function
Monocytes	Tissue macrophages	Phagocytosis
Neutrophil polymorphs	Migrate from blood into tissues as part of inflammation (q.v.)	Phagocytosis
Basophil polymorphs	Mast cells	Secrete defensive substances such as heparin and histamine (roles in inflammation, q.v.)
Eosinophil polymorphs	Migrate from blood during immune reaction (q.v.)	Phagocytic for antigen–antibody complex
Lymphocytes, including plasma cells	Circulate in tissue fluid and lymph	Produce antibodies
—	Fibroblasts	Produce collagen
Platelets	—	Not strictly cells but play an important role in blood clotting

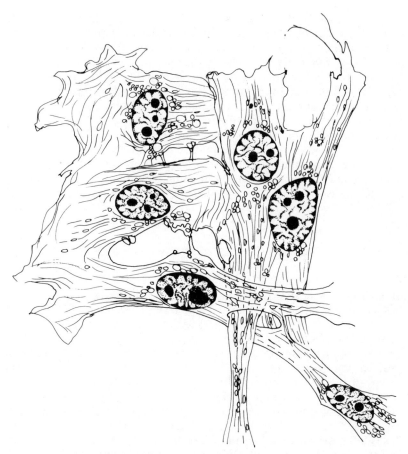

FIG. 4.6. Fibroblasts showing elongated form and prominent nuclei; the granules are associated with collagen formation, discussed in later chapters. Redrawn, with acknowledgements, from *Gray's Anatomy.*

of fibroblasts is to lay down new collagen fibre in order to form mechanically-strong scar tissue. This defensive function is described in Chapter 15.

The Origin of Defensive Cells

The origin of the individual types of defensive cells is still far from clear. However, it is clear that many defensive cells have their origin (if traced back far enough) in *stem cells* of the bone marrow. It is useful to visualize the bone marrow as a factory producing defensive cells. The life span of most defensive cells is quite short; consequently, the 'bone-marrow factory' is in constant production simply maintaining the body's stocks of defensive cells. Damage to the factory, if it reduces output of defensive cells, puts the body at risk especially from micropredators. Hence, effects on bone marrow of ionizing radiations and certain chemical substances can be regarded as immobilization of bodily defence mechanisms. As will be stated later, the reason why the bone marrow is so vulnerable to attack by substances is because of its very activity in producing new cells. In this respect, a parallel can be drawn with the process whereby sperm is produced from seminiferous epithelium. Here, too, new generations of cells are continually being produced; consequently, sperm production may be interfered with by ionizing radiation.

Radiation Damage to White Blood Cells

In Figs. 4.2, 4.3 and 4.5, shown alongside the normal cells, are drawings of cells damaged by ionizing radiations. The drawings are based on photographs published by Hempelmann *et al.* (1952) in connection with a study of nine cases of acute radiation syndrome. These authors reported on the accidents which occurred during the Los Alamos development of the atomic bomb. For example, within a year of the first military use of the bomb, two nuclear accidents occurred at the laboratory. These accidents involved temporary uncontrolled fission reactions during which ten persons were exposed to ionizing radiations. Of these persons, two died as a result of their exposure. The other eight received smaller and in some cases harmful doses of radiation. Among the harmful effects of the radiation are included morphological changes in blood white cells. Some of these are illustrated in the figures.

In passing, it should be pointed out that Britain's legislation concerned with protection against ionizing radiations requires blood cell investigations as a matter of routine. Features such as those illustrated in the figures are looked for among the cells of the exposed population. (Peripherally, it should be noted that the value of such tests is questionable in the protection of persons against ionizing radiations. There is no doubt that following exposure to radiation cells show damage, but it is debatable whether the statutory tests provide information of practical value which could not be inferred from measurements made directly of radiation by means of physical measurement instruments.)

CONCLUSIONS

Defensive cells are crucially important in defending the living body against harmful inputs of all kinds, of which those associated with work probably make up only a small proportion of the total of harmful inputs with which the body has to contend. Micropredators, that is living things, generally represent a much greater danger than inputs of occupational origin. That the human body survives for so long without disease being induced by micropredators is a tribute to the effectiveness of, *inter alia*, the body's defensive cells.

REFERENCES AND FURTHER READING

HEMPELMANN, L. H., LISCO, H. and HOFFMAN, J. G. (1952). The acute radiation syndrome: a study of nine cases and a review of the problem, *Annals of Internal Medicine*, **36**, 279–510.

LA VIA, M. F. and HILL, R. B. (eds) (1975). *Principles of Pathobiology*, 2nd Edition, New York: Oxford University Press.

WARWICK, R. and WILLIAMS, P. L. (eds) (1973). *Gray's Anatomy*, 35th Edition, Edinburgh: Longman.

Chapter 5

Inflammatory Response and Immune Response

OBJECTIVES OF THE CHAPTER

This chapter:
(1) provides an outline of inflammatory response, and its natural history,
(2) provides an outline of immune response and certain of its functions,
(3) prepares the ground for Chapters 13, 14 and 15 which deal with inflammation, altered sensitivity, and disorders of repair, and
(4) presents certain key concepts in outline: acute inflammation, chronic inflammation, surveillance, disposal of antigens, self-disposal, types of immune response, link between immune response and inflammation, steps in the immune response, antigen–antibody reaction, immunoglobulins, and the role of the complex in phagocytosis.

INFLAMMATORY RESPONSE

Inflammatory response (inflammation) is the reaction of a tissue to harm which is insufficient to kill the tissue. It follows the entering of the body of any one of a large number of different types of harmful input. This entrance is effected by way of inhalation, ingestion, absorption, pervasion, implantation, surface penetration, certain kinds of trauma, and energy transformation. The pattern of inflammation varies to some extent according to the type of input, but there are important aspects of the inflammatory response which are common to all types of inputs.

A distinction is drawn in clinical pathology between acute and chronic inflammation. Acute inflammatory responses are of short duration, whereas chronic inflammatory responses may last weeks, months or years. Some authorities define three forms of inflammation: acute, sub-acute, and chronic. The three are distinguished according to the types of cell predominating. Table 5.1 summarizes the cellular pattern.

Acute inflammation may subside entirely or form pus which is then discharged, following which there is making good by scar tissue. Inflammation may spread rapidly to involve large areas of tissue, or it

TABLE 5.1
INFLAMMATION: PREDOMINANT CELLS

Acute inflammation	Neutrophils
Sub-acute inflammation	Lymphocytes and monocytes with granulation tissue*
Chronic inflammation	Lymphocytes, monocytes and plasma cells with overstimulation of collagen*

*These topics are discussed in later chapters.

may spread sufficiently to obstruct the blood supply, resulting in death of tissue and the subsequent invasion of the dead tissue by bacteria (gangrene).

Inflammation is a defensive process of great importance. But if called upon to act for too long it can, itself, result in disease. The dividing line between the defensive and the harmful aspects of inflammation is not always easy to draw; that there are beneficial and harmful effects of inflammation is a concept which needs to be kept in mind in any consideration of inflammation and related processes.

Acute Inflammation

Acute inflammation, which may or may not turn into chronic inflammation, is typified by the following sequence of events:

(1) the small vessels in the area of tissue affected by the harmful input constrict briefly,

(2) the same blood vessels then dilate,

(3) coincidentally with the dilation the vessel walls increase in permeability,

(4) as a result, protein-rich fluid exudes from the blood vessels into the tissues, causing them to swell,

(5) white cells migrate through the vessel walls and are drawn towards the harmful input. The migrant cells phagocytose the harmful input and any destroyed tissue, and

(6) tissue-dwelling macrophages join the white cells and take part in the phagocytosis, and the area of inflammation may be localized by a boundary of fibrin (a protein associated with the clotting of blood).

The cells which congregate in the inflamed area include lymphocytes and plasma cells as well as microphages and macrophages. Pus resulting from acute inflammation contains polymorphs in abundance.

Figure 5.1 depicts events (4) and (5).

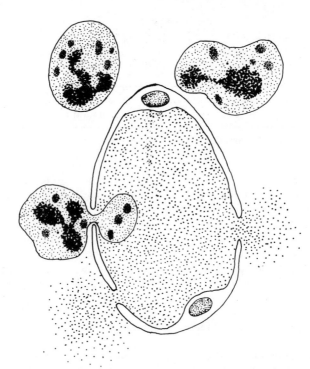

FIG. 5.1. Protein-rich fluid exuding from a capillary blood vessel into the tissues; and a neutrophil in the process of migrating through the capillary wall. Redrawn, with acknowledgements, from Carr, I. (1972), *Biological Defence Mechanisms*, Oxford: Blackwell.

Chronic Inflammation

The typical pattern of chronic inflammation is an accumulation of macrophages, lymphocytes and plasma cells, the virtual absence of polymorphs and the stimulation of fibroblasts to produce excessive amounts of collagen. In certain chronic inflammatory conditions macrophages die and other macrophages phagocytose their dead brethren, the combination of living and dead macrophages forming structures termed giant cells (Fig. 5.2). Chronic inflammation may develop as a sequel to acute inflammation, but it may also arise without an apparent intervening period of acute inflammation.

Scarring is a reparative process by which gaps in tissue are made good. But in certain patterns of chronic inflammation the repair

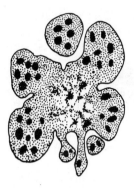

FIG. 5.2. Impression of a giant cell drawn from chronically inflamed tissue.

process becomes disordered and there is excessive production of collagen. The overgrowth of scar tissue shrinks and contracts, tearing and distorting adjacent tissues (discussed in Chapter 15). In the lung this tearing and distortion results in the clinical condition emphysema. Extensive scarring and fibrosis typify certain categories of pneumonoconiosis, an important category of occupational dust diseases described later.

IMMUNE RESPONSE

Immune response describes a group of mechanisms concerned with defence and the preservation of normal body integrity. Originally, 'immune response' was used to describe certain mechanisms involved in the living body's response to micropredators, particularly bacteria and viruses. Subsequently, immune response has acquired a broader meaning. It now covers defensive mechanisms falling under the following general headings:

(1) *Surveillance.* The mechanisms involved in the immune response may provide constant surveillance for genetic mutation. Occasionally dividing cells give rise to mutants, that is deviants from the normal cell pattern. The multiplication of these mutants could produce abnormal tissue, though not always since deviation may be an adaptive process. Surveillance by the immune response may result in mutants being recognized as alien and their being destroyed. Some authorities look upon surveillance as being the most important function of the immune response.

(2) *Disposal of antigens.* A wide range of chemical substances and micropredators provoke an immune response, whatever route they

enter the body. In its classical form this immune response involves the production of antibodies over the space of a few days. When next the antigen enters the body it reacts with the antibody and the combination of antigen and antibody is then phagocytosed or dealt with by some other defensive mechanism.

(3) *Self-disposal.* Redundant blood and tissue cells are dealt with by phagocytosis (by fixed phagocytes in spleen, for example). There are mechanisms involved in this process which ensure that the macrophages phagocytose only the redundant cells and not the functional cells, a differentiation which may well be an aspect of immune response.

The structures principally involved in the immune response are the lymph nodes, the spleen and the thymus. The cells principally involved in the immune response are the large and small lymphocytes and plasma cells.

Types of Immune Response

Some elements of the immune response take place in the body fluids—the humoral response. Other elements of the immune response take place in cells—a process known as the cell-mediated response.

The immune response is continually being activated. It is convenient to recognize three categories of immune response: (a) advantageous, (b) neutral, and (c) disadvantageous, of which a further classification is given in Chapter 14.

People who suffer from asthma, hay fever, or allergic dermatitis experience the disadvantageous category of the immune response. We are normally unaware of the neutral and advantageous immune responses taking place within our bodies, although occasionally the advantageous responses are accompanied by a transient illness which may be temporarily uncomfortable but which lays down a defensive process able to cope with potentially far more serious disease.

The immune response and disorders of the immune response are crucial considerations in the understanding of disease of non-occupational as well as occupational origin. The pathogenesis of much disease involves aspects of the immune response which, clearly, is a subject of immense importance in the understanding of disease generally. The following outline deals with those aspects of the immune response which are of greatest importance to the student of occupational health and safety. What follows is thus a summary of selected material and not the topic as a whole.

It is important to link the immune response with inflammation, since the two can be conveniently regarded as stages in a continuous process. Inflammation represents the final phase of the process in which the harmful input and the damaged tissue are removed prior to the repair process. The immune response can be regarded as the mechanisms concerned with identifying and preparing the harmful inputs or the damaged tissues for the inflammatory process. Because immune response is such a large subject it stands alone, implying perhaps its separation from inflammation and inflammatory processes. For our purposes, this compartmentalization needs to be avoided.

As previously mentioned, a wide range of harmful inputs can provoke the immune response. Micropredators and chemical substances have already been identified. Inorganic as well as organic chemical substances can produce the immune response. Certain of the energies cause tissue damage which in turn provokes the immune response. In addition, it is widely recognized that there is a psychological component to diseases such as asthma and allergic dermatitis. On these grounds the psycho–social climate can also be regarded as an input contributing to though not necessarily causative of immune response. It is recognized that adverse climatic conditions such as high temperature and humidity can exacerbate certain immune responses, so that these also can be regarded as interactive factors.

Some occupational diseases are disorders of growth, occupational cancer for example. It is an interesting question for research whether the occupational cancer involves a failure or disorder of the immune reaction against neoplastic cells.

Certain occupational diseases represent a disorder of repair. Whether these disorders have their origins prior to the repair processes, that is in the inflammatory or even the immune phases of the response, will be examined later.

Steps in the Immune Response

The three features which characterize immune responses generally are:
  (1) specificity,
  (2) memory, and
  (3) self-recognition.
The first focus for attention can conveniently be the antigen–antibody reaction. Antigens may be extrinsic—that is entering the body from outside—or intrinsic—that is originating within the body. Both extrinsic and intrinsic antigens stimulate antibodies. Antibodies are proteins (gamma globulins) and are named immunoglobulins, of which there are several categories and components, each coded according to their molecular weight and other properties.

Human serum immunoglobulins are synthesized in plasma cells. Five types of immunoglobulins, all with defensive functions are most commonly identified. These are: IgM, IgG, IgA, IgD, and IgE. Of these IgM and IgG form the bulk of circulating antibodies. IgA is found in secretions from the gut and the respiratory pathway. IgE is found in association with cell membranes in, for example, release of histamine from mast cells following the immune response, and probably IgE is associated with phagocytosis by macrophages. The role of IgD is uncertain.

Immunoglobulins constitute the principal component, perhaps the only component, of the humoral immune response; they are produced by plasma cells.

When the body first experiences an antigen the following chain of events takes place. Soon the antigen appears in the lymph nodes, stationed in the system draining the tissue through which the antigen has entered. The appearance of the antigen within the lymph node stimulates cell division among the lymphocytes. After about 5 days antibodies (that is immunoglobulins) are detectable in the lymph nodes. At the same time large numbers of lymphocytes start to pass from the lymph nodes into the lymphatic circulation and thence into the blood stream. From the blood stream they gather in the spleen and bone marrow, continuing producing the antibody, which like the lymphocytes, now begins to circulate freely in the blood.

What happens when the body is exposed to the antigen for a second time, perhaps a matter of months after the first exposure? The antibodies produced by the initial reaction are still available; the body 'remembers' the antigen—hence the characteristic of memory. It is believed that antigenic memory resides in a portion of the lymphocyte population. The memory is highly specific, that is the reaction takes place only with an antigen identical to the one which stimulated the formation of the antibodies in the first place—hence the characteristic feature of specificity. The antibodies react only with an antigen which is identical to that which provoked their formation. Thus antigen and antibody fit together like lock and key. This second entry leads to an immediate combination of the antigens with the antibodies, a combination called the complex, which is associated with the following elements of the inflammatory processes:

(1) The apparent rendering sticky of the antigen. If the antigen is a particulate the acquired stickiness will cause particulates to clump together, apparently making them easier prey for the macrophages and microphages. This stickiness also causes the particulates to adhere to the macrophage, hence facilitating the engulfing process.

(2) The complex appears to underlie the chemotaxis, which is fundamental to phagocytosis.

(3) In certain circumstances, in combination with other factors, the antibody can exercise a destructive effect on the antigen.

(4) The destructive effect of the antibody on the antigen can liberate substances which themselves provoke the inflammatory reaction.

(5) In combining with the antigen the antibody can reduce the antigen's potential for biological activity.

(6) Antigen and antibody combined together in a complex may then combine with further proteins—known as complement—which play a part in facilitating phagocytosis, neutralization or destruction of antigen. The antigen–antibody reaction can occur to an excessive degree. Certain antigens are harmful to certain people because of the violence of the antigen–antibody reaction which they provoke. Harmful excessive reactions are considered further in Chapter 14, which presents a classification of disadvantageous (that is, harmful) reactions.

Manifestly, the body's own cells should not act as antigens causing the formation of 'anti-self' antibodies. In fact, *self-recognition* (the third defining characteristic of immune response) can become disordered but consideration of this topic is beyond this book's scope.

CONCLUSIONS

Inflammatory response and immune response are vital defence processes. Both, however, can be the locus of disease. If the inflammatory response subsides reasonably quickly all is well; if it persists then health of the individual will be undermined. The immune response is responsible for safeguarding the body against invasion by a wide range of micropredators. There are, however, important instances of the immune response acquiring harmful characteristics.

BIBLIOGRAPHY

BURNET, M. (1970). *Self and Not-Self*, Melbourne: University Press.
WEIR, D. M. (1973). *Immunology for Undergraduates*, 3rd Edition, Edinburgh: Churchill Livingstone.

Chapter 6

Stress, Stress Responses and Resistance, and Homeostasis

OBJECTIVES OF THE CHAPTER

This chapter:
  (1) outlines the role of homeostatic mechanisms and identifies some of the occupationally significant inputs likely to activate homeostatic mechanisms,
  (2) identifies difficulties in the definition of stress,
  (3) identifies a definition of stress which links it with cortisol secretion, and
  (4) shows that, for cortisol-linked stressors at least, there is a plausible, though as yet unproven, mechanism whereby stress might affect the pathogenicity of chemical substances.

INTRODUCTION

'Stress' differs from other inputs/modes of action discussed in this book: unlike the others it is difficult to identify let alone quantify. Stress is a word which is widely used and it has many meanings. Of relevance here are those concepts of stress which relate to inputs and modes of action of occupational significance. Two examples, both from 1976, illustrate the use of 'stress', and 'stress response':

(1) the American Conference of Governmental Industrial Hygienists (ACGIH, 1976) stated in the preface to their list of Threshold Limit Values (see Chapter 24) that physical factors such as heat, ultra-violet and ionizing radiation, humidity, abnormal pressure (altitude) may place *added stress* on the body so that effects from exposure to chemical substances may be increased.

(2) Kryter (1976) in discussing the effects of noise on people, included the conclusion that:

> Autonomic system stress responses could conceivably be a contributing factor to ill-health in some persons as the result of noise in their living environment directly interfering with auditory communications and sleep, and, thereby, creating the feelings of annoyance and anger that serve as the direct cause of the stress responses.

STRESS AND RESPONSES TO STRESS

There are important implications in both the examples quoted. The ACGIH is drawing attention to the point that certain energies and climatic inputs may interact with chemical substances in such a way that the harmful effects of the chemical substances are increased. Kryter concluded that peoples' feelings towards noise as a factor in the external milieu could conceivably lead to disease. Both views are in line with the 'common sense' view of stress. But the problem with the common sense view is that there is no firm scientific basis for wholly accepting it although some experimental evidence points in its direction. There are still important gaps in knowledge and, in certain areas, the evidence is contradictory. For example, an editorial in the *British Medical Journal* (1977), discussing stress and heart disease, wrote that although 'stress' is widely believed to cause coronary heart disease, the evidence put forward to support the belief cannot be accepted uncritically and neither is there any clear definition of stress (the editorial, by repeated use of quotation marks for the word 'stress', showed quite clearly the belief that the very concept of stress is a matter for scepticism).

While it is certainly true that there are problems with the definition of stress, scrutiny of research findings in certain fields, at least, supports the view that stress is a concept worthy of further consideration.

Part of the problem with definitions arises because the concept is applied in at least four quite distinct areas of research, and all with different meanings. In the field of psychiatry, 'stress' and 'stressful situations' are equated with those capable of arousing anxiety in individuals. Thus a stress reaction in psychiatrists' conceptualization is related to or identical with anxiety. (For discussions of this aspect of stress see Lazarus, 1966 and 1976; Levitt, 1968.)

In the fields of physiological and endocrinological research, especially where animals are used for experimental purposes, stress is conceived in more traumatic terms. Three examples taken more or less at random from reported studies illustrate the point: Greer *et al.* (1970) reported studies into traumatic stress in which the stimulus was a 'leg break' performed on experimental animals; Cote and Yasumura (1975) reported a study on stress in immature rats in which histamine was used in 'stressful' doses (histamine causes capillary dilation and increased capillary permeability). Even where data from human subjects are used, the stressful stimuli used in endocrinological research tend to be extreme. For example, Yalow *et al.* (1969) studied the effects of a variety of stressful stimuli including insulin-induced hypoglycaemia (reduced blood-sugar), electroconvulsive shock therapy and surgical procedures (unspecified). At the other extreme of the scale of severity are to be found researches in experimental psychology. For example, Simpson *et al.* (1974) used white noise as a stimulus, labelling 80 dB(A) as 'stressful' and 50 dB(A) as 'non-stressful' noise. Other stressful stimuli commonly used in psychological researches include difficult mental tasks of various kinds.

In the field of accident research stress has been equated with emotional distress and related to disturbing events of life (see, for example, Selzer and Vinokur, 1974).

As is apparent from the studies just cited, there is variation from context to context in what authorities regard as stressful. Although the pioneer of stress research, Hans Selye, put forward a unifying hypothesis through which all forms of stress could be linked, it is doubtful whether he visualized such a wide range of stimuli being used under the heading of stress. Certainly, since Selye's first report of the unifying theory in 1936, research has produced some evidence of unifying factors. But, as yet, no single process, or group of processes, has been identified to link the entire range of stimuli at present labelled stress.

Nelson (1976) has summarized an important area of research, pointing to the uniqueness of the adrenal cortex in the endocrine system. It alone produces hormones which are necessary for life, and in accelerating quantities in response to, and as protection against, a variety of stimuli of a stressful nature. Significantly, excessive secretion by the adrenal cortex for a prolonged period causes harm to physiological mechanisms. The adrenal cortex can be labelled, justifiably, the organ of 'stress response'.

The stresses associated with adrenal cortical secretion include lowered blood-sugar, damage to tissues, and oxygen deficiency (anoxia). Within a few minutes of exposure to any one of these, the hormone released from the pituitary gland called adrenocorticotrophic hormone (ACTH) rises sharply. The effect of ACTH on the adrenal cortex is to increase the secretion of hormones from the adrenal cortex, notably a specific hormone called cortisol. Vander *et al.* (1970) described the principal functions of cortisol in relation to stress. Most, but not all, of the actions of cortisol appear to fit the hypothesis that it increases the body's resistance to stress in the short term. These actions include:

(1) antagonism to insulin—insulin reduces blood-sugar, thus antagonism of insulin maintains or increases blood-sugar;

(2) stimulation of protein catabolism—protein breakdown will liberate energy-containing substances to fuel body metabolism; and

(3) stimulation of gluconeogenesis—the formation of glucose from 'reserve' sources.

Cortisol increases the ability of skeletal muscle to maintain prolonged contraction, and it delays the onset of fatigue. Cortisol has a complex action in resisting that effect of certain types of stress which induces widespread dilation of small arteries. Unless this dilation is resisted, the body rapidly goes into circulatory collapse. It appears that increased secretion of cortisol is necessary in order to combat this effect.

The stress-adaptive roles of certain other effects of cortisol are not so readily explicable; for example, cortisol inhibits inflammation by preventing neutrophil emigration and it impairs wound healing.

In the long term, excessive formation of cortisol may block the formation of antibodies and reduce the synthesis of structural body proteins. Excessive cortisol is known to stimulate abnormal mental activity but it is not clearly established whether this effect occurs in connection with stress-induced over-secretion of cortisol.

THE EMERGING CONCEPTS OF STRESS

The studies cited, and others, lead to the concept that stress is any input or combination of inputs the mode of action of which is to increase the secretion of cortisol. A number of factors have been identified with this capability, but what is not yet known is the extent to which the comparatively mild 'stresses' such as those used in psychological research are also capable of provoking increased secretion of cortisol. If it turned out that the mild stimuli were not cortisol-linked this would not prove that such stimuli were without effects. Nevertheless, a distinction might usefully be drawn between cortisol-linked and non-cortisol-linked stressors.

The link with cortisol fits in with one of the central concepts in this book, namely that defensive processes called upon to act to excess or for prolonged periods may themselves become harmful. Such is certainly the case with cortisol of which prolonged, excessive secretion has damaging effects. The effect of cortisol in excess on people simultaneously exposed to harmful inputs, such as chemical substances, unrelated to the cortisol-provoking stimulus is a matter for speculation and for research. However, there is plainly a possibility that cortisol-linked stimuli might undermine the body's medium and long-term resistance to pathogenicity by chemical substances.

As has been pointed out, it is by no means certain that all so-called stressors necessarily induce the excessive secretion of cortisol.

Whilst it is useful to link the concept of stress with cortisol secretion it is also necessary to consider other inputs which, though they may not activate cortisol secretion, nevertheless call into play the body's *homeostasis*, because the function, or malfunction, of these mechanisms is also a matter of importance in human safety.

HOMEOSTASIS

Homeostasis can be regarded as the actions of the restorative processes which maintain the constancy of the body's internal milieu. Numerous inputs of occupational significance will tend to alter the internal milieu. The absence of any alteration despite a measurable 'dose' of input is indicative of the effective operation of one or more of the homeostatic mechanisms. It is a matter of widely held belief in physiology that, within limits, the activation of homeostatic mechanisms is not, *per se*, harmful. But it is also recognized that the restorative capabilities of the homeostatic mechanisms are not infinite. That is to say there will be magnitudes of inputs which overload the homeostatic mechanisms.

Stress is sometimes used to describe the input (or its mode of action) when the magnitude is such that homeostatic mechanisms are overburdened. Whether the overburdening is invariably associated with excessive production of cortisol is not clear from present knowledge. It is reasonable, however, to assume that some (possibly most) of the inputs which overburden homeostatic mechanisms are those which also stimulate excessive production of cortisol. This assumption raises the hypothesis that it is the very overburdening of the homeostatic mechanisms which acts as the stimulus for cortisol production. Exploration of that hypothesis, interesting challenge though it is, is beyond the scope of this book.

Homeostatic mechanisms are described in connection with a host of physiological variables including blood pressure, blood flow to various organs, volume of body fluid, blood temperature, blood pH, blood and tissue fluid osmotic pressure, and oxygen and carbon dioxide concentrations in circulating fluid. It is clear that by one or more mechanisms a range of inputs originating in the external milieu are likely to initiate the homeostatic mechanisms for these variables, whether separately or collectively. A high air temperature, for example, may initiate the homeostatic mechanisms to temperature. High humidity activates homeostasis because it affects the heat loss from the body. Thermal homeostatic mechanisms are also likely to be activated by radiations of various kinds because these are often transformed so as to produce a rise in tissue temperature (see Chapter

17). An abnormal atmospheric pressure may affect the gas exchange between the lungs and the external milieu and thus initiate relevant homeostatic mechanisms.

On the input–output model of Chapter 2, most if not all energies and climates, including psycho–social climate can be considered as possible activators of one or more of the body's homeostatic mechanisms.

It is convenient to regard the body fluids, especially blood and tissue fluids, as being the milieu of which the constancy is maintained. Thus, the function of homeostatic mechanisms is to be seen as the maintenance of the stability of the body fluids.

In Chapter 11 transport systems are discussed and diffusion is mentioned as a process of transport. Diffusion relies on the existence of a gradient such that molecules or ions pass from areas of high to low concentration. Thus, within the overall stability of the body's internal fluid milieu, gradients must exist naturally. These gradients exist, however, between the interiors of cells and the fluid which surrounds them. The fluid represents the fixed point; the gradient exists because the interior of the cell differs in some physical or chemical respect from the fluid which surrounds it. The body cells live by commerce with the tissue fluid which surrounds them, aided by gradients in one direction or the other.

In discussing the 'environmental' problems of cells functioning in the human body, Robinson (1975) drew a comparison with free-living unicellular organisms inhabiting an ocean. He likened their world to one in which the physical, and chemical properties vary very little; the organisms' own activities have no effect on these properties. Living cells in the human body, about 30 kg of them, inhabit an 'ocean' of about 20 litres. Unlike real ocean-dwelling organisms, theirs is a desparately overcrowded world, highly susceptible to change induced by their activities, and by the influence of factors outside the body. As Robinson commented, it is not surprising there are numerous mechanisms cooperating to maintain the stability of the internal milieu of the human body.

In the next chapter thermoregulation is outlined; this is a clear example of a homeostatic mechanism. By this means the temperature of the blood supply to vital organs is maintained constant against rise or fall. Within certain ranges of air temperature and humidity the body copes with constant or varying exposure, repeatedly, without adverse effect—or not to an extent which is apparent. Exposure beyond those ranges, but within certain limits, can be coped with by the body for a while but eventually harm results until the exposure is discontinued or other steps are taken. Where the limits are exceeded, damage is inevitable. The pattern of response to adverse thermal environment just outlined is consistent with the presence of homeo-static mechanisms able to maintain the internal milieu against fluctua-tions in the external milieu indefinitely, so long as the external fluctuations are within certain bounds. Beyond these bounds homeostatic mechanisms can cope with short exposures but damage results if they are called upon to operate for too long. Beyond certain limits, the homeostatic mechanisms cannot cope at all.

Homeostatic mechanisms involve the nervous system and the endocrine system as instrumental parts of the control processes of homeostasis. Indeed, homeostasis is one of the central subjects of physiology and as such embraces some of the most complex areas of physiology. What follows is strictly an outline, therefore, of a vast area of knowledge which in any case is the subject of intensive research.

Broadly speaking, homeostatic mechanisms comprise three neces-sary mechanisms:

(1) *Sensors*—to detect beginnings of the departure from stability of a physiological variable in the internal milieu. The sensors are nor-mally highly specific.

(2) *Effectors*—these are organs or systems, such as muscles, blood vessels, and glands which by their action effect the changes required for re-stabilization.

(3) *Integrating mechanisms*—these link sensors to effectors, and integrate or co-ordinate signals from sensors for a range of variables so that the signals relayed to the effectors are appropriate.

Integration is carried out in the brain. The focal point for the signals from all sensors is a part of the brain called the hypothalamus. This receives signals from the other parts such as the specialized organs of sensory reception (eyes and ears for example). Also received by the hypothalamus are signals from those parts of the brain in which the emotions are centred. From the hypothalamus, signals may pass via the autonomic nervous system to the effector organs. Responses mediated in this way will normally be very rapid. Somewhat slower are the endocrinal (hormonal) responses also in-itiated by the hypothalamus, but mediated by a different route.

The hypothalamus is linked by short blood vessels with the pitui-tary gland situated close to it. The pituitary gland liberates certain hormones which act on further glands such as the adrenal cortex, which in turn produce hormones which influence the behaviour of the effectors.

CONCLUSIONS

When homeostatic mechanisms are overburdened, harm is likely to result. In several instances, overburdening of the homeostatic mechanisms by an input will be associated with excessive secretion of cortisol. Cortisol, at first, increases the body's resistance but beyond a certain point, excess of cortisol becomes harmful. Thus, we have an example where a defensive process can itself, after prolonged action, become harmful.

The point made in the previous paragraph appears to be reasonably well established; rather more speculative is the effect of cortisol in excess on people simultaneously exposed to harmful inputs such as chemical substances. Circumstances could be visualized where an individual is exposed to chemical substances and simultaneously exposed to a cortisol-provoking input of, say, climatic origin. Plainly, the possibility exists that the cortisol-linked input might act to undermine the body's medium and long-term resistance to the chemical substance, thereby increasing its pathogenicity. Here, of course, we have the very point expressed by the ACGIH preface, outlined at the start of this chapter. Certainly, several hypotheses for research could be put forward linking stress, stress responses, resistance to stress, homeostasis, and pathogenicity of chemical substances—but not much can be advanced by way of established knowledge about these relations.

REFERENCES

ACGIH (1976). American Conference of Governmental Industrial Hygienists TLVs: *Threshold Limit Values for Chemical Substances and Physical Agents in the Work Room Environment with Intended Changes for 1976*, Cincinnati: ACGIH.

British Medical Journal, editorial (1977). Stress, coronary disease and platelets behaviour, **1**, 408.

COTE, T. E. and YASUMURA, S. (1975). Effect of ACTH and histamine stress on serum corticosterone and adrenal cyclic AMP levels in immature rats, *Endocrinology*, **96**, 1044–7.

GREER, M. A., ALLEN, C. F., GIBBS, F. P. and GULLICKSON, C. (1970). Pathways at the hypothalamic level through which traumatic stress activates ACTH secretion, *Endocrinology*, **86**, 1404–9.

KRYTER, K. D. (1976). Extra-auditory effects of noise, in *Effects of Noise on Hearing*, D. Henderson, R. P. Hamernik, D. S. Dosanjh and J. H. Mills, eds, New York: Raven Press.

LAZURUS, R. S. (1966). *Psychological Stress and the Coping Process*, New York: McGraw-Hill.

LAZARUS, R. S. (1976). *Pattern of Adjustment*, New York: McGraw-Hill.

LEVITT, E. E. (1968). *The Psychology of Anxiety*, London: Staples Press.

NELSON, D. H. (1976). Regulatory mechanisms of the pituitary and pituitary–adrenal axis, in *Pathophysiology*, E. D. Frohlich, ed., Philadelphia: Lippincott.

ROBINSON, J. R. (1975). *A Prelude to Physiology*, Oxford: Blackwell.

SELYE, H. (1936). Thymus and adrenals in the response to the organism to injuries and intoxications, *British Journal of Experimental Pathology*, **17**, 234–48.

SELZER, M. L. and VINOKUR, A. (1974). Life events, subjective stress, and traffic accidents, *American Journal of Psychiatry*, **131**, 903–6.

SIMPSON, G. C., COX, T. and ROTHSCHILD, D. R. (1974). The effect of noise stress on blood glucose level and skill performance, *Ergonomics*, **17**, 481–7.

VANDER, A. J., SHERMAN, J. H. and LUSIANO, D. S. (1970). *Human Physiology: The Mechanisms of Body Function*, New York: McGraw-Hill.

YALOW, R. S., VARSANO-AHARON, N., ECHEMENDIA, E. and BERSON, S. A. (1969). HGH and ACTH secretory responses to stress, *Hormone Metabolism Research*, **1**, 3–8.

Chapter 7

Thermoregulation

OBJECTIVES OF THE CHAPTER

This chapter:
  (1) outlines the biological need for thermal stability in key areas of
      the body by inference from effects of heating and cooling on
      living tissues,
  (2) identifies disorders of thermoregulation and links these with the
      concept that excessive responses on the part of defensive
      processes can be harmful,
  (3) outlines thermoregulatory mechanisms, including routes by
      which the body gains and loses heat, relating these mechanisms
      to homeostasis as discussed in the previous chapter,
  (4) considers briefly the biological phenomena which standards for
      hot and cold working milieux must contend with,
  (5) identifies effects of variation of blood flow in skin,
  (6) presents key data relating to human 'thermal' biology, and
  (7) mentions one example of a limit intended to safeguard people at
      work from heat stress.

EFFECTS OF HEAT

The body's vital biochemical processes are sensitive to changes in
temperature. These processes produce heat; and relatively small rises
in temperature increase their speed of reaction. Hence, living cells
cannot withstand large rises in temperature because a rise in temperature causes loss of control over the biochemical processes. In the
event of a temperature rise occurring, enzymes may be inhibited or,
as a more severe effect, inactivated permanently. Enzymes are proteins and the effect of temperature rise on proteins is to cause them to
become denatured. This is an irreversible change familiar as the
difference between hard and soft-boiled eggs.

Cellular biochemistry is sensitive to decreases in temperature,
which cause it to slow down. The slowing down of cellular biochemistry is innately less damaging than its speeding up. Unless the
temperature falls so low that ice crystals form within cells, the
changes from 'moderate' cold are largely or wholly reversible. This

explains why, at the cellular level, rises in temperature are essentially more damaging than decreases.

Changes in cell biochemistry are likely to have the most generally disturbing effects on highly specialized cells such as those of brain or liver. Thus the existence of a thermoregulatory mechanism could be predicted. This is required to maintain the operating temperature of cells, the normal biochemical functioning of which is vital.

Heavy work in hot, moist external milieux could affect the temperature of cells in key tissues if there was no defensive process of thermoregulation. But harm can also result from prolonged activity of the thermoregulatory defence. Disorder of thermoregulation underlies the following observation made by J. B. S. Haldane in 1927:

> Perhaps the hottest place in England is about a mile underground in a well-known Lancashire coal-pit, where the miners work in boots and bathing-drawers, and empty the sweat from their boots at lunch. One man sweated eighteen pounds in the course of the shift, and it is probable that even this figure had been exceeded. This sweat contained about an ounce of salt—twice what the average man consumes in all forms per day. The salt loss was instinctively made up above ground by means of bacon, kippers, salted beer, and the like. And as long as they did not drink more than a quart of water under ground, no harm came to the miners. But a man who has sweated nearly two gallons is thirsty, and coal-dust dries the throat, so this amount was often exceeded and the excess occasionally led to appalling attacks of cramp, often in the stomach, but sometimes in the limbs or back. The victims had taken more water than was needed to adjust the salt concentration in their blood, and the diversion of blood from their kidneys to their muscles and skin was so great that they were unable to excrete the excess. The miners in question were offered a solution of salt in water which was of about the composition of sweat, and would be somewhat unappetising to the average man. They drank it by quarts and asked for more. And now that it has become their regular beverage under ground there is no more cramp and far less fatigue. It is almost certain that the cramp of stokers, and of iron and glass workers, which is known to be due to excessive water-drinking, could be prevented in the same way.

It is now recognized that there are three categories of disorder of the thermoregulatory processes:

(1) heat cramps (as described by Haldane) caused by disturbances

in the water–electrolyte balance. Heat cramps result from excessive intake of water, combined with an inadequate intake of sodium chloride to make up the balance. The decline in the blood's osmotic pressure in relation to muscle cells causes the various symptoms. The condition, as Haldane pointed out, responds readily to correction by increased salt intake.

There is an important acclimatization factor. People doing hard work in hot climates for the first time lose salt in their sweat in the same concentration as they would in temperate climates. However, the acclimatization results in relatively less salt being lost in sweat. The adapted state is marked by ready sweating without excessive salt (electrolyte) loss.

(2) Heat exhaustion, a condition where the individual is totally or almost totally incapacitated by heat. This is a potentially serious condition which, if not treated, can result in death. It is to be distinguished from the torpor and lassitude induced as a normal reaction to relatively high temperatures in the external milieu.

Heat exhaustion results from loss of blood volume. This results when fluid intake is not commensurate with the sweating induced by the body's need to lose heat. In order to prevent the electrolyte concentration rising in the blood as the water from it is lost, the body's homeostatic mechanisms cause electrolytes to be dumped in the sweat or urine, as in heat cramps. Osmolarity is more or less maintained even where fluid intake and not only electrolyte intake are inadequate. The dumped electrolytes take with them, so to speak, a certain amount of water additional to that being lost by sweating. The net result is progressive dehydration resulting in a severe reduction in the total volume of fluid available for circulation in the blood.

The effect of the loss of circulating blood volume will ultimately be the same as that induced by direct loss of blood volume by haemorrhage—circulatory failure. Heat exhaustion is readily corrected by adequate fluid and electrolyte intake, death only occurring where the victim does not receive these in sufficient time to avoid the irreversible effects of circulatory failure.

(3) Heat stroke is the most acute of the three conditions. This is regarded as evidence of a failure in the brain's control over thermoregulation (see later); the nature of the failure is not precisely understood. It is as if the body's thermostat had failed so that control over body temperature is lost, resulting in core temperatures in the region of 42°C (core temperature is normally 37°C). Once this failure is established, the body temperature spirals upwards and death follows rapidly.

EFFECTS OF COLD

The biochemical processes in the key tissues slow down with a decrease in operating temperature. Prolonged exposure of brain cells to temperatures below about 21°C is fatal because the biochemical processes have been reduced to a rate insufficient to maintain vital functions.

Trench-foot and frost-bite can be regarded as local failures in the thermoregulatory mechanisms of the tissues. Trench-foot results from prolonged exposure to cold—at a temperature above the freezing point of water. The damage occurs to the arteries of the affected part and to the tissues of that part generally. Both are affected by the prolonged oxygen deficiency which results from the local shutting down of circulation, a change in the thermoregulatory mechanism discussed below.

When cells are cooled sufficiently, ice crystals form within the cells' contents and disrupt their structure (by destroying membranes, for example) resulting in the cells' death. When the tissues containing the dead cells are rewarmed, bacterial micropredators invade. If the cell death is very extensive there will also be disruption of the blood supply to the tissues and consequently severe disruption of macrophage and microphage defensive activities. The disablement of the defences allows the micropredators to flourish once rewarming has taken place; the result is gangrene.

In some respects frost-bite and burns are similar. Burns, resulting from the application of very high temperature to living tissues, represent the death and destruction of cells within the tissues and disruption of the blood supply bringing impairment of defensive processes. The dead tissue provides an excellent culture medium for invading micropredators.

Thermoregulatory mechanisms cannot cope with local temperature changes of great magnitude, so cell death and tissue destruction result.

THERMOREGULATORY MECHANISMS

Man belongs to the homeothermic group of animals. These animals have homeostatic mechanisms the duty of which is to maintain the internal temperature at key points against variations in heat input and output. Such variations result from changes in internal processes as well as the external milieu. All biochemical processes involve the production of heat and, within limits, the speeding up of biochemical processes results in increased production of heat. Moreover, an increase in temperature results in the speeding up of biochemical processes. These three conditions acting together would produce a positive feedback loop, resulting in a runaway biochemical reaction and self-destruction. Clearly, homeotherms must have elaborate thermoregulatory mechanisms.

Other members of the animal and plant kingdoms are poikilothermic. These organisms have less elaborate mechanisms to maintain temperature control in tissues. Consequently, the tissues ultimately stabilize near the temperature of the external milieu (although some poikilotherms do, in fact, store heat). The rate of biological activity varies accordingly.

Figure 7.1 outlines the interconnections of the thermoregulatory processes. The heat output is adjusted by the various regulatory processes so that the input of heat does not result in temperature rise in sensitive tissues. Heat inputs may come from external heat received by radiation and conduction. However, the contribution of these is normally small compared with the internal heat generated by biochemical processes. Thermoregulation appears to be 'designed' to combat biochemical rather than external heat. It should be emphasized that all biochemical processes produce some heat, but certain mechanisms (such as those occurring in muscle activity and liver) produce large quantities of heat.

The blood acts as both a transport and a storage medium for heat. The body's 'thermostat' is situated in the hypothalamus which, though it contains more than one nerve centre concerned with thermoregulation, for present purposes can be regarded as a single centre. The sensing receptors are in the hypothalamus itself, possibly in other organs and in the skin. The hypothalamic centres stimulate certain responses when the input indicates 'too hot', and other responses when the indication is 'too cold'. In diseases accompanied by fever substances known as pyrogens are formed in tissues. Pyrogens disturb the setting, so to speak, of the thermostat.

The hypothalamic thermostat is widely accepted as a concept but there is discussion about its functional characteristics (Bligh, 1973 and 1975). One school of thought holds that the thermostat compares signals from thermal sensors with a reference or set-point signal; this is the set-point theory. Another school of thought holds that the thermostat is the controlling element in a servo-mechanical system maintaining the balance between heat production and heat loss. Current observations appear to favour the set-point rather than the servo-mechanical theory. Certainly, the set-point theory is more readily reconciled with the concept that the thermoregulatory homeostasis is aimed at maintaining the constancy of temperature *locally* in the tissues. Obviously, if heat production and heat loss are

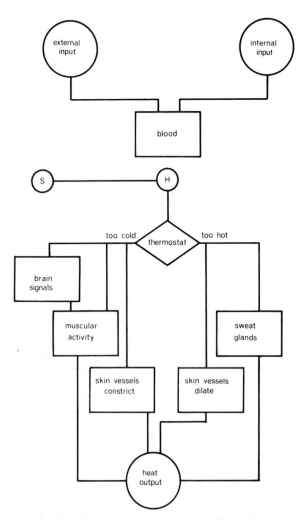

FIG. 7.1. The interconnections of the thermoregulatory processes. The blood acts as a transport system as well as a storage compartment for body heat; heat is gained by the body, and therefore the blood, from inputs from the external milieu, and inputs from the internal milieu arising from biochemical processes. Blood temperature is sensed by the hypothalamic thermostat (H); there is also sensing of the skin at (S). The hypothalamic thermostat sends a 'too hot' or a 'too cold' signal to the various organs which respond accordingly. The responses result in an increase or decrease in heat output from part or whole of the body.

not balanced body temperature will rise; but an overall balance between heat production and heat loss does not preclude the possibility of excessive temperature *locally* in tissues. So, even if the servo-mechanical theory is accepted, it would still be necessary to consider a set-point hypothesis for local thermal homeostasis.

It will be recalled from the previous chapter that homeostatic mechanisms, typically, comprise three elements: sensors, effectors, and integrating mechanisms. In Fig. 7.1 the sensors are situated in the hypothalamus and also in the skin (consideration of the possibilities of other sensors is not within this chapter's scope); the effectors bring about the changes identified in the figure, for example, shivering and sweating; and the integrating mechanism can be considered to lie within the hypothalamus itself.

The 'too hot' responses consist mainly of dilatation of the arterioles of the skin and stimulation of the sweat glands. The effect of the arteriolar dilatation is to increase the quantity of blood close to the surface of the body, the result of which is an increase in heat loss from the blood to the external milieu. The dilatation of the skin arterioles is visible in the flushing of the skin and in the warm glow which exercise produces.

The amount of heat lost by this route is dependent on the temperature gradient, which will be zero when the temperature of the external milieu matches that of the skin. No heat transfer can then take place. Under these circumstances heat can only be lost by means of evaporation. Evaporation takes place in the lung and as a result 'insensible' heat loss also occurs in the lung, a loss which clearly is increased as the respiratory rate rises. Respiratory evaporative cooling is, however, less important than that of the skin. The many sweat glands in the skin are capable of producing large quantities of sweat. Cooling of the skin results from the sweat's evaporation, since sweat dripping off the body in fluid form removes only a little heat. Hence, the humidity of the external milieu is the all-important factor. If the humidity is high, evaporative cooling is relatively small. Consequently, a large quantity of sweat is produced and heat cramps and other heat-induced conditions often result. The problems identified by Haldane in the Lancashire miners were exacerbated by the high humidity as well as by the high temperature of the milieu in which they worked.

Standards aimed at the thermal aspect of the external milieu therefore have to take into consideration heat input–output from radiation, convection and conduction. Other factors to be considered include the possibilities of heat loss by evaporation—controlled by relative humidity and air movement—and the amount of internal heat generated, which is largely determined by the activity of the individual in the external milieu. It is also clear that thermal standards should be different for acclimatized and non-acclimatized subjects. Appendix 7.1 identifies an example of a thermal environmental standard, taken from ACGIH (1976).

When the input to the hypothalamus indicates 'too cold', involuntary activity is set up in the skeletal muscles, familiar as shivering. There is also increased voluntary activity initiated by the cerebral cortex (see Fig. 7.1) which takes the form of such activities as hand clapping and foot stamping. These behavioural mechanisms are of overwhelming importance in non-occupational settings, but occupation may limit opportunities for behavioural responses to too hot or too cold. There is also involuntary activity in which the skin arterioles are caused to constrict, reducing the heat exchange with the external milieu and shutting down sweating.

In cold conditions the arteriolar constriction in the skin ensures the minimum loss of core heat although the skin itself cools. There can be a substantial temperature gradient between the core and the skin, but there can also be a temperature gradient from the trunk to the tips of the limbs. This is possible because the arteries to the limbs have veins lying closely alongside them. Heat being carried away from the centre in the arterial blood warms the venous blood returning from the periphery. This explains why hands and feet can be chilled 'all through' while the core temperature remains within the correct operating limits. It also explains how trench-foot and frost-bite can occur, since the heat bypass resulting from the transfer of heat between the arteries and veins deprives the periphery of the limbs of the heat. The associated shut-down of peripheral circulation results in lack of oxygen in the peripheral parts of the tissues, and their being cooled to a point where the biochemical activity of the cells is insufficient to maintain vital functions.

Table 7.1 sets out certain data relating to human heat production and loss. These data are useful in that they give an indication of the capability of the thermoregulatory processes.

A recurring theme throughout human safety is that defensive processes operating under extreme conditions may themselves become the cause of disease and injury. Just such an example is seen in the thermal bypass which deprives the periphery of vital heat. Another example is seen in milieux in which the 'too hot' mechanisms are extensively activated. Under these conditions there will be a considerable increase in blood flow through the skin compared with normal. This extra flow is in parallel with the ordinary circulatory requirements and it must impose extra pumping requirements on the heart if muscles and other organs are not to be starved of blood. Thus hard work in hot milieux leads to greater demands on the heart.

TABLE 7.1

Base for metabolic heat production, per hour for a 70 kg man	300 kJ
Heat output per m² (total surface area of body about 1.8 m² specific heat of body approximately same as water)	167 kJ
Latent heat of vaporization	2.4 kJ per ml of water
Evaporation from skin in a temperate climate per day (this is different from sweating in that it represents simply the evaporation of the skin's moisture)	900 ml
Evaporation of water from lungs per day	400 ml
Sweating maximum	12 litres per day
Temperature gradient core to big toe in unclothed subject with limb exposed to temperature of 19°C—and without effects of air movement or humidity:	
core temperature	37°C
big toe temperature	20.5°C
skin temperature at groin	34.2°C
skin temperature at knee	27.2°C
skin temperature at ankle	25.2°C

Conditions which make heat loss from the skin ineffective cause increased skin circulation and correspondingly increased loads on the heart. The end of this process is complete circulatory failure. Manifestly the limits up to which the circulation can cope will be determined by the age and fitness of the individual, together with all the factors governing the loss of heat from the body.

A number of authorities speak of heat stress and heat strain. Stress represents the degree of applied thermal disadvantage to the individual and strain represents the response—in some instances pathogenicity—to the thermal stress. Both are useful terms, but I have avoided them in order that stress may be used in its broader meaning as explained in the previous chapter. Excessive heat applied by whatever mode of input may cause extra production of cortisol, as may excessive cold. However, it is not clear on present knowledge whether excessive heat and cold always produce cortisol-linked stress. Because of this uncertainty the term 'heat stress' is, in my opinion, best avoided unless the mode of action involved is clearly understood and stated.

This chapter has been concerned with defence processes against excessive thermal inputs. Although there is very little evidence upon which to draw, it is commonly supposed that heat illness of any kind is uncommon in temperate climates. Much more common are complaints about thermal discomfort. It is likely that the sensations which monitor thermal comfort/discomfort come primarily from the thermal receptors in the skin. There is a tendency to use the same systems of measurement for thermal comfort/discomfort as are used for the prediction of heat overload and heat illness, a parallelism which is, to a point, appropriate. At root, the approach to assessment of situations where the defence processes are brought into play is, however, quite different from that used to assess and predict the sensation of comfort and acceptability.

APPENDIX 7.1: AMERICAN CONFERENCE OF GOVERNMENTAL INDUSTRIAL HYGIENISTS— THRESHOLD LIMIT VALUES FOR HEAT STRESS

The ACGIH Threshold Limit Values (1976) in relation to heat stress condition is expressed in terms of the wet bulb–globe temperature index (WBGT), considered by ACGIH to be the most appropriate technique currently available.

The TLVs rely on the assumption that nearly all workers who are acclimatized, fully clothed, and who have adequate water and salt intake should be able to function effectively under the working conditions specified without their core temperatures exceeding 38°C. Methods for calculating WBGT values depend on the presence or absence of solar (radiant) energy.

In outdoor conditions with solar energy load, the following expression is used:

$$WBGT = 0.7WB + 0.2GT + 0.1DB$$

and for indoor or outdoor conditions without solar energy, the following expression is used:

$$WBGT = 0.7WB + 0.3GT$$

where: WBGT = wet bulb–globe temperature index, WB = natural wet-bulb temperature, DB = dry-bulb temperature, GT = globe thermometer temperature. The instrumentation for measuring WBGT includes a black globe thermometer, a wet-bulb thermometer, and a dry-bulb thermometer.

Under the ACGIH TLVs for heat exposure, values exceeding those shown in Table 7.A1 are permitted if the workers have been undergoing medical surveillance and it has been established that they are

TABLE 7.A1
ACGIH THRESHOLD LIMIT VALUES FOR PERMISSIBLE HEAT EXPOSURE
(VALUES IN DEGREES C WBGT)

| | Work load | | |
Work Pattern	Light	Moderate	Heavy
Continuous work	30.0	26.7	25.0
Each hour 75% work/25% rest	30.6	28.0	25.9
Each hour 50% work/50% rest	31.4	29.4	27.9
Each hour 25% work/75% rest	32.2	31.1	30.0

more tolerant to work in heat than the average worker. However, workers should not be permitted to continue their work when their core temperature exceeds 38°C. Work load is defined in the ACGIH document, which should be consulted for further information.

REFERENCES

AMERICAN CONFERENCE OF GOVERNMENTAL INDUSTRIAL HYGIENISTS (1976). *Threshold Limit Values for Chemical Substances and Physical Agents in the Work Room Environment with Intended Changes for 1976*, Cincinnati: ACGIH.

BLIGH, J. (1973). *Temperature Regulation in Mammals and Other Vertebrates*, Amsterdam: North Holland.

BLIGH, J. (1975). Physiological responses to heat, in *Fundamental and Applied Aspects of Nonionizing Radiation*, S. M. Michaelson, M. W. Miller, R. Magin and E. L. Carstensen, eds, New York: Plenum Press.

HALDANE, J. B. S. (1927). *Possible Worlds and Other Essays*, London: Chatto and Windus.

Chapter 8

Metabolic Transformation and Conjugation; Toxicity and Pathogenicity

OBJECTIVES OF THE CHAPTER

This chapter:
(1) outlines metabolic transformation and conjugation,
(2) distinguishes between toxicity and pathogenicity,
(3) shows how metabolic transformation and conjugation can render non-pathogenic a substance which is otherwise toxic,
(4) shows how metabolic transformation and conjugation cause a relatively non-toxic substance to be pathogenic,
(5) identifies factors which determine whether a toxic substance is pathogenic or not, and
(6) identifies the concept of biochemical pathway.

INTRODUCTION

In this chapter the main concern will be with certain key defence processes which operate at the molecular and sub-cellular level. A glossary has been included at the end of the chapter defining certain of the terms used.

The earliest theories about disease generally were derived from observations of distortions of living tissues. Early observers had limited optical magnification at their disposal. Their theories of disease could be derived only from interpretation of the structural distortions they could observe. The crucial advance was made by Virchow (1859) with his cellular theory. As we have already seen, changes in location, structure, and function of cells can account for disease processes in the occupational as well as the non-occupational sphere. The development of electron microscopy combined with advancing biochemical knowledge, has focused attention at the sub-cellular level. Increasingly, disease processes are being understood at the molecular level. It has been clear for some decades that studies of injury at the sub-cellular and molecular levels promise important progress in the understanding of disease generally. Recent years have seen the development of a new study, pathobiology, in which the knowledge gained in recent years in fields such as molecular biology is applied to problems in abnormal biology and, in particular, the pathology of human disease (see, for example, La Via and Hill, 1975).

Disease (in plants and animals as well as human beings) is being researched increasingly on the basis of hypotheses linking injury, at the sub-cellular level, and disease. There is a movement away from viewing disease as 'spontaneous' disorder of biological processes. The point will be made again in Chapter 16 with specific reference to cancer. In the past cancer was regarded as being largely spontaneous in origin; the current view holds that a high proportion of cancers have environmental causes. Environmental causes produce injuries at the molecular and sub-cellular level.

Cancer apart, a number of occupational diseases can now be explained in terms of injury at the sub-cellular or molecular level. However, as has been stated previously, whether exposure to inputs capable of causing disease actually results in disease depends on a number of factors, including the effectiveness of defence processes. That proposition is true, too, at the molecular and sub-cellular level: whether inputs capable of inflicting damage to cells and their components actually result in disease depends, *inter alia*, upon the effectiveness of the defence processes at the *molecular level and the sub-cellular level.*

TOXICITY AND PATHOGENICITY

At this stage it is useful to distinguish between toxicity and pathogenicity*. Toxicity should be looked upon as a laboratory term useful in the description of the injury-producing characteristics of chemical substances. Various systems have been developed to evaluate the toxicity of chemical substances. There have been classifications based on the sources of toxic substances, whether synthetic or naturally occurring, and functional classifications based on the type of injury most characteristic of the substances. But there are two principal difficulties with this approach.

(a) toxicity of a substance cannot always be predicted from knowledge of its chemical composition and physical properties. Certain generalizations can be made, but there are numerous examples of chemically dissimilar substances producing similar toxic effects. There are equally numerous instances of substances which are seemingly closely related chemically but which have quite different toxicological properties, and

*The definitions used here are not universally accepted. They suffice for the purposes of this book but the reader will find alternative terminology in use elsewhere.

(b) the nature and pattern of disease arising from exposure to a chemical substance, if any, cannot be reliably predicted solely from knowledge of the chemical and physical properties of a substance.

'Pathogenicity' can be used to describe the potential of a chemical substance for producing disease in people exposed at work to that substance. *Whether disease results depends upon the effectiveness of defence processes as well as upon the toxicity of the chemical substance.*

The study of pathogenicity thus involves knowledge of epidemiology, anatomy and physiology, and pathobiology and pathology as well as chemistry and biochemistry. This chapter is concerned—briefly—with one part of the whole subject of pathogenicity: metabolic transformation and conjugation.

METABOLIC TRANSFORMATION AND CONJUGATION

The first step in the understanding of metabolic transformation and conjugation is to identify their location in the biochemical network. Figure 8.1 shows a diagram dealing with nutrients and non-nutrients. Both can be exogenous—that is derived from outside the body. The diet is the major source of nutrients and non-nutrients. The process of ingestion is partially selective and some non-nutrients, such as cellulose fibre, are not absorbed from the gut. However, many non-nutrients are ingested along with nutrients, later to be removed from the body along certain biochemical pathways. Nutrients contribute to anabolism and catabolism (see Appendix 8.1: Glossary), processes which produce *residual* substances of a non-nutrient character. These are removed from the body, generally speaking, along the same biochemical pathways as the ingested non-nutrients. Biochemical pathways, in general, essentially consist of a series of biochemical steps represented by enzyme-mediated reactions.

In the occupational context, nutrients as well as non-nutrients enter the body by means of the various modes of input. Nutrients will join other nutrients on the biochemical pathways. The non-nutrients will follow any pathway capable of coping with them. This underlines an important anxiety which must always be in the minds of chemists and biochemists considering new chemical substances, namely, is there a biochemical pathway capable of dealing with the new substance? If not, then there is a risk of the substances not being removed or, worse still, accumulating in the tissues. Whether the accumulation results in harm will depend on the biological activity of the substance.

According to Parke (1968), only a few chemical substances so far identified appear to have no appropriate biochemical pathways. It is

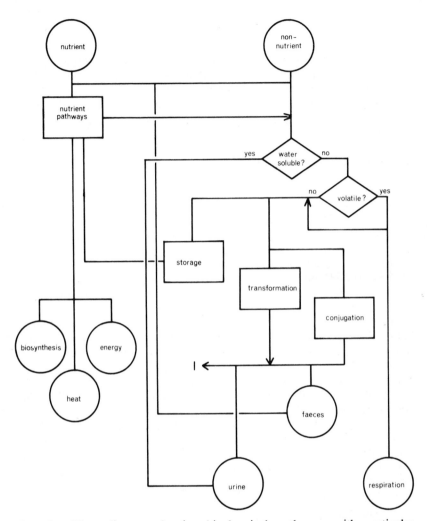

FIG. 8.1. Flow diagram showing biochemical pathways with particular reference to metabolic transformation and conjugation. *Nutrients* and *non-nutrients* represent inputs, both of which can enter the nutrient pathways. The normal outputs of the nutrient pathways are biosynthesis (synthesis of new tissue or repair of old tissue), energy for muscular activity and other energy-consuming body functions, and biochemical heat of reaction. Residual substances from the nutrient pathway enter the non-nutrient pathways.

Water soluble non-nutrients, extrinsic and intrinsic, are bio-dumped in the urine. If not water soluble they may be volatile, in which case bio-dumping takes place by means of respiratory exchange. If not volatile, non-nutrients are acted upon by transformation or conjugation and the resultant substances appear in urine or faeces. Occasionally, conjugation and transformation result in substances which are more rather than less bioactive so that intoxication (I) occurs. Nutrients and non-nutrients both enter storage compartments from which they can, under suitable circumstances, be released thereupon to enter both the nutrient and non-nutrient pathways.

interesting that some of these substances occur naturally, for example the South African plants *Dichapetalum cymosum* and *D. toxicarium*. A further source of anxiety, usefully highlighted at this point, is that one chemical substance may have effects on the biochemical pathways by which *other substances* are inactivated or removed from the body.

From Fig. 8.1 it can be seen that outputs from the biochemical pathways appear in the urine, faeces, and expired air. These three routes of output account for most of the body's biochemical output. Metabolic transformation and conjugation are groups of biochemical processes involved in the biochemical pathways which deal with non-nutrients of any origin. These reactions take place extensively in liver but they also occur in lung, the cells of the gut wall, and the kidneys.

Metabolic transformation, in terms of biochemical reactions, describes reactions which generally increase the polarity (see Glossary) and the water solubility of the non-nutrients being transformed. The substances resulting from these reactions may be water soluble enough to be excreted directly, or they may become the intermediates in further biochemical reactions. Metabolic transformation relies on basic chemical reactions such as oxidation, reduction, and hydrolysis.

Conjugation describes a group of reactions in which substances endogenous to the body combine with the non-nutrient. The resulting conjugates are usually more polar, and more soluble. They too may be excreted directly, or they may form the intermediates for further chemical reactions.

Certain chemical groupings, such as —CN, —COOH, —NH₂, —SH, are biologically very active. They are deactivated by conjugation in which the active group is masked by the conjugant. Thus the conjugate is biologically less active than its precursor, as well as being more water soluble.

Normally the metabolic transformation and conjugation reactions lead to substances which are less harmful than their precursors. Sometimes, however, the intermediate is more biologically active than its precursor. *Detoxication* describes those reactions which reduce biological activity, *intoxication* those which increase biological activity.

Intoxication is of considerable interest in pharmacology. Therapeutic drugs are eliminated along the biochemical pathways for non-nutrients. By judicious chemistry therapeutic drugs can be synthesized so that the intermediates from transformation or conjugation retain their biological activity or, better still, increase biological activity. It is not surprising, therefore, that research in phar-macological toxicology is concerned with intoxication as well as detoxication. From the standpoint of occupational toxicology, know-ledge of possible causes of intoxication is important because it helps explain why a seemingly biologically inactive substance can be so pathogenic. A striking example of intoxication can be seen in occu-pational cancer of the bladder, which is discussed in Chapter 18. Table 8.1 gives examples of transformation and conjugation reactions of significance in an occupational context; for a much fuller list see Norton (1975).

FURTHER COMMENTS ON THE DIFFERENCE BETWEEN TOXICITY AND PATHOGENICITY

Metabolic transformation and conjugation may be crucial factors in determining whether a substance identified in the laboratory as toxic is, in fact, pathogenic. The reverse is also true, since metabolic transformation and conjugation may result in the formation of a highly pathogenic substance from a substance which in the test-tube appears biologically inactive and harmless.

Instances of pathogenicity can be found where the biochemical pathways for dealing with nutrients are insufficiently specific to enable discrimination between nutrients and non-nutrients which closely resemble the nutrients. The incorporation in the biochemical pathways of non-nutrients as well as nutrients of similar features may result in *lethal synthesis* when the non-nutrients have pathogenic effects.

A number of trace elements which at low concentrations are nutrients—and essential nutrients—are harmful at high concentra-tions. One example is manganese, which may be a co-factor in a number of enzyme reactions, particularly those involved in phos-phorylation and fatty acids synthesis, both key reactions in anabolic and catabolic metabolism. But excess of manganese due to chronic inhalation of manganese dioxide produces a serious disease of the central nervous system, not unlike Parkinson's disease.

There are still other factors to be considered in the pathogenicity of chemical substances. It may be, for example, that one substance inhibits or stimulates the biochemical pathway affecting a nutrient. A number of metals, of which lead is a well-known example, impair enzyme systems. The resulting anaemia represents a disturbance in the metabolism of iron, a nutrient substance.

TABLE 8.1

EXAMPLES OF CHANGES IN SUBSTANCES BROUGHT ABOUT BY METABOLIC TRANSFORMATION AND CONJUGATION
The examples given in the table are taken from Norton (1975); references are given to sources of further information about the particular reactions.

Substance	Structure	Changes	Type	References
Acetone	CH_3COCH_3	Unchanged + carbon dioxide	Oxidation	Williams (1959) Sakami and Lafaye (1951)
Benzene		Unchanged + —O conjugate	Oxidation, conjugation	Parke and Williams (1950), (1953)
Benzyl chloride	—CH_2Cl	—CH_2 mercapturic acid	Conjugation	Stekol (1938)
Carbon disulphide	CS_2	$R_2NCS—S^- \rightarrow SO_4^{2-}$	Oxidation	Browning (1965)
Carbon tetra-chloride	CCl_4	Unchanged + carbon dioxide + chloroform	Hydrolysis (not demonstrated)	McCollister *et al.* (1951)
Ethanol	C_2H_5OH	$CH_3CHO \rightarrow CH_3COOH \rightarrow$ metabolic pool + CO_2	Oxidation	Williams (1959)
Methanol	CH_3OH	Unchanged + HCOOH + CO_2	Oxidation Oxidation	Williams (1959)

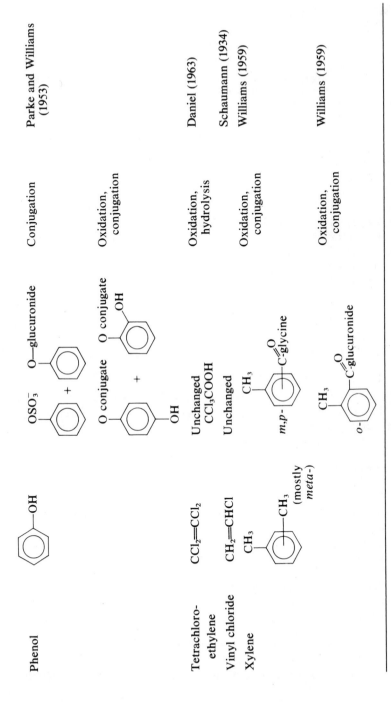

Compound	Structure	Metabolite(s)	Process	Reference
Phenol	—OH (phenol)	OSO_3^- (O-sulphate) + O—glucuronide + O conjugate + O conjugate OH (quinol) + OH (catechol)	Conjugation	Parke and Williams (1953)
Tetrachloro-ethylene	$CCl_2{=}CCl_2$	Unchanged CCl_3COOH	Oxidation, conjugation	Daniel (1963)
Vinyl chloride	$CH_2{=}CHCl$	Unchanged	Oxidation, hydrolysis	Schaumann (1934)
Xylene	CH_3 / CH_3 (mostly meta-)	CH_3 —C(=O)-glycine (m,p-); CH_3 —C(=O)-glucuronide (o-)	Oxidation, conjugation	Williams (1959)
			Oxidation, conjugation	Williams (1959)

CONCLUSION

Metabolic transformation and conjugation are important biochemical defence processes which are recognizable in a number of the biochemical pathways. A grasp of these two processes provides a good basis for first approaches to the study of pathogenicity.

Unfortunately no science of pathogenicity exists *per se*. Instead, use has to be made of knowledge in toxicology. Hodge (1975) described toxicology as the science of poisons which combines biochemistry and pharmacology. These two subjects, extensive though they are, do not cover wholly the pathogenicity of chemically induced occupational disease; for comprehensive understanding of that topic knowledge is needed of matters such as biological defence processes and the relation between disease and occupational exposure which, in turn, draws on subjects such as epidemiology (the study of disease in populations), and matters of pathobiology and pathology such as those already mentioned and others to be dealt with in following chapters.

APPENDIX 8.1: GLOSSARY

Anabolism: the building-up of tissues; synonymous with biosynthesis.

Catabolism: the breakdown of carbon compounds to form energy and intermediates for biosynthesis. There is an intermediate category between anabolism and catabolism, known as amphibolism, which combines certain of the characteristics of the other two. Certain nutrients are thus broken down into intermediates and then recombined in different combinations for biosynthesis.

Conjugation (metabolic conjugation): the process of chemical combination that takes place between a substance having a functional group with a polar moiety. The result of the combination is the formation of a substance which is more water soluble than the substance from which it was formed.

Detoxication: results from the processes in which non-nutrients and waste products from nutrients are broken down or inactivated in some way.

Hydrolysis: in inorganic chemistry this refers to the formation of an acid and a base from a salt by action with water; it is caused by disassociation of water into its constituent ions. In organic chemistry, hydrolysis refers to the decomposition of organic compounds by interaction with water resulting in the addition of an —OH group.

Intoxication: results from the processes in which intermediates are produced which are biochemically more active than their precursors. This is the opposite of detoxication in which the biochemical activity of the intermediates is progressively reduced with each stage of the detoxication process.

Median lethal dose (LD_{50}): defined as the dose at which 50 per cent of a population of one species will die within a specified time and under stated conditions of an experiment; the population might be animals, microorganisms, or living cells.

Metabolism: the complex of biochemical processes concerned with the building-up of tissues or the production of energy, and dealing with non-nutrients and waste products from biosynthesis or energy production. (Other definitions of metabolism may be broader than that given here.)

Oxidation: the addition of oxygen to a compound. Generally, oxidation is any reaction involving the loss of electrons from an atom.

Polarity: a property of certain covalent bonds. Covalent bonds involve the sharing of electrons between two adjacent atoms. Frequently, the two nuclei of the bonded atoms do not have equal shares of electrons. Hence, the electron cloud is relatively dense round one of the nuclei. The result is that one end of the bond is negative in charge with respect to the other. There is thus a negative and a positive pole; the bond is said to be a polar bond, that is to say it possesses polarity.

Molecules are said to possess polarity when the centre of the negative charge is not coincidental with the centre of the positive charge.

Polarity largely determines the solubility characteristics of substances. Polar substances dissolve readily in polar solvents, such as water, and non-polar substances dissolve readily in non-polar solvents. Many, but not all, of the organic solvents are non-polar.

Reduction: any chemical process in which an electron is added to an atom or an ion.

Transformation: the splitting of organic substances thereby producing substances with increased water solubility.

REFERENCES

BROWNING, E. (1965). *Toxicity and Metabolism of Industrial Solvents*, Amsterdam: Elsevier.

DANIEL, J. W. (1963). The metabolism of ^{36}Cl-labelled trichloroethylene and tetrachloroethylene in the rat, *Biochemistry and Pharmacology*, **12**, 795.

HODGE, H. C. (1975). Foreword, in *Toxicology*, L. J. Casarett and J. Doull, eds., New York: Macmillan.

LA VIA, M. F. and HILL, R. B. (1975). *Principles of Pathobiology*, 2nd Edition, New York: Oxford University Press.

MCCOLLISTER, D. D., BEAMER, W. H., ATCHISON, G. J. and SPENCER, H. C. (1967). Absorption, distribution and elimination of radioactive carbon tetrachloride by monkeys, *Journal of Pharmacology*, **102**, 112.

NORTON, T. R. (1975). Metabolism of toxic substances, in *Toxicology*, L. J. Casarett and J. Doull, eds., New York: Macmillan.

PARKE, D. V. (1968). *The Biochemistry of Foreign Compounds*, Volume 5, Oxford: Pergamon.

PARKE, D. V. and WILLIAMS, R. T. (1950). Studies in detoxication. 30: The metabolism of benzene, *Biochemical Journal*, **46**, 236–43.

PARKE, D. V. and WILLIAMS, R. T. (1953). Studies in detoxication. 54: The metabolism of benzene; the metabolism of C^{14}-phenol, *Biochemical Journal*, **55**, 337–40.

SAKAMI, W. and LAFAYE, J. M. (1951). The metabolism of acetone in the intact rat, *Journal of Biological Chemistry*, **193**, 199–263.

SCHAUMANN, O. (1934). *Medizin und Chemie*, **2**, 139. (Cited by Norton, 1975.)

STEKOL, J. A. (1938). Mercapturic acid synthesis in animals. IX: The conversion of benzyl chloride and S-benzylcysteine into benzylmercapturic acid in the organism of the dog, rabbit and rat, *Journal of Biological Chemistry*, **124**, 129–33.

VIRCHOW, R. (1859). *Die Cellularpathologie in Ihrer Begründung auf Physiologische und Pathologische Gewebelehre*, Berlin.

WILLIAMS, R. T. (1959). *Detoxication Mechanisms*, 2nd Edition, London: Chapman & Hall.

Section 3:
NORMAL BIOLOGICAL PROCESSES

Chapters 9–11 deal with a number of important vital processes. As is explained in Chapter 9, a distinction can be made between vital life processes and defensive processes. Growth, cell division, cellular genetics, boundaries and transport systems are the vital processes outlined in this section.

Chapters 9 and 10 contain information necessary for an appreciation of the processes involved in cancer and other disorders of growth of occupational significance.

The discussion of cellular genetics leads to a recognition of the roles of protein synthesis within cells. Such roles include the (a) 'structural'; (b) regulatory, because as enzymes, these play a crucial role in the body's biochemistry; and (c) others such as hormones.

The synthesis occurs within cells from nutrients which come, ultimately, from the diet. An outline of how such substances enter the body indicates some of the ways in which harmful substances can enter the body and be transported within it.

Chapter 9

Growth

OBJECTIVES OF THE CHAPTER

This chapter provides:
(1) concepts of vital processes and defensive processes,
(2) an outline of how growth takes place in living tissues and the central role of cell division,
(3) the concept that living tissue is in a dynamic state throughout normal life,
(4) an outline of cell division, explaining how chromosomes become visible and how they behave during cell division,
(5) an identification of certain important structures within cells, and
(6) an indication that some patterns of disease result from disorders in the processes of growth.

INTRODUCTION

In this chapter various aspects of growth are dealt with. Replication and division of chromosomes* are an essential part of growth by cell division. Growth and, in particular, the processes of growth need to be considered because chromosomes are the repositories of genetic information which is the target of many chemicals and energies. Chromosomes' condensation and formation is a useful starting point for discussion of the genetics of the cell.

Previous chapters have dealt with those processes which have a recognizable defensive function of special importance in the context of occupational health and safety. This chapter is concerned, as are subsequent chapters, with certain other processes, such as growth. These are also defensive, in the sense that they are essential to the maintenance of life. They are, moreover, the subject of disorder in a number of occupational injuries and diseases. The defensive processes described in previous chapters can, however, be distinguished

---

*The term 'chromosome' is commonly used to describe structures visible only during certain phases of the cell life cycle (described later). However, some authorities use 'chromosome' to describe the general arrangement of the genetic complement within the cell nucleus. In this book the former terminology is followed.

from those described here in the sense that defence only becomes a
factor when there is a harmful input. On the other hand, those
processes which form the subject of this and succeeding chapters are
in continuous operation during life. Inevitably, to distinguish in this
manner is only partially valid but the distinction applies in a sufficient
number of instances for it to be useful for present purposes. Such a
categorization is admittedly imposed and artificial, since in the living
body the processes of defence and the processes of life function
inseparably.

GROWTH OF LIVING TISSUES

Knowledge of how living tissues grow is necessary to the understanding of occupational cancers, because these are disorders of
growth. Such knowledge is, however, also important in the understanding of other occupation-linked diseases. Examples are
mutagenesis and teratogenesis (see Chapter 16), both of which can be
induced by chemical substances.

The everyday concept of growth centres on increase of bodily
dimensions. This concept can be said to be erroneous, in that growth
continues in most living tissues throughout life. As new tissue is
formed by cell-division, the old 'spent' tissue is removed by processes
such as phagocytosis. Dimensional stability represents a dynamic
balance between the growth processes on the one hand, and the
tissue-destructive processes on the other.

Cell division, known as mitosis, is a central process in tissue
growth. However, dimensional increase of tissues also follows the
accumulation of water, of fat and other organic substances, and of
inorganic substances. There are some accumulations which take place
in static stores isolated from the main stream of physiological activity
but these instances are to be regarded as exceptions since most
accumulations take place exclusively as a result of a shift in the
dynamic balance, in the tissues, between input and output. Accumulation is normally, therefore, not a static deposit but, rather, the result
of physiological 'work in progress'.

A key concept here requires emphasis: living tissues, generally,
should not be perceived as static structures. They are more appro-
priately seen as an expression of the current net balance of a
multitude of processes. Much occupational disease—and disease
generally—can be explained as a disturbance in the equilibrium of
growth. Equilibrium is a concept which is frequently encountered in
the physical sciences and which explains numerous phenomena in
those sciences. It is essential, in order to explain living structures, to
extend the concept of equilibrium to growth. Le Gros Clark (1975)

stated the concept and its significance:

> In all studies of the tissues of the body and of the functional units in which they are organized, it is of the greatest importance to realize that the fabric of which they are composed is in a continual state of flux during life. Yet they maintain their morphological identity in spite of the unceasing removal and substitution of their material elements which occur in the process of metabolic interchange. Strictly speaking, there is nothing static about any anatomical units in the living body ... even the inorganic constituents of bone are being continually removed and replaced. The material of a bone in life, therefore, is not the same from one month to another. Notwithstanding this fact, however, in its contour and architecture the bone outwardly preserves its identity. Again, the skin of the hand, which appears to be fixed in position with a permanent pattern of lines and creases, is in a continual process of removal and replacement. Thus a cross-section of skin as seen under the microscope is also a cross-section of time—it represents an arbitrary point in an uninterrupted process of change. But not only does this process of change affect anatomical units as a whole, it involves the very molecules out of which living matter is constituted. In the phenomenon which has been called 'transamination', nitrogen is being continually shifted on in living tissue from one organic molecule to another in an everlasting chain of 'hand-to-hand' transport. Thus, not even the ultimate elements of living substance are stable. Yet the pattern to which they conform remains the same. It seems almost to demand the postulation of a pre-existent system of physical forces with a stable spacial organization to whose pattern the molecules of organic substances must perforce adapt themselves as they become assimilated into living matter. It is as though, when we look at the living body, we look at its reflection in an ever-running stream of water. The material substratum of the reflection, the water, is continually changing, but the reflection remains apparently static. If this analogy contains an element of truth, if, that is to say, we are justified in regarding the living body as a sort of reflection in a stream of material substance which continually passes through it, we are faced with the profound question—what is it that actually determines the 'reflection'? Here we approach one of the most fundamental riddles of biology—the 'riddle of form' as it has been called, the solution of which is still entirely obscure.*

*From *The Tissues of the Body* by W. E. Le Gros Clark (6th Edn., 1975), p. 9, reproduced by permission of Oxford University Press.

One would add that at least some disease represents ripples on the surface of the stream or turbulence in its flow.

CELL DIVISION

Figure 9.1 is a three-dimensional diagram of a cell. From this can be seen the principal features of living cells generally. Various components of the cells are identified, only some of which are involved in cell division. The remaining cellular components identified in the diagram are involved in other cellular processes and are mentioned in other chapters.

The following description applies generally to all cells capable of cell division, and excludes certain highly specialized cells, such as nerve cells, which are not capable of cell division. After cell division is complete, a single cell will have produced two daughter cells with genetic complement normally identical in every way to the parent. The morphological and functional identity of such cells is maintained through the generations by the passing of genetic information from one generation to the next. This process is described in the next chapter. Changes in the genetic information result in changes in the morphology and function of the daughter cells. If generation upon generation of cells are changed the result, ultimately, will be change in tissue. For the purposes of this book, interest is in changes in tissues with harmful consequences. However, the possibility should be borne in mind that changes in the genetic information that passes from one cell generation to another may also be the means by which living organisms adapt to changing circumstances. Hence, normal growth is dependent upon faithful and organized replication.

Figure 9.2 depicts the life cycle of living cells. Cells have a life span which varies from days to months. Many cells in an adult live out their existence in the absence of any further growth and division. Other cells, however, are required to go through cycles of division. Such cells include those of skin. Those parts of the cell cycle involving division and growth have been most studied in cells in culture (that is outside the living body) and four major phases have been identified. The phase of cell division (mitosis), M-phase, lasts for the shortest time and is succeeded, in each daughter cell, by a period of growth G1. Each daughter cell then begins the replication of its genetic complement. Replication of genetic complement takes place by DNA synthesis (see Chapter 10); replication is denoted by the S-phase. At the end of S-phase a second phase of growth takes place denoted G2. Each cell then proceeds to another round of cell division and the cycle is repeated. Mitosis can be divided into several stages, to be described shortly.

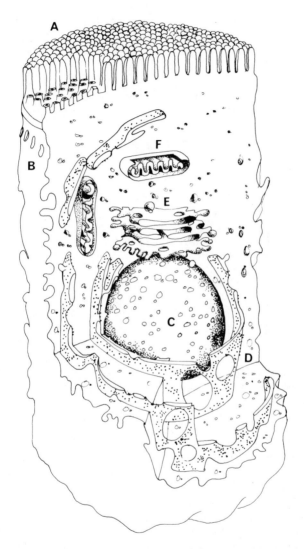

FIG. 9.1. Cut-away drawing of a cell from the small intestine: simplified.
(Certain of the structures shown are discussed in later chapters.) (A) The
absorptive surface of the cell with finger-like protuberances called villi in
contact with the intestinal contents; (B) the cell wall in contact with neighbouring cells with areas for cellular intercommunication; (C) cell nucleus,
with apertures in the nuclear membrane; (D) the endoplasmic reticulum,
studded with ribosomes; (E) the Golgi apparatus; and (F) mitochondrion.
          Modified, with acknowledgements, from *Gray's Anatomy*.

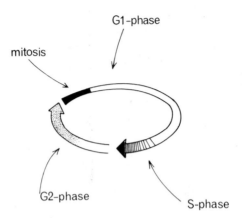

Fig. 9.2. Diagram of the cellular life cycle. Mitosis is the phase of cell division. G1 is a period of growth; G2 is a second period of growth. During the S-phase, DNA synthesis takes place (see text).

During the period between cell divisions (G1, S and G2) the cell is active in synthesis of proteins and all other components. Only during mitosis is synthetic activity suspended. Genetic complement is not visible as discrete structures during the G1-, S- and G2-phases. The discrete structures, once they become visible, are known as chromosomes. These can only be distinguished during mitosis when they are in a condensed state. It is the condensed state which permits the physical separation of the identical copies produced as a result of DNA synthesis.

The phases of the cell cycle mark differences, not only in cellular activity, but also in cellular sensitivity to harmful inputs.

G2 can be looked upon as the premitotic phase when the chromosomes condense from material called *chromatin* dispersed through the nucleus. The cell's sensitivity to injury varies with the phase it happens to be in when the injury occurs. Cells in the early G1- and G2-phases are least sensitive, whereas cells in mitosis, late G1- or early S-phase are most sensitive. Variation in sensitivity is particularly marked towards ionizing radiations or the so-called radiomimetic agents such as the nitrogen mustards (these are also known as the alkylating agents).

The differential sensitivity is exploited in the therapy of cancer by ionizing radiations; the therapeutic radiations applied to cells in mitosis, late G1- or early S-phases will be most harmful to cells in those phases. Because cancer cells are dividing frequently, the radiations are effective against them. However, the same effect will occur in tissues where a good deal of mitosis normally takes place, such as bone marrow or the spermatogenic tissues. Hence, sensitivity to radiation and certain chemical substances will be greater in these tissues than in those where mitosis is relatively uncommon. Differential sensitivity within the cellular life cycle provides one important explanation of differences in radiosensitivity from one tissue to another.

It is convenient to describe the process of cell division chronologically, and for this purpose to subdivide the subject into a series of steps (see Fig. 9.3). The word 'interphase' is used to describe the state of the cell when it is executing its non-mitotic activities, which are suspended when mitosis starts. In a nucleus not currently undergoing division there are granules and threads of chromatin, a nucleoprotein which, even when distributed throughout the nucleus, contains the genetic information. At the start of the cell division the chromatin threads shorten and thicken into rod-shaped bodies called visible chromosomes. These carry the genetic information in *theoretical* units called genes, arranged along the length of the chromosome. The number of chromosomes is constant from one species to another, humans having 46.

The first step of mitosis is the *prophase*, in which the chromosomes are visible in the form of convuluted threads loosely bundled together. The cell cytoplasm contains a tiny structure called the centriole, which has a pivotal role in cell division. It divides into two in early prophase (if it is not already double; in some cells the centriole appears always as doublet). The two daughter centrioles separate and move to opposite sides of the nucleus. The cytoplasm lying between the centrioles is then traversed by very fine spindle fibres—the asta—which run between the two centrioles.

The *metaphase* begins when the nuclear membrane starts to break down. The chromosomes arrange themselves in the equatorial plane mid-way between the centrioles. Each chromosome then splits longitudinally, forming a pair of chromatids. Chromatids separate and migrate to the centrioles. If the centrioles can be regarded as the poles then one chromatid from each pair will move 'north' whilst the other moves 'south'.

Anaphase begins when the chromatids separate. When the chromatids have reached the centrioles the spindle fibres elongate in preparation for the final stage of mitosis. This is the *telophase* in which the chromatids become daughter chromosomes. They become enclosed by the newly formed nuclear membrane. In the sense that they become invisible, the chromosomes then seem to 'dissolve' into chromatin, to be distributed throughout the new nucleus. At the same time a line of fission appears in the cellular cytoplasm and the parent cell finally splits into the two genetically identical daughter cells. This

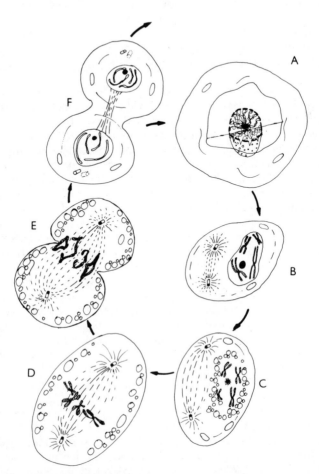

FIG. 9.3. Phases of mitosis (cell division). (A) *Interphase* in which the cell is in its functional state, that is its non-mitotic state; (B) *prophase* in which centrioles separate and spindles and asters form; the chromosomes shorten and divide into chromatids; (C) *prometaphase* (end of prophase) in which the nuclear membrane has disintegrated; (D) *metaphase* in which the chromosomes line up at the equator; (E) *anaphase* in which the chromatids which reform into chromosomes, divide, and migrate 'north' and 'south'; and (F) *telophase* in which the nuclear envelopes are reformed and cleavage takes place between the two cells which then continue in the life cycle as depicted in Fig. 9.2 (modified, with acknowledgements, from *Gray's Anatomy*).

entire process has been observed in liver cells to average 49 min (Brues and Marble, 1937) while in other cells mitosis ranges from 30 min to several hours.

'Mitotic figures' refer to the stages of mitosis in which the chromosomes or chromatids are visible. 'Mitotic rate' refers to the frequency with which cells undergo mitosis.

Where cell division takes place as part of the ordinary growth of tissues the term mitosis is used. *Meiosis* describes the process of cell division in the reproductive cells before fertilization, but disorders of meiosis as a specific topic are beyond the scope of this book.

As previously stated, important classes of disease can be understood in terms of disorder of growth; the growth becomes disorganized. Processes by which growth is organized are incompletely understood but an important clue was provided in 1967 when Bullough reported finding a substance capable of inhibiting mitosis. The substance was isolated from superficial layers of skin. It was sought there because observation had shown that when the superficial layers of skin are removed, the basal layer of skin (see Fig. 13.1) increases its mitotic activity. This observation led to the postulation that the outer layer of skin produced an inhibiting effect on the basal layer; Bullough showed that the inhibitor was, in fact, a chemical substance. A number of other substances, including cortisol (see Chapter 6) also have inhibitory effects. Obviously the role of growth inhibitors generally in disorders of growth is the subject of important research.

CONCLUSIONS

The essential process of growth is cell division. Chromosomes form during mitosis. During this and other phases of the cellular life cycle, cells are most sensitive to injury. This fact explains the apparent paradox that inputs such as ionizing radiations or certain chemicals can be both inhibitors and initiators of disorders of growth.

REFERENCES AND BIBLIOGRAPHY

BRUES, A. M. (1954). Ionizing Radiations and Cancer, *Advances in Cancer Research*, **2**, 177–95.

BRUES, A. M. and MARBLE, B. B. (1937). An analysis of mitosis in liver restoration, *Journal of Experimental Medicine*, **65**, 15–27.

BULLOUGH, W. S. (1967). The vertebrate epidermal chalone, *Nature*, **214**, 578.

LE GROS CLARK, W. E. (1975). *The Tissue of the Body*, 6th Edition, Oxford: University Press.

Chapter 10

Genetics of the Cell

OBJECTIVES OF THE CHAPTER

This chapter provides a brief outline of:
(1) two important processes included, for the purposes of this chapter, under the heading *genetics of the cell*: synthesis of protein, and the passing of genetic information from cells to their offspring,
(2) the relation between structure and function of cells, protein synthesis, and the transfer of information from one generation of cells to the next,
(3) the roles of protein produced in cells, and
(4) mutation.
In addition, a brief glossary of terms commonly encountered in biochemical biology is included.

INTRODUCTION

'Genetics of the cell' is used here as a label for two important processes which take place within cells:
(1) synthesis of protein, and
(2) the passing of genetic information from cells to their offspring.
'Genetics of the cell' is here confined in meaning to the two processes, but it should be noted that there are others which could also be included under the heading. Selection of the two for the purposes of this chapter has been influenced by their biological importance and their being the sites of the harmful actions of numerous inputs (input here has the meaning given to it in Chapter 2) which are encountered in the occupational milieu.

CELL STRUCTURE AND FUNCTION

Figure 10.1 (a modification of Fig. 9.1) depicts some cell structures and identifies certain associated cell functions connected with the two processes with which this chapter is concerned. It is convenient to see the cell as a container for a vast number of processes, each of which could be characterized with an input–output model. Nutrients,

118

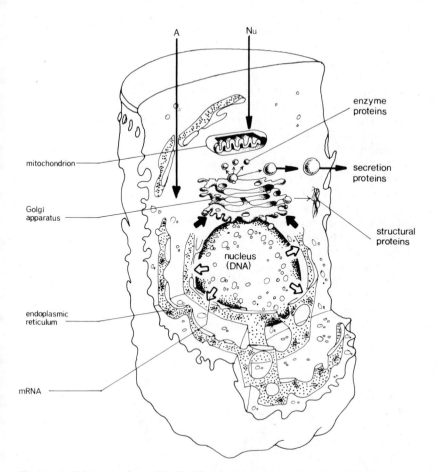

FIG. 10.1. Diagram of a cell indicating structures and functions. Nutrients (A) in the form of amino acids enter the cell by modes of transport such as those described in Chapter 11. The nutrients diffuse through the cellular cytoplasm and are available for protein synthesis at the ribosomes. Other nutrients (Nu) are processed at the mitochondrion (normally, cells have many mitochondria) supplying energy for the various biochemical reactions taking place within the cell. The nucleus contains DNA, in the structure of which is the genetic information. RNA forms messsenger RNA (mRNA—see Appendix 10.1) within the nucleus. The messenger RNA passes to the ribosomes attached to the endoplasmic reticulum, and also to ribosomes lying free within the cytoplasm. At the ribosomes, from the RNA template, and with the aid of other RNAs, protein synthesis takes place from the amino acids located in the cytoplasm. Proteins synthesized at the endoplasmic ribosomes pass to the Golgi apparatus. Here, secretory vacuoles form; these move to the cell surface. Non-endoplasmic ribosomes produce enzyme proteins and structural proteins which may, or may not, leave the cell. (Based on, with acknow-ledgements to, *Gray's Anatomy*.)

for example, enter the cell by one or more of the means identified in Chapter 11. Some of the nutrients reach the mitochondria (of which there may be many in the typical cell). At a mitochondrion nutrients are broken down to smaller molecules in a series of biochemical steps. The breakdown of nutrients liberates energy which is used in the synthesis of an energy storage molecule called ATP (adenosine triphosphate). Waste products, water and carbon dioxide are formed during the nutrient breakdown, and the smaller molecules produced are used as the building blocks for the synthesis of larger molecules such as proteins.

The outcome of these metabolic steps, which correspond to catabolism and anabolism, can be summarized:

  (a) the formation of energy-rich molecules which can take part in cellular activities requiring energy,
  (b) the formation of small molecules which become the building blocks of the larger molecules of the cell, and
  (c) the formation of waste products which are subsequently removed from the cell.

The Golgi apparatus, the endoplasmic reticulum, and the nucleus can be visualized as a hollow interconnecting system intimately

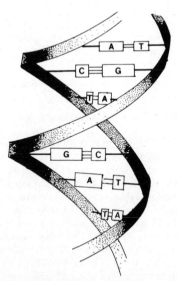

FIG. 10.2. The double-helical molecule of DNA showing an impression of the way in which the bases link the two strands of the double helix. (A) adenine; (C) cytosine; (G) guanine; and (T) thymine. Note that in DNA, (A) always pairs with (T) and (G) always pairs with (C).

connected with protein synthesis. For the purposes of this chapter the hollow system as a whole will be looked upon as the principal site of protein synthesis although, in fact, the process of protein synthesis actually takes place at the ribosomes. These are small structures attached to the endoplasmic reticulum and also found dispersed throughout the cell cytoplasm.

The nucleus contains deoxyribonucleic acid (DNA) which has a stable molecular structure suitable for the storage of genetic information. DNA never leaves the nucleus (not, at least, for the purposes of this chapter). Figure 10.2 shows the structure of DNA in simplified form.

Mitosis, as previously described, involves division of the cell to give two daughter cells. The genetic information is copied *prior* to mitosis. During mitosis, the identical copies are *physically* separated, each daughter cell acquiring one of the copies.

ROLES OF PROTEINS PRODUCED IN CELLS

Proteins' roles are identified here under three headings: (a) as enzymes, (b) as structural components, and (c) as secretory products.

(a) Proteins as Enzymes
All enzymes are proteins. Enzymes are catalytic molecules that control most of the chemical reactions of cells. They differ in the kinds of molecules they affect and the kinds of reactions they catalyse. Many enzymes produced in the cell are required for processes internal to the cell. However, other enzymes may be needed for processes external to cells, or connected with cells remote from the cells responsible for synthesis. Such enzymes are secreted by the cell. Not all secretory proteins are enzymes—hence, secretory proteins may be considered under a separate heading.

(b) Proteins as Structural Components
All cells contain substantial quantities of 'structural' protein. 'Structural' proteins (in membranes, for example) are defined by a negative characteristic: that no enzymatic regulatory role is known for them at present. Sometimes the structural protein has special mechanical properties (muscle cells, for example, have large quantities of contractile structural protein). Other cells produce structural proteins whose ultimate destination lies outside the cell. In Chapter 15, the importance of collagen will be highlighted in connection with repair, and disorder of repair: collagen is a structural protein secreted by fibroblasts.

(c) *Proteins as Secretions*
It has been mentioned that cells may secrete enzymes, and that cells may secrete structural proteins. In addition, cells may also produce secretory products which are neither enzymes nor structural proteins. For example, certain hormones are proteins. One example is adrenocorticotrophic hormone which triggers the secretion of cortisol, the hormone so closely identified with stress, discussed in Chapter 6. Other secretory proteins are antibodies mentioned in Chapters 5 and 14.

PROTEIN SYNTHESIS

A cell's morphology, organization, and function are critically dependent upon the nature and quantity of proteins synthesized within the cell. Structural proteins produced should be appropriate in type and quantity; and the cell's response in terms of its production of

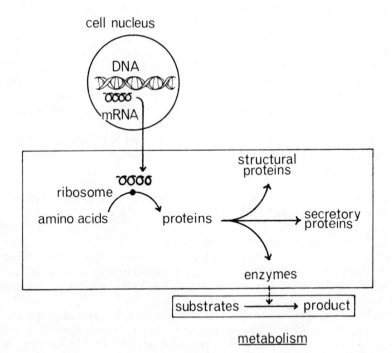

FIG. 10.3. Outline of protein synthesis. mRNA is messenger RNA; substrate is a general term for substances converted to products with the mediation of enzymes.

secretory proteins should also be appropriate to meet demands placed on the cell. For metabolism to function adequately the metabolic enzymes should be appropriate in nature and extent. Control of protein synthesis is thus a crucial matter. Control may be exercised at a number of points in the protein synthesis pathway when, for example, the cell responds to external and internal demands for changes in the formation of structural proteins, secretory proteins or enzymes. Protein synthesis follows of a blueprint, so to speak, taken off the master copy carried in the nuclear DNA. However, the blueprint may not be a complete copy of the master copy. Through the mediation of control processes, the information may be modified.

Figure 10.3 identifies certain of the processes involved. As stated, the cell nucleus contains within the DNA structure the cells' genetic information. Messenger RNA forms within the nucleus, the genetic information being encoded in its structure (the formation of RNA from DNA is known as transcription). Messenger RNA travels to the ribosomes via the hollow system consisting of the endoplasmic reticulum linked to the nucleus. At the ribosomes the messenger RNA (in company with other forms of RNA not discussed here) acts as the template for the assembling of amino acids into proteins (this process is known as translation).

As previously mentioned, at every stage control is exercised by enzymes (not only are these produced by the processes but they also are the agents by which control is exercised over the processes).

THE PASSING OF GENETIC INFORMATION FROM CELLS TO THEIR OFFSPRING

In the last chapter it was seen that cell division normally results in the production of two daughter cells, each containing genetic information identical to the parent cell from whose division they resulted. The genetic complement, that is to say the essential information which must pass from parent to daughter cell, is located in the nucleus. There, replication of genetic information takes place. The migration of the chromatids 'north' and 'south' during cell division carries to each future nucleus a complete complement of genetic information. All this is assuming, of course, that no abnormal processes have intervened, such as mutation—outlined below. Replication of genetic information involves the separation of the strands of the DNA double helix and the formation of complementary strands from which are then formed new, identical, DNA molecules. The process of replication is outlined in Fig. 10.4.

After a cell has divided there is a time of growth when no

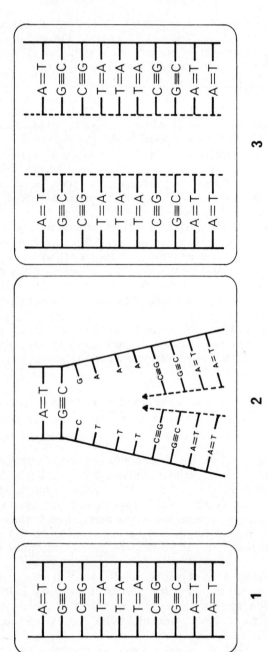

FIG. 10.4. Diagram showing replication of DNA. In frame 1 is shown a portion of a double-stranded DNA molecule. In frame 2 the strand is separating. New (complementary) single-strands are synthesized. The separation of the double-strand and the subsequent synthesis of complementary strands are mediated by enzymes. The two halves of the parent double-strand act as a template for the synthesis of the complementary single strands.

In frame 3 is shown the two 'daughter' double-strands which have been produced by the process of replication. They are exact replicas of the parent double-strand. Each contains a half of the parent double-strand. (N.B. for simplicity, synthesis of complementary strands is shown to proceed in the same direction. Strictly speaking, this is incorrect since DNA strands are antiparallel, but this need not concern us here.)

detectable DNA synthesis occurs, followed by a period of DNA synthesis (the whole DNA complement is replicated) and finally there is another period of growth when no DNA synthesis is detectable. The cell then divides, and the cycle is repeated. During all stages except cell division the DNA is available for messenger RNA synthesis.

MUTATION

Mutation refers to any alteration in the genetic information carried by DNA. It is a mechanism by which genetic variation can be introduced; mutation and selection are the processes which form the basis of the Darwinian theory of evolution. However, in the context of human exposure to environmental inputs of which many are mutagenic (that is capable of inducing changes in genetic structure) mutation can be treated as a mode of harmful action.

Mutation describes effects which range from the alteration of a single base pair to deletions, insertions or rearrangements of large sections of a genome. An alteration in DNA necessarily results in a change in information content of the affected DNA. Depending on the type, position and magnitude of this change in information content, a number of effects may be observed. For instance if the change occurs in a gene which codes for an enzyme (the *one gene one enzyme* hypothesis is adopted in this book) the change may be such that the synthesis of the enzyme is no longer possible, or, an enzyme may be produced but it will be malfunctional. Sometimes the mutation may have little effect and an enzyme is produced which functions adequately despite the mutation. Another factor is the role of the enzyme itself which may not be vital to the organism.

DNA codes not only for proteins such as enzymes but also for other molecules important to the cell. Furthermore, mutation may affect the genome in such a way that the physical behaviour of structures such as chromosomes is affected.

Mechanisms have been identified by which the cell can repair errors introduced into DNA. Unless this repair occurs, the error is inherited by subsequent generations of cells. Inheritance may further be prevented by mechanisms such as immunosurveillance. If the mutation induces a change in a cell and the immune system then treats it as 'foreign', the cell may be destroyed before it passes on its mistakes to the daughter cells.

Included among the ways in which mutations may be induced is chemical change in the nucleotide bases, induced, for example, by chemical substances or ionizing radiations. Figure 10.5 shows the

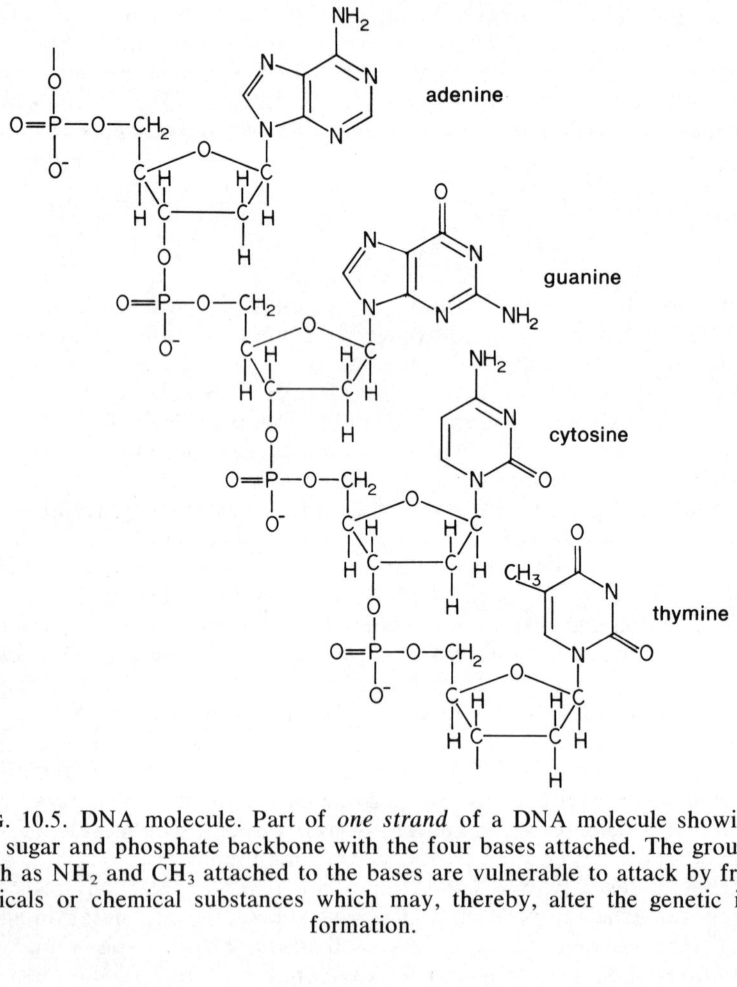

FIG. 10.5. DNA molecule. Part of *one strand* of a DNA molecule showing the sugar and phosphate backbone with the four bases attached. The groups such as NH_2 and CH_3 attached to the bases are vulnerable to attack by free radicals or chemical substances which may, thereby, alter the genetic information.

chemical structure of DNA and Fig. 10.6, for comparison, that of RNA. Chemical change in, for example, —NH_2 radicals could be produced by a range of chemical reactions. Such chemical alterations could result in mutation.

SIGNIFICANCE OF MUTATION; CHROMOSOME ABERRATIONS

It is not possible to generalize about the significance of mutation. Possible effects have to be considered for individuals and their offspring, for cells and their offspring, and for cells not about to

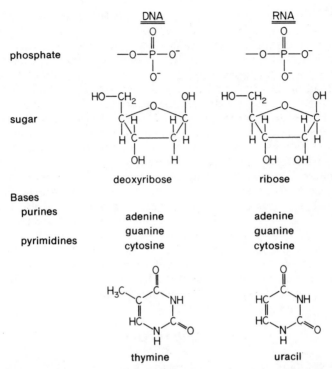

FIG. 10.6. Comparison of DNA and RNA. Apart from the information shown, DNA has a double-helical molecule whereas RNA has a single-helical molecule.

divide. Mutation may appear as an abnormality transmitted from one generation to the next. Obviously, this possibility has been greatly feared in connection with ionizing radiation. In general, however, human mutation is more a theoretical than an actual risk of occupation. Mutation may or may not result in abnormalities in cells' offspring. If any abnormalities do occur, there is also considerable variability in the effect. Some are of vital consequence to the offspring cells whereas others appear to cause no great disruption. In cells not about to divide, mutation may not manifest itself because the mutated part of the DNA molecule is not involved in processes currently underway in that cell. The abnormality induced by the mutation would be expressed only in later generations of cells.

Mutation may be manifest in visible abnormalities of chromosomes (it will be recalled that the chromosomes are only visible at certain phases of the cells' life cycle). It is generally agreed that where abnormalities of chromosomes can be seen, this can be taken as evidence that genetic injury has taken place. However, the *absence* of chromosome aberrations is no guarantee of the *absence of mutations*. This is because the visible chromosomes each contain a vast amount of genetic information in the DNA of the chromosomes. Chromosomes can appear, under the ordinary microscope at least, to be quite normal and yet be carrying altered DNA, the manifestation of which could be abnormality of a significant kind in the daughter cells.

In Chapter 16 disorders of growth are discussed and the point is made that cancer cells often display chromosome abnormalities. However, not all chromosome abnormalities are necessarily linked with cancer. Hence the observation that an input, whether a chemical or an energy, is capable of inducing chromosome aberrations is not in itself sufficient evidence for the conclusion to be drawn that the substance is carcinogenic. There are, moreover, difficulties in the interpretation of experimental data. Many of the experiments are done on cells cultured 'in the test-tube'. However, abnormalities induced by a particular input in the test-tube do not necessarily occur to identical cells in their normal milieu within living tissues. One of the reasons, of course, is that in the living tissues cells are defended by numerous homeostatic and defensive processes.

APPENDIX 10.1:   GLOSSARY

Note: Not all of the terms included here appear in Chapter 10.

Codon: in nucleic acid, a sequence of three adjacent nucleotides. A codon contains the code for one specific amino acid.

Gene: a mutable segment of DNA carrying genetic information. Genes are theoretical units.

Gene expression: refers to the assembling of genetic information in the replicated DNA.

Genetic code: the set of triplet code words used in DNA to specify the various amino acids and proteins.

Genetic information: the hereditary information contained in a sequence of nucleotide bases in DNA.

Genome: the total complement of genetic information within a cell.

Heritable material: DNA or RNA containing genetic information, carried by the chromosomes.

Messenger RNA: RNA molecules which carry the genetic message from DNA within the nucleus to the ribosomes; abbreviated to mRNA.

Microsomes: text-books on toxicology refer to microsomes, or microsomal fractions as the sites in which some metabolic transformations and conjugations (see Chapter 8) take place. The toxicologists' microsomes consist of parts of cells separated from cytoplasm by centrifugation. Microsomes are ribosomes attached to fragments of endoplastic reticulum. The production of the microsomes by centrifugation is thus a toxicologists' artifact. Nevertheless, it provides important opportunities for toxicological study.

Nucleoside: a substance consisting of either a purine or pyrimidine base linked covalently to a pentose (sugar).

Nucleotide: this is a nucleoside which is phosphorylated at one of its pentose hydroxyl groups.

Purine: a nitrogenous base which is a component of nucleotides but which differs in composition from pyrimidine.

Pyrimidine: a nitrogenous base which is a component of a nucleotide.

Replication: synthesis of a daughter RNA molecule complementary to a parental DNA molecule.

Transcription: the process whereby the genetic information contained in DNA is used to specify a complementary sequence of bases in an RNA chain.

Translation: the process in which the genetic information present in an mRNA molecule codes the sequence of amino acids during synthesis of proteins.

FURTHER READING

AUERBACH, C. (1976). *Mutation Research*, London: Chapman & Hall.

LOEWY, A. G. and SIEKEVITZ, P. (1970). *Cell Structure and Function*, 2nd Edition, London: Holt.

SMITH, A. (1976). *Protein Bio-Synthesis*, London: Chapman & Hall.

WATSON, J. D. (1970). *Molecular Biology of the Gene*, 2nd Edition, New York: Benjamin.

Chapter 11

Transport Systems: Input, Distribution and Storage, and Bio-dumping

OBJECTIVES OF THE CHAPTER

This chapter provides:
(1) an outline of some of the processes by which substances gain entrance to the body,
(2) an explanation of 'boundary' as a concept connected with modes of input,
(3) a brief outline of cell membrane,
(4) a discussion of certain principles involved in substances crossing cell membranes,
(5) an outline of intra-cellular transport,
(6) a brief outline of the processes involved in the distribution of substances within the body,
(7) a note on the quantitative aspects of accumulation and retention, together with an outline of the concepts of biological half-time, and
(8) a brief outline of bio-dumping via the kidney.

INTRODUCTION

This chapter is concerned with the processes by which harmful substances reach the body's organs on which they exert their harm. Mention is also made of certain quantitative aspects of accumulation of substances in tissues. The processes about to be described can be visualized, for descriptive purposes, as a series of stages through which the input substances move.

The first stage involves a substance being presented to one of the body's boundaries; the second stage involves passing that boundary; and the third is the substance's being taken by means of an intermediate transport process to another boundary. Passing that boundary is the next stage and the final stage is reaching the target organ, by means of still other transport processes. Figure 11.1 identifies the various stages.

130

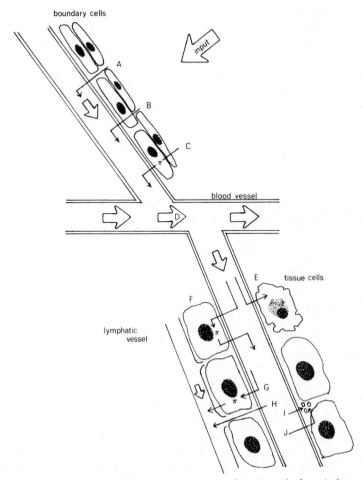

FIG. 11.1. Diagram showing substances crossing boundaries, being transported in blood, leaving blood to enter tissues of target organ, and being drained from tissues by lymphatic vessels. (A) Substance passing between cells of boundary; (B) substance passing into boundary cell, across it, and leaving the other side; (C) substance entering boundary cell, entering biochemical pathways, and leaving cell, perhaps in changed form; (D) blood transport processes such as binding to plasma proteins or red blood cells, or simple water solubility; (E) substance leaving blood, entering cell of target organ, and causing injury to that cell; (F) substance entering cell of tissue, entering biochemical pathway, and returning to blood circulation in a changed form; (G) substance entering cellular biochemical pathways but leaving via the lymphatic vessels; (H) substance passing directly from the blood stream through the vessel wall, normally capillary wall, into the lymphatic vessels; (I) substance leaving vascular system and entering storage compartment; (J) substance entering tissue cell with or without entering biochemical pathway and passing to tissue storage compartment; and (π) biochemical pathways.

BOUNDARIES AND THE INSIDE OF THE BODY

Each of the modes of input identified in Chapter 2 involves the crossing of one or other boundary between the body and the external milieu. Hence, in order to reach the inside of the body, a substance must penetrate at least one of the boundaries which separate the body from the external milieu.

The skin can be referred to as a boundary between internal and external milieux. Its structure is such that any substance presented to it must pass through a large number of cells before reaching the inside of the body. Furthermore, the outer layers of cells in the skin are tough and inactive, and as long as these qualities are retained the skin provides an effective barrier to polar, water-soluble substances. It is also reasonably effective as a barrier against lipid-soluble substances, although some pervasion undoubtedly takes place; various experiments have shown that it takes place in measurable quantities. For example, Dutkiewicz and Tyras (cited by Wahlberg, 1976) reported the rate of percutaneous absorption of toluene as 14–23 mg/cm^{2}h and that of benzene, 0.4 mg/cm^{2}h.

Apart from skin, another obvious boundary is the membrane which forms the gas exchange surface of the lungs (see Chapter 3). Yet another boundary is that formed by the gut lining, though here the term 'inside the body' is less clearly defined. In some instances substances cannot be regarded as being 'inside the body' unless they have actually penetrated the intestinal lining. Thus cellulose (roughage) in the diet does not penetrate the intestinal lining, and in this sense is never 'inside the body'. However, the colonization of the gut and its contents by aggressive micropredators can set up functional disturbances, ranging from trivially uncomfortable to lethal. These micropredators need not pass all the way through the gut lining in order to cause such disturbances. In that sense, therefore, they are not 'inside the body', even though the effect of their activities may result in the widespread disturbance of systems very definitely inside the body. There are other examples: carcinogens may do no more than gain access to a few cells of the intestinal lining in order to initiate carcinogenesis, yet they are scarcely inside the body.

CROSSING BOUNDARIES

Boundaries are composed of cells. There are two routes across the boundary:

(1) By substances passing *between* the cell in the small spaces which separate cells; (A) in Fig. 11.1. A possible example of inter-

cellular penetration of boundaries appeared in Chapter 3 where mention was made of lung macrophages carrying particles through the gap between adjacent cells in the alveolar membrane. (It should be noted that the gap size varies from tissue to tissue. Gaps may be important as an explanation of leakage from blood vessels.) The passing of a substance between constituent cells of boundaries is most likely when the boundaries are only a few cells thick, as is the case with the boundaries presented by lung and gut. Moreover, the surface area in both organs is very high, a fact which favours boundary crossing.

(2) By substances entering boundaries' cells, being in some way transported across those cells, and leaving them on the opposite side; this process can be referred to as intra-cellular transport; (B) and (C) in Fig. 11.1. A number of important cellular processes will be considered under this heading.

Intra-cellular Transport

Intra-cellular transport can be described under two headings: (1) transport across cellular membranes; and (2) transport across cells, and transport out of cells.

(1) *Transport across Cellular Membranes*
Cell membranes show an outer and inner layer of protein molecules, sandwiched between which are phospho-lipid molecules. (The nature of the cell membrane's microstructure has been the subject of detailed research. Following Davson and Danielli's seminal study in 1952, important reports have included those by Lucy (1968), Hendler (1971), Wallach and Fisher (1971), and Colacicco (1972).)

The outer layer is normally sticky, a fact which encourages adhesion of cells. The immune response is closely associated with this stickiness. Malignant cells in cancer possess little or no stickiness and can therefore spread easily.

In Fig. 9.1 are structures called microvilli, finger-like processes protruding from the surface of the cell. They are recognized to be mostly concerned with transport systems across the cell membrane. Plainly, the microvilli increase the surface area. The deep clefts between the microvilli appear to be the site at which substances become entrapped in pinocytosis. The microvilli are, presumably, functionally similar to the filopodia and to other protuberances of the cell membrane of phagocytic cells.

Observation of the diffusion of molecules into living cells shows that large, polar, water-soluble molecules do *not* readily diffuse into living cells. Small, polar, water-soluble molecules do enter living cells, supposedly by means of pores postulated in the cell membrane. These pores, however, permit only the passage of the small molecules. Lipid-soluble, non-polar molecules of all sizes permeate readily into living cells by solution in the phospho-lipid layer of the cell membrane.

The central role of water as a solvent in living processes might, at first sight, lead one to expect that water solubility is a key to the passive entry of molecules into living cells. But it is lipid solubility not water solubility which provides a key, and acceptance of this fact provides the first step towards an explanation of the behaviour of many lipid-soluble substances, such as the aromatic hydrocarbons. They enter living cells readily and accumulate in tissues with high lipid content, for example brain, liver, and adipose tissue. Non-lipid-soluble substances, such as sugars, amino acids and also ions, all enter cells, but not by diffusion. Non-lipid-soluble substances enter cells by means of facilitated and active transport systems.

The cell membrane is relatively impervious to the transport of ions, except by active transport systems. Potassium ions are kept within cells whereas sodium ions are actively 'pumped' out. This ionic separation is responsible for a number of vital biological phenomena. For example, an electrical potential difference can exist between the inside and outside of the cell, a phenomenon which plays a role in both brain and nerve function, and in the process of hearing. By means of ionic pumping, the cells' internal milieu is maintained in a different chemical state from that of the external milieu, represented by the tissue fluid. The differential is maintained by life processes at a cost of energy consumption. The death of the cell, or the impairment of its energy utilization, or the puncturing of the cell membrane destroys this vital differential.

Chapter 4 outlined the process of phagocytosis, in which the macrophage engulfs the particle. However, phagocytosis is a means whereby *all* particles find their way into cells. Phagocytosis of alien particles by macrophages can take place, but so also can phagocytosis of nutrient particles from, for example, the intestine. Pinocytosis is a term used to describe a process similar to phagocytosis, though it involves smaller particles, such as molecules and droplets of fluid.

Transmembranosis is a term, used in some writings, descriptive of processes in which molecules of a substance are transported across the cell membrane without the apparent involvement of phagocytosis or pinocytosis. Transmembranosis presumably includes diffusion of small molecules through the pores in the cell membrane, which involves no energy expenditure on the part of the cell. Transmembranosis can conveniently be used to describe all the 'energy-free' processes by which substances enter cells.

Diffusion results from the random movement of molecules of liquids (and also gases but consideration of these is not relevant here). The movement, called Brownian movement, results in molecules becoming dispersed within a solution. Fick's law states that the rate of transport of molecules by diffusion is proportional to the concentration gradient which provides the driving force for diffusion. Mathematical considerations, mentioned by Robinson (1975), showed that the distance travelled and the time taken to travel the distance are related by an inverse square law. This leads to the important conclusion that diffusion takes place rapidly over short distances, but only slowly over long distances. This observation helps to explain the role of diffusion: it is important for the movement of molecules over short distances but quite inadequate for the movement of molecules over long distances for which bulk flow of fluid is necessary.

Osmosis is the movement of water through a semi-permeable membrane (to be visualized as a sieve with small holes in it structured so as to allow through water but not the solute molecules). Osmosis can be described as follows: Suppose a semi-permeable membrane is used to separate pure water from sucrose solution and that the semi-permeable membrane prevents the sucrose from diffusing into the pure water. It is a fact that diffusion always takes place from high to low concentration. The concentration of water in the sucrose solution is *lower* than that of the pure water; consequently water passes through the semi-permeable membrane into the solution so that it becomes increasingly dilute.

Suppose now that the experiment is extended so that the passage of water into the solution is resisted by the application of pressure to the solution. If the pressure is steadily increased a point would be reached where the water ceased passing into the solution—this point marks the osmotic pressure. If the pressure is then further increased water would start to flow back from the solution into the pure water.

Ultra-filtration is the name given to the 'reverse' flow brought about by hydrostatic pressure in excess of osmotic pressure.

Permeation by solution—where a substance is soluble in the stuff of the cell-structured boundary or cell membrane, the substance will permeate by means of its solubility across the boundary or membrane.

(2) *Transport across Cells, and Transport out of Cells*
In the previous paragraphs various energy-free processes have been described whereby substances enter cells by crossing a membrane. But these processes do not apply solely to cellular membranes; indeed, a boundary formed by a sheet of cells can be regarded as a 'membrane'. Hence, the various processes previously described are

applicable to the body boundaries, especially where these are composed of layers only two or three cells deep.

Unfortunately, 'membrane' is commonly used to denote both cellular membranes and boundaries composed of cells, such as the alveolar membrane. For the purposes of this chapter 'membrane' is used exclusively in the context of cellular membrane, and 'boundary' is reserved for tissue membranes composed of sheets of cells. However, this restriction of terminology is limited in usefulness because 'membrane' is so often used to describe both cellular membranes and multi-cellular boundaries. Consequently, 'membrane' has to be used with a dual meaning elsewhere in this book.

Quite another view can be taken of the cells which constitute boundaries. Substances which gain access to cells become involved with cellular processes and by this means cross the cell. They leave it again by re-crossing the cellular membrane. It is postulated that carriers exist within the cellular cytoplasm and that these serve to carry substances from one side of the cell to another. Also, the processes described in the previous chapter (the synthesis protein or the energy metabolism at the mitochondria) can also be visualized as intra-cellular transport processes.

Trump and Arstila (1975) described exotropy and esotropy. Esotropy refers to the 'turning in' of a membrane so that the hitherto outer surface becomes the inner surface of a newly formed vacuole (Fig. 11.2). Esotropy takes place at the cellular membrane during phagocytosis; it also takes place from the Golgi apparatus into the cellular cytoplasm (see Chapter 10).

Exotropy refers to the 'turning out' of a membrane. On the outer surface of a cell this would involve the 'budding off' of a fragment of cytoplasm. Trump and Arstila used mitosis as a caricature of exotropy. It also occurs when buds of endoplasmic reticulum form within the endoplasmic channel taking with it the fragment of cellular cytoplasm.

In esotropy and exotropy we see two processes which play a fundamental role in intra-cellular transport; by recalling the processes of protein synthesis and also energy metabolism we can visualize intra-cellular transport as an input–process–output model, where the processes are represented by events at the mitochondria and the ribosomes. Manifestly, these events can be disordered by factors which interfere with cellular functioning at any point in the input–output model.

Certain substances which have large molecules and which are not lipid soluble appear to be able to cross cellular membrane relatively rapidly. Yet esotropy does not appear to be involved. Nor can diffusion, osmosis, or permeation by solution provide the explanation

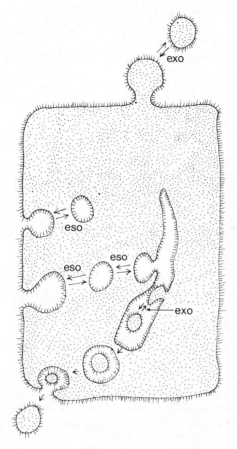

FIG. 11.2. Exotropy and Esotropy. Diagram showing exotropy (exo) and esotropy (eso). The two terms are described in the text. Redrawn, with acknowledgements, from Trump and Arstila (1975).

of the means whereby those substances enter cells. In order to provide an explanation 'facilitated transportation' is postulated. Facilitated transportation implies that the substances enter into temporary combination with an enzyme within the cellular membrane which acts as a ferry taking it from one side of the cellular membrane to the other. Facilitated diffusion is important because it provides an explanation of how cellular membranes can be selective in a way which would not be possible with diffusion or osmosis.

Facilitated transportation describes methods of transport across cell membrane where the quantities and rates of substances transported are not consistent with the purely passive process of diffusion. The existence of various supplemental processes is postulated; for example carriers within cell membrane, and enzymes active at the surface of the cell.

The characteristics of a facilitated transport system are:

(1) transport kinetics which are not wholly consistent with diffusion,

(2) specificity and selectivity which arise because a facilitated transport system tends only to carry substances possessed of specific chemical properties,

(3) competition between substances with similar chemical properties (a possible cause of selectivity),

(4) limited capability of the system—reflecting the possibility that a given facilitated transport process can become 'saturated' and act as a 'bottle-neck' for the system (this is a partial explanation of why the transport kinetics are not wholly consistent with diffusion), and

(5) the susceptibility of the transport system to being poisoned or inhibited in a variety of ways.

DISTRIBUTION

Substances presented to the cell-structured boundaries of skin, lung, and gut are transported before gaining access to the inside of the body. As already indicated, this applies to all substances except those which penetrate between cells. Now we look briefly at the systems by which substances are transported round the body to their target organs or to their storage sites.

The boundary membranes (lung, gut, and skin, etc.) are all closely associated with a network of capillaries. Substances transported across barrier cells therefore appear quickly in the circulating blood. Those substances deposited in tissue fluids by the membrane cell transport are 'drained' via the lymphatic system, ultimately to be brought into the main blood circulation. Chapter 3 contained a brief examination of the lymphatic drainage of the lung, and noted that the concept of lymphatic drainage, and of circulation, applies to all soft tissues of the body. Thus by one route or another most substances entering the body reach the blood, which provides the body's principal transport system.

Blood is a complex substance possessed of a number of important functions, one of which is the distribution of nutrients and non-nutrients. Water is a constituent of plasma (the fluid part of blood), so water-soluble substances circulate readily in the blood.

The cellular components of blood (mentioned in Chapter 4) act as vehicles for transport within the blood transport system. Phagocytic cells can pick up substances; both cells and the carried substances can then be swept round the circulation. The red cell contains haemoglobin which, by loose binding, carries the necessary quantities of oxygen to the tissues in which oxidation takes place. Haemoglobin allows the transport of more oxygen than would be possible on the basis of simple solubility. Unfortunately, however, haemoglobin can also act as the transport system for harmful substances such as carbon monoxide.

Plasma contains molecules of protein as well as water. These fall into two broad classes, known as albumins and globulins (globulins have already been encountered in the context of immunoglobulins, see Chapter 5). Both albumins and globulins act as carriers, as a result of their attachment to a wide range of nutrients and non-nutrients which find their way into the circulating blood. In some cases the attachment can be so tight that an otherwise harmful non-nutrient is prevented from reaching the target organ; in other instances the attachment can be loose and impermanent. If any factor interferes with the binding, the harmful non-nutrient may be released in sufficient quantities to cause damage, perhaps long after original exposure to the harmful substance. The blood circulation also carries particles which have eluded the defence processes. Micropredators are moved round the body and cancer can also spread in this way.

In previous chapters it has been implied that the biochemical pathways take place within cells and, in many instances, within the organelles of cells. Though this is correct, steps in biochemical pathways may also involve a change of site from one organ to another. The transport from one site to another is often provided by blood. Hence, the intermediates can sometimes be extracted from blood samples and provide the basis for biological measurement.

STORAGE

Substances accumulate in the body, but the sites of these accumulations may not be particularly susceptible to their biological activity. For example, in aromatic hydrocarbons the lipid solubility, which favours their entry to the body, also favours their accumulation in adipose tissue. Equilibrium is maintained between the adipose tissue and the circulating blood, and the former may be able to mop up relatively large quantities of lipid-soluble substances. In this way the adipose tissue acts as a protective reservoir. Here there may be a link between nutritional status and pathogenicity. People with little or no reserve of adipose tissue may be relatively susceptible to harmful lipid-soluble non-nutrients.

The brain and liver possess relatively large quantities of lipids. In the case of certain substances, such as the halogenated hydrocarbons, this may result in accumulation which is selective in its effect. The well-known narcotic effects of halogenated hydrocarbons reflect the brain's particular affinities for lipid solvents. The toxic effects of certain halogenated hydrocarbons on the liver may reflect particular lipid affinities with these hydrocarbons.

The fate of inhaled particles was discussed in Chapter 3, in which the lung's interstitial spaces were identified as a site for storage of certain categories of inhaled particles. In the case of metals such as iron and tin, the storage in the interstitial spaces takes place apparently without any harmful effect on the lung structure.

ACCUMULATION AND RETENTION: QUANTITATIVE ASPECTS

Nordberg (1976) has discussed the quantitative aspects of accumulation and retention for certain metals. However, his observations have general application. He pointed out that accumulation in an organ takes place when uptake exceeds elimination (bio-dumping). 'Biological half-time' is commonly used to express the elimination rate. It is defined as the time taken for the concentration in an organ to fall by one half and is calculated according to the following equation:

$$C = C_0\, e^{-bt} \tag{1}$$

where C is the concentration in the organ at time t, C_0 the concentration in the organ at time 0, b the elimination constant, and t the time.

Elimination constant and biological half-time are related:

$$T = \frac{1}{b} \log_e 2 \tag{2}$$

where T is the biological half-time, and $\log_e 2 = 0.693$.

On the assumption that eqn. (1) is valid, the Task Group on Metal Accumulation (cited by Nordberg) proposed the following expression of the amount accumulated in an organ during continuous exposure:

$$A = \frac{a}{b}(1 - e^{-bt}) \tag{3}$$

where *A* is the accumulated amount, *a* the fraction of daily intake taken up by the organ, *b* the elimination constant, and *t* the time of exposure. At steady state $a = b$ (by definition) and therefore $A = a/b$.

A number of important assumptions attach to these three equations. It is assumed that intake and absorption are constant, and that the exponential function adequately represents elimination. In the case of eqn. (3) it is also assumed that uptake exceeds elimination and that a constant fraction of the intake is taken up by the organ. The overall assumption is that the movement of substances is dependent only on concentration gradient. Clearly, active transport systems which can move substances against concentration gradients invalidate this assumption. Moreover, the involvement of more than one storage compartment would considerably complicate the equations. As Nordberg rightly pointed out, data need to be established in order to broaden the predictive possibilities of the three equations. Meanwhile, they represent the foundation—however inadequate—for a quantitative approach. This is exemplified by the findings of the Task Group on Metal Accumulation that biological half-time for lead in bone varied from 64 days, for spines of rats, to 7500 days for the skeleton of a dog. The biological half-time for humans' skeletal lead is said to be 18 years.

BIO-DUMPING

Water-soluble substances are eliminated from the body (bio-dumping) via the kidney; see Fig. 11.3. An outline of kidney function, moreover, provides a glimpse of how a number of harmful agents cause harm not only to bio-dumping processes but also to vital homeostatic processes.

The kidney is a selective *two-way* (see below) filter, across one face of which blood is continually flowing. The very large surface area is provided by the capillary blood vessels within the kidneys' glomeruli (the filter units). Filtration takes place so that water-soluble substances pass readily into the urine's side of the filter. High molecular weight substances, such as proteins, are normally held back. Damage to the glomeruli, for example by cadmium or mercury, causes the filter to leak protein. Among the water-soluble substances filtered at the glomeruli are substances which the body cannot lose in quantity without being harmed. Such substances are absorbed back into the blood stream from the kidney tubules (see Fig. 11.4) and in this way vital substances are conserved.

The renal tubular system represents processes by which homeo-stasis is exercised over, for example, water balance, the electrolyte

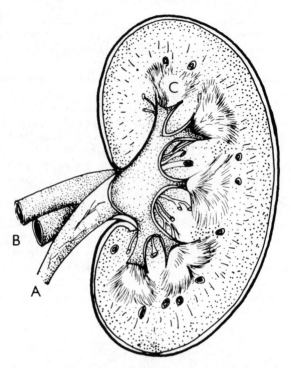

FIG. 11.3. Drawing of the kidney showing, in section: (A) ureter, the duct through which the urine flows, (B) renal artery and vein carrying blood to and from the kidney, and (C) area of the kidney shown in diagrammatic form in Fig. 11.4.

concentration, and the pH of the blood. The homeostasis is achieved by selective secretion and absorption of water and electrolytes in such a way that the body water balance and chemical constancy of the blood are maintained. Maintenance of water balance by the kidney against a concentration gradient is a vitally important homeostatic mechanism. The epithet 'two-way' reflects the ability of the kidney to absorb and secrete water, electrolytes, and other substances.

It is not within this book's scope to consider kidney function in any greater detail but it should be pointed out that this function, and its study, represents a profoundly important field in human physiology. It should also be emphasized that kidney function is vulnerable to attack by inputs encountered in the course of occupation, sometimes with fatal results as the following case history shows.

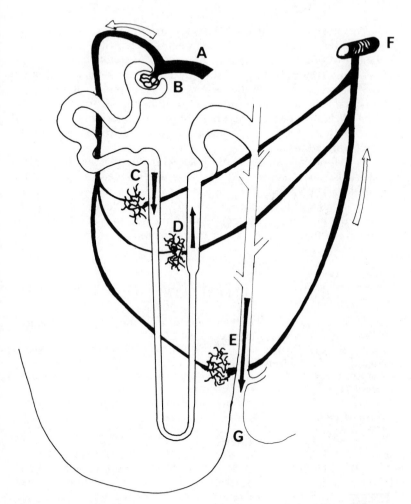

FIG. 11.4. Simplified diagram of renal function. (A) Branch of renal artery carrying blood to capillaries in glomerulus, (B) glomerulus, (C, D, E) capillary network at which absorption and secretion takes place, acting as a homeostatic mechanism for substances in the blood such as electrolytes, (F) renal vein, taking blood from kidney, and (G) the renal papilla where the 'fully processed' urine leaves the kidney to start its journey down the ureter.

Trump and Arstila (1975) described a clinical case which exemplifies the vital role of the kidney:

> An unspecified amount of mercuric chloride solution was drunk by a 42-year-old laboratory worker. She felt sick and began to vomit. Following admission to hospital she was unsuccessfully treated with a substance (dimercaprol) aimed at binding the heavy metal. On the second day following admission to hospital her urine was found to contain 8 g of protein per 24 h (the normal value is in the range 10 to 150 mg per 24 h) and blood. The output of urine then ceased and she died on the fifth day following admission.
>
> Post-mortem examination of her kidneys revealed widespread necrosis (see Chapter 13) of the epithelial cells of the kidney tubules. The raised level of protein in her urine on the second day indicated that damage to the kidney had occurred; the subsequent failure of urine output indicated that the damage was so severe that even the water transportation systems were out of action.

REFERENCES

COLACICCO, G. (1972). Surface behaviour of membrane proteins, *Annals of the New York Academy of Sciences*, **195**, 224–61.

DAVSON, H. and DANIELLI, J. F. (1952). *The Permeability of Natural Membranes*, Cambridge: The University Press.

HENDLER, R. W. (1971). Biological membrane ultrastructure, *Physiological Review*, **51**, 66–97.

LUCY, J. A. (1968). Ultrastructure of membranes: micellar organization, *British Medical Bulletin*, **24**, 127–9.

NORDBERG, G. F. (ed.) (1976). *Effects and Dose–Response Relationships of Toxic Metals*, Amsterdam: Elsevier.

ROBINSON, J. R. (1975). *A Prelude to Physiology*, Oxford: Blackwell.

TRUMP, B. F. and ARSTILA, A. U. (1975). In *Principles of Pathobiology*, 2nd Edition, M. F. La Via and R. B. Hill, eds, New York: Oxford University Press.

WAHLBERG, J. E. (1976). Percutaneous toxicity of solvents. A comparative investigation in the guinea pig with benzene, toluene and 1,1,2-trichloroethane, *Annals of Occupational Hygiene*, **19**, 115–19.

WALLACH, D. F. H. and FISCHER, H. (eds) (1971). *The Dynamic Structure of Cell Membranes*, Berlin: Springer-Verlag.

Section 4: MODES OF ACTION, PATHOLOGICAL PROCESSES AND DISEASE FOLLOWING HARMFUL INPUTS

Chapters 12–16 deal with non-specific pathological changes arising as a result of harmful inputs of one type or another. It should be recalled that 'pathological' includes changes which may be wholly or partly beneficial, and that certain harmful outputs represent the hyperaction or malfunction of pathological processes ordinarily beneficial in their operation.

Although these chapters are dealing with non-specific processes, examples are given of responses to specific agents of relevance to occupational health and safety.

Chapter 12

Dose, Effects, Quantitative Relations and Target Organs

OBJECTIVES OF THE CHAPTER

This chapter provides definitions and descriptions of certain concepts important for a quantitative approach to human safety:
  (1) effects,
  (2) dose–response relations, dose–effect relations, dose–output relations and dose,
  (3) critical effect, critical organ, critical organ concentration, critical concentration, lethal concentration, sub-critical effects, and sub-critical concentrations, and target organs,

and

  (4) outlines functions of certain target organs, and
  (5) discusses certain quantitative problems in specifying dose.

DEFINITIONS

The previous chapter examined how input substances cross barriers, how they move about once inside the body, and how an approach is made to the quantitative aspects of accumulation. This chapter will examine some general aspects of harmful chemical substances acting as inputs to the body, as defined in Chapter 2. Further consideration will be given to the quantitative aspects.

As a preamble it should be pointed out that a chemical substance may become pathogenic (see below) for different reasons:
  (a) because it is present in such quantity or its nature is so toxic that the bodily defences are of no avail,
  (b) because the defences are defective and therefore unable to cope with attack which might otherwise be overcome (it should be recalled that defences can be impaired by factors apart from whatever chemical substance is acting as input),
  (c) because there exists no biological defence process against the attacking substances, or
  (d) because the defence process produces an excessive reaction which becomes a disease process.

This may happen when defence processes function for an excessive time or to an excessive extent, or when the individual is, for some reason, especially sensitive to the manifestations of a particular defence process.

'Pathogenic' was used in Chapter 8 to describe the potential of a chemical substance for producing disease in people. But there are changes induced by chemical substances for which 'disease' is too strong a word; indeed, there are changes caused by chemical substances which the body *appears* to be able to tolerate without any harmful outcome.

The term 'effects' is used here to cover the range of changes from the pathogenic to the apparently harmless, and includes those resulting from overreaction on the part of bodily defence processes. The term 'effects' and others to be described in this chapter are based on definitions which have been put forward by the Task Group on Metal Accumulation (1973, cited by Nordberg 1976) in their consideration of the problems of quantitative relations between toxic metals and biological effects. Although the terms used were derived to apply to metals, they are applicable beyond the boundaries of metal toxicology, as will be demonstrated.

Knowledge of the quantitative relation between any input and its effect is sought for purposes such as standard setting, as described in Chapter 24. The terms 'dose–response relation' and 'dose–effect relation' are used to signify the quantitative relations.

The Task Group drew a distinction between *dose–response relations* and dose–effect relations, response being used to signify that proportion of a population which experiences a specific effect following exposure of the total population to specified harmful input. The correlation of the response with estimates of the *dose* (defined later) provides a dose–response relation, which is normally expressed as a graph, with percentage of population affected on the ordinate and estimated dose on the abscissa.

A *dose–effect relation* is the correlation between *dose* (defined later) and the magnitude of effect in a specified proportion of the population.

Figures 12.1 and 12.2 are included in order to exemplify the concepts of dose–response relation and dose–effect relation. Both relate to noise and its effect on hearing. Dose of noise can be defined as the quantity of frequency-weighted sound energy received during a specified period of time. The effect of noise on hearing can be expressed in decibels of hearing loss attributable solely to noise (see Appendix 12.1). Noise dose is expressed, in decibels, as *noise immission level* denoted E_A. *Hearing loss* is expressed in decibels. Figure 12.1 shows the dose–response relation for noise immission

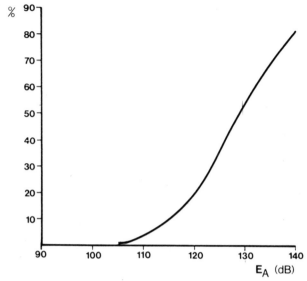

Fig. 12.1. Dose–response relation for noise and its effect on hearing. Dose of noise is given as noise emission level in decibels (abscissa). The response is the percentage of an exposed population experiencing noise-induced hearing loss of 40 dB or more, measured as the average of three test frequencies, 1000, 2000 and 3000 Hz.

level and a noise-induced hearing loss of 40 dB or more, measured as the average of three test frequencies 1000, 2000, and 3000 Hz. (This frequency combination and the value of 40 dB have been chosen by way of examples because these correspond to the values used in connection with the British National Insurance Scheme for Occupational Deafness.)

Figure 12.2 displays the dose–effect relation for noise and noise-induced hearing loss. The abscissa is noise dose as in Fig. 12.1. The ordinate is hearing loss, measured in decibels as the average loss at the three test frequencies 1000, 2000, and 3000 Hz. The parameters in Fig. 12.2 represent the centiles of the population. The 50th centile is, of course, the median value. Thus the 50th centile curve represents the dose–effect relation for the median individual in a population exposed to noise. However, it should be noted that *the curve of the 50th centile does not necessarily represent the progress of any one individual as his/her noise dose increases with time.* Rather, the curve should be regarded as a collection of points linked by a curve where each point represents the median for one population.

Empirical knowledge relating noise and hearing loss is now exten-

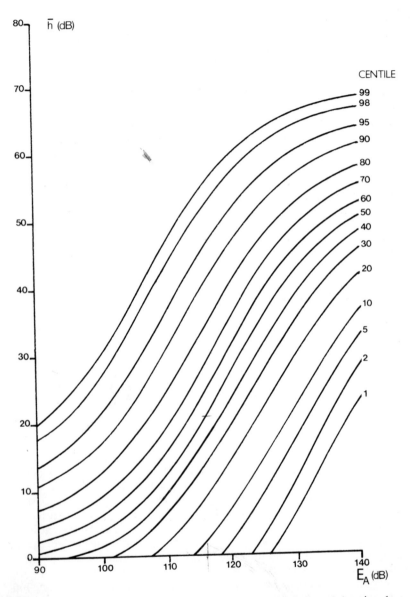

FIG. 12.2. Dose–effect relation for noise and noise-induced hearing loss. Abscissa: noise dose, as in Fig. 12.1. Ordinate: hearing loss, measured in decibels as the average loss at the three test frequencies, 1000, 2000 and 3000 Hz.

sive; consequently, a family of curves of dose–effect relation can be drawn. With many other harmful inputs, knowledge is nothing like as extensive and therefore it would not be possible to draw up curves in as much detail and with as much apparent precision.

As stated, _effect_ can be any observable biological change associated with the input concerned, though ideally it should be quantifiable. It is implicit in dose–effect relations that effect is related to and caused by the dose. Effect, therefore, is carefully defined so as to _include_ observable biological changes and to _exclude_ outputs which imply no biological change. This point can be clarified by reference, for example, to the excretion of cadmium in urine, which may reflect no more than the absorption, transport and excretion of cadmium ions by ordinary processes; and which may be without any accompanying observable biological effect in the individual. On this pattern, measurement of cadmium in urine may be regarded, assuming no retention, as a measurement of the _dose_ of cadmium input. A quantitative statement of the relation between, for example, urinary concentration of cadmium and the daily input of cadmium in the diet would be a '_dose–output relation_', and not a dose–effect relation, because no biological change is observable.

Superficially, the dose–output relation appears to be justification of biological monitoring in which the output by biological dumping is used as an indicator of exposure. However, use of a dose–output relation in this way requires the presumption that, over prolonged exposure periods, no effect (as defined above) is likely to occur within the range of the dose–output relation in use for the biological monitoring. For many substances no such presumption can safely be made because relevant knowledge does not exist. The key issue is observability; plainly, as knowledge advances more and more effects will be recognized.

It is now necessary to consider in more detail what is meant by 'effect' in the context of dose–effect relation. The term 'effect' is used to denote a biological change which is caused by exposure to a dose of a stipulated input. Used thus, effect does not necessarily denote adverse biological change, but embraces any biological change. Certain effects become adverse, however, but only if they are present to a sufficient degree or for a sufficient period of time. Hence the importance of the adverse effect for the individual may or may not be immediate.

Critical effect is the term used to define that point at which an effect can be said to be adverse. Effects, particularly adverse effects, are generally manifestations of the change in an organ and particularly the cells of the organ. The term _critical organ_ is that used to define the organ which first attains the _critical organ concentration_ (defined below). Such a condition only applies under specified circumstances of exposure. (This definition of critical organ is that reported by Nordberg and not that of the International Commission on Radiological Protection.)

The function of an organ is controlled by the behaviour of the cells within that organ. It should be stressed that organs normally contain more than one type of functional cell, and in discussion of adverse effects the key consideration is, thus, the concentration for a cell.

The *critical concentration* for a cell has been defined as that at which adverse changes, whether reversible or irreversible, occur in the cell. *Lethal concentration* has been defined as the cellular concentration necessary to cause the death of the cell, while *critical organ concentration* is defined as the stage at which the mean concentration in *any* of the organ's cells reaches critical concentration.

Critical organ concentration seems, given the present state of knowledge, to be the parameter of greatest utility in estimating dose. Whole body concentration provides a less useful criterion, because the organs in which greatest accumulation occurs may not be critical organs. Bone, for example, accumulates lead, but the critical organ is bone marrow, which is functionally separate from the bone which surrounds it. At some time in the future it will, no doubt, be possible to estimate dose in terms of critical cell concentration—or subcellular concentration—but at present this is impracticable.

There are complexities in the specification of effect, since certain effects, such as death, are of an all-or-none character, while others are of a graded nature, such as occupational deafness. Specification is further complicated by the fact that certain all-or-none effects (cancer, for example) require only a trigger. Once triggered they continue by self-propagation or by other processes independent of the dose of the triggering input. On the other hand, many observable and gradable effects are both trivial and reversible.

However, the complexities do not end here. The specification of dose needs to take account of all possible modes of input, and the non-occupational as well as the occupational possibilities. For example, in the case of metals like lead, in most if not in all countries, input by ingestion from the normal diet is inevitable. Any occupational exposure, probably by inhalation, will be supplemented by the non-occupational dose. Combination of the two may cause a critical organ concentration to be reached in the bone marrow or in other organs.

At an exposure level below that which would cause critical concentration in a critical organ there may be effects which are, apparently, without adverse consequences. These may be detectable, for example by biochemical tests. Such effects are known as *sub-critical effects*, and the concentrations needed to produce them are termed *sub-critical concentrations*.

Ideally, *dose* should be defined as the concentration of substance (or of other form of input) at the site of effect, regard being made to the time for which the concentration is maintained. Because of practical difficulties of measurement, it is seldom possible to specify the dose actually at the site of effect. In the instance of asbestos the dose in the critical organ—lung—is inferred from measurements of the milieux to which the organ is being exposed. This inference is complicated by the activities of the respiratory defence processes. Dose is, in fact, a quantity which is less easy to specify than the definition would appear to suggest; some of the problems involved in the concept of dose and its relation with the time over which the dose is distributed are identified later in this Chapter.

In the case of substances which are transported in the blood, blood concentration is sometimes used as an indicator of dose. On the other hand, blood concentration results from a number of factors, such as accumulation and retention, and hence may not reflect dose nor relate quantitatively to effect.

Target organs are defined, for the purposes of this book, as organs in which critical effects are observed as the result of exposure to a harmful input. There are many identifiable instances of inputs which affect a number of critical organs. Which they affect depends upon the circumstances of exposure, the interplay of defence processes and the susceptibility of the individual, as well as the tissues of the target organ. Thus, in discussing effects it is required that all possible target organs are considered.

The definition of 'target organs' must, necessarily, be wide, and must include, where appropriate, systems and tissues as well as organs. For example, the target organ of hydrogen sulphide (see Chapter 22), which attacks the nerve tissue and causes respiratory paralysis, might be categorized as the central nervous system. Crocidolite (see Chapter 20) induces serious disease of the pleura (the tissue lining in the inner surface of the chest wall, the lungs, the inner surface of the abdominal cavity and the abdominal organs). In this instance the pleura is the target organ.

Obviously there are as many target organs as there are organs, systems, tissues and different types of cell in the body. It is intended that in this study only certain target organs will be discussed. These are listed, together with a very brief outline of their function, in Table 12.1. Readers should have outline knowledge of the principal elements of the structure as well as the function of these organs. Those

TABLE 12.1

TARGET ORGANS, WITH AN OUTLINE OF THEIR PRINCIPAL FUNCTIONS

Target organ	Principal functions
Skin	Protects against friction, water/fluid loss, entry of harmful inputs; thermal insulation; self-greasing by means of sebaceous glands; thermoregulatory by means of sweat glands; receives afferent information
Respiratory tract	Oxygen and carbon dioxide exchange; defence against aerosols; warming and moistening of incoming air; excretion of gases, vapours Metabolism: transformation and conjugation
Blood, plasma, blood-forming organs; circulatory system	Chief transport system for oxygen, carbon dioxide nutrients, heat and fluids
Kidney, urinary tract	Excretion: water, salts and nitrogenous wastes (includes homeostasis as well as bio-dumping) Secretion: hormones for controlling blood pressure and production of red blood cells Metabolism: transportation and conjugation
Liver	Secretory: (a) bile—contains waste non-nutrients, aids digestion (b) heparin—anti-coagulant for blood Storage: (a) vitamins (b) iron (for haemoglobin) (c) glycogen—energy store substance Metabolism: transformation and conjugation
Brain and nervous system	Information processing and control of bodily activities
Bone	Support framework for movement and protection (certain bones house blood-forming organs; but those are functionally separate from bone)
Gut	Input of nutrients; digestion; excretion of non-nutrients; defensive processes of gastric-acid barrier
Lymphoid system and lymphatics	Tissue drainage; filtration; site of defensive processes such as immune response and phagocytosis
Ductless glands	Such as thyroid, parathyroids, adrenals (suprarenals) produce hormones—substances exercising key control over function and morphology

requiring further information are referred to the reference section, which contains examples of introductory books from which the necessary outline knowledge can be obtained.

In the following chapters the four categories of non-specific effects (irritation, inflammation and degeneration; altered sensitivity; disorder of repair; and disorder of growth) are discussed briefly in relation to their relevant target organs.

PROBLEMS WITH SPECIFYING DOSE FOR CARCINOGENS: QUANTITATIVE ASPECTS

The specification of dose for carcinogens presents a number of theoretical difficulties which need to be discussed. For the purposes of the present discussion it is assumed that the quantitative aspects of a carcinogen are being studied. Further, it is assumed that individuals receive, each day, a measurable daily dose of carcinogen, d. It is observed that victims develop cancer after T years of exposure; T may be termed the induction period.

The total quantity, D, of carcinogen received by a victim up to the time the cancer develops will be given by

$$D = 240 \ T \times d$$

on the assumption that there are 240 working days per year.

Summation Law

Carcinogenesis could obey a *summation law* as discussed by Süss *et al.* (1973). Under a summation law, cancer would develop at a constant value for the *total dose D* no matter how D was delivered. Süss *et al.* found that a summation law is, in fact, valid for certain carcinogens administered to experimental animals.

Stepping outside the field of carcinogenesis for a moment we can see an example of a summation with occupational deafness. Underlying the mathematical relation between noise and occupational deafness is a summation law: occupational deafness is related to the total quantity of frequency-weighted sound delivered to the ear.

Time Dependency for Carcinogenesis

To return to carcinogenesis: Süss *et al.* found that the summation law did not hold in the case of a number of carcinogens given to experimental animals. They noted that the data in some experiments showed a better fit with the following relation:

$$dT^2 = \text{constant}$$

They went on to explore data from several experiments involving different animals and different carcinogens and they found that the relation between d and T could be written in a general form:

$$dT^n = \text{constant}$$

From the experimental data they noted that n lay in the range between 1 and 4. For human data, so far as these were available, it appeared that n could be as high as 6. The relation $dT^n = \text{constant}$ gives time an important role in carcinogenesis: with a long latent period only very small doses of carcinogen would be required in order to initiate the disease where n is greater than 1. The relation implied by the equation postulated by Süss *et al.* may be called a time related law.

The Implications of a Time Related Law for Dose Relations

The relation postulated by Süss *et al.* can be applied to a particular carcinogen, asbestos. For the purposes of the present discussion the constant can be labelled K, hence

$$dT^n = K$$

In Chapter 20 mention is made of the British Occupational Hygiene Society's Hygiene Standard for Chrysolite Asbestos Dust (1968). The Standard stated:

> It is probable that the risk of being affected [*by asbestos, my note*] will be less than 1% for an accumulated exposure of 100 fibre years per cubic centimetre. This is, for example, a concentration of 2 fibres per cubic centimetre for 50 years, 4 fibres per cubic centimetre for 25 years or 10 fibres per cubic centimetre for 10 years.

It is evident that the Hygiene Standard assumed a summation law for asbestos and asbestosis. The Hygiene Standard specifically excluded cancer from its scope but subsequent evidence has linked asbestos exposure to lung cancer. *As a matter of speculation* it is interesting to explore possible quantitative relations between asbestos and the initiation of lung cancer.

Let K be 100 fibre years per cubic centimetre; let T, in this instance the induction period of lung cancer from asbestos, take any value between 1 and 50 years; and let n have any value between 1 and 6.

Now

$$dT^n = K$$

and

$$\log K = \log d + n \log T$$

therefore

$$\log d = -n \log T + \log K$$

Thus, $\log d$ can be plotted against $\log T$ with n as the slope. Figure 12.3 shows the results. Twenty years is a plausible value to take for T; at this value it can be seen that the daily dose of fibres required to initiate lung cancer varies considerably with n. With the higher values of n remarkably small doses of asbestos would be required for daily

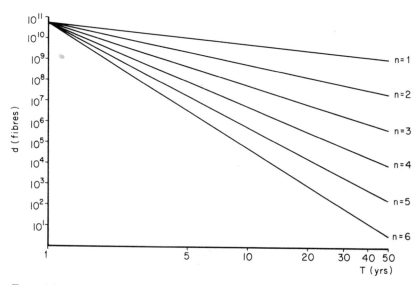

FIG. 12.3. Asbestos and the induction period for lung cancer. Graphs showing relation between T, n and d for $K = 100$ fibre years per cubic centimetre of air. d is the dose of fibres assumed to be inhaled during the course of an 8 h working day assuming an average lung ventilation of 5000 cubic centimetres per minute; T is the induction time measured in years; and n is the exponent in the equation $dT^n = K$.

Notes: (a) It is assumed that all fibres inhaled are retained and that all retained fibres play a full part in the initiation of the lung cancer.

(b) 'Fibres' are defined here in exactly the same way as in the BOHS Hygiene Standard (see text); d represents the total number of fibres, inhaled each day within the BOHS size range as defined in 1968.

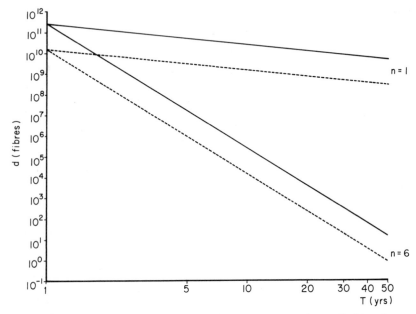

FIG. 12.4. Asbestos, the latent period of lung cancer, and two limiting doses of asbestos. Graph showing the influence of K compared with n. (-----) $K = 25$ fibre years per cubic centimetre; (———) $K = 400$ fibre years per cubic centimetre. Other data as shown in Fig. 12.3.

exposure to initiate lung cancer. Here is a plausible explanation for suggestions that lung cancer can arise following exposure to apparently trivial quantities of asbestos.

As a further step in speculation, an exploration can be made of the effect of using values for K other than 100 fibre years per cubic centimetre. Figure 12.4 shows data for 25 fibres per cubic centimetre and 400 fibres per cubic centimetre for $n = 1$ and $n = 6$, demonstrating the importance of n compared with K. If *in reality n* for lung cancer caused by asbestos were shown to have a value in excess of 1 then a reduction in the 'limiting dose' of asbestos from 100 fibre years per cubic centimetre to 25 fibre years per cubic centimetre would not extirpate the risk of lung cancer from asbestos.

Dose-Rate Dependency

In Chapter 17, in the discussion on energies, it is observed that certain effects of ionizing radiations are dose-rate dependent. One way of expressing such a relation is:

$$d \times t^n = \text{constant}$$

where d is the daily dose, as before, and t is the number of hours in a day over which the daily dose is delivered. Although no such relation has been demonstrated empirically for carcinogenic substances the possibility cannot be ruled out that some, at least, behave in this way. Indeed, there are hints of dose-rate dependency for effects other than carcinogenesis caused by both energies and substances, such as thermal changes in tissue from applied heat, and inflammation caused by irritant substances.

The purpose of the preceding discussion has been to bring forward the possibility that other factors besides dose may have to be considered in the establishment of dose–response and dose–effect relations. Such factors include the value of the exponent n in the general relation $dT^n = \text{constant}$, the induction period of the disease, and dose-rate dependency.

In Chapter 24 mention is made of 'time-weighted average' as a method of calculation. This method of calculation, and others used in practical occupational hygiene, rely on an assumption that a summation law, or something akin to it, is valid generally in human safety. In many instances such validity has yet to be established. This comment is not to be seen as a criticism aimed at undermining the practice of occupational hygiene but, rather, a reminder that 'dose–response relations' and 'dose–effect relations' are sometimes complicated matters requiring knowledge of the temporal pattern of exposure as well as the total dose received during a stipulated period.

APPENDIX 12.1: CALCULATIONS RELATING NOISE AND NOISE-INDUCED HEARING LOSS

Noise can be measured using the A scale of frequency weighting (see Fig. 12.A1). Such measurements are expressed in dB(A) and denoted by L_A.

Noise emission level, E_A, is given by the following expression (Robinson, 1970):

$$E_A = L_A + 10 \log T/T_0$$

where T is the period of exposure in years to the noise level L_A, and $T_0 = 1$. (The purpose of T_0 being 1 is to achieve a dimensionless ratio for the logarithm.)

Hearing loss attributable to noise in an individual of age N is given

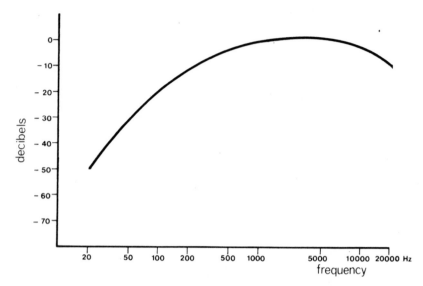

FIG. 12.A1. A-weighted curve.

by the following expression:

$$H - F(N) = h$$

where H is the hearing level at a stipulated frequency, expressed in decibels relative to a standardized zero. The hearing level is found by asking a subject to listen to a pure tone at a stipulated frequency, the tone then is adjusted in level until the individual is just able to hear it. $F(N)$ is the hearing level supposedly of an average male or female of age N; h is the noise-induced *hearing loss*. It is assumed that the effects of age and noise are additive.

Noise level, nose emission level and noise-induced hearing loss can be related mathematically by means of the following expression (Robinson, 1970):

$$H(P) = 27.5 \left\{ 1 + \tanh \frac{E_A - \lambda(f) + u(P)}{15} \right\} + u(P) + F(N)$$

where N is the age in years,

$$F(N) = \begin{cases} 0 \text{ when } N \leqslant 20 \\ c(f).(N - 20)^2 \text{ when } N > 20 \end{cases}$$

$c(f)$ has the values given in Table 12.A1,
$\lambda(f)$ has the values given in Table 12.A1,

$$u(P) = 6\sqrt{2} \, \mathrm{erf}^{-1} \left(\frac{P}{50} - 1 \right)$$

P is the centile of population for which $H \geqslant H(P)$

TABLE 12.A1
VALUES OF $c(f)$ AND $\lambda(f)$ (FROM ROBINSON, 1971)

Frequency, Hz (f)	1	2	3
c	0.004 3	0.006	0.008
λ (dB)	126.5	120.0	114.5

Average hearing loss; suppose an individual's noise-induced hearing loss at 1000 Hz is h_1, h_2 at 2000 Hz, and h_3 at 3000 Hz, respectively; the average hearing loss $\bar{h}$ is then given by:

$$\bar{h} = \frac{h_1 + h_2 + h_3}{3}$$

REFERENCES

BRITISH OCCUPATIONAL HYGIENE SOCIETY (1968). *Hygiene Standard for Chrysotile Asbestos Dust*, Oxford: Pergamon.

NORDBERG, G. F. (ed.) (1976). *Effects and Dose–Response Relationships of Toxic Metals*, Amsterdam: Elsevier.

ROBINSON, D. W. (1970). In *Hearing and Noise in Industry*, W. Burns and D. W. Robinson, eds, London: HMSO.

ROBINSON, D. W. (1971). In *Occupational Hearing Loss*, D. W. Robinson, ed., London: Academic Press.

SÜSS, R. KINZEL, V. and SCRIBNER, J. D. (1973). *Cancer Experiments and Concepts*, New York: Springer-Verlag.

Other Reading
For outline knowledge of structure and function of bodily organs:

GREEN, J. H. (1976). *An Introduction to Human Physiology*, 4th Edition, London: Oxford University Press.

ROWETT, H. G. Q. (1973). *Basic Anatomy and Physiology*, 2nd Edition, London: John Murray.

VINES, A. E. and REES, N. (1973). *Human Biology*, 2nd Edition, London: Pitman.

For Reference
WILLIAMS, P. L. and WARWICK, R. (eds) (1973). *Gray's Anatomy*, 35th Edition, Edinburgh: Longman.

Chapter 13

Inflammation as a Harmful Process

OBJECTIVES OF THE CHAPTER

This chapter:
(1) emphasizes the point that inflammation may be defensive or harmful,
(2) indicates what is understood by cell death, necrosis, and cell degeneration,
(3) emphasizes the point that much disease (occupational and non-occupational) represents the harmful manifestations of inflammation,
(4) explains the terminology of disease resulting from harmful inflammation in the target organs skin, respiratory pathway, blood, blood forming organs and circulatory system, kidney and urinary tract, liver, gut, and the special senses,
(5) outlines the mode of action of certain harmful inputs causing inflammation, cell death, necrosis, or degeneration, and
(6) outlines the structure of skin.

INTRODUCTION

In Chapter 5 a look was taken at inflammatory response and immune reaction. Inflammation, it will be recalled, is the reaction of tissues to injury (of almost any type) which is insufficient to kill the tissue. Inflammation is labelled 'acute' or 'chronic' according to its duration. 'Acute' implies hours or days whereas 'chronic' implies months or years. Inflammation can also be classified on the basis of the types of cells predominating.

Injury to cells may be sufficient to cause their death. *After* the cells are dead changes take place within them which, collectively, are called necrosis. One further pattern of response by cells to injury needs to be noted here: degeneration. This refers to a condition where, in response to injury, the cells do not die but show marked changes in structure and function. It is these changes which are known as degeneration. Degeneration is separate from chronic inflammation, although the two may co-exist.

As stated in Chapter 5, inflammation is a defensive process but if it persists for too long it becomes the cause of disease. A number of important occupational diseases manifest the harmful features of inflammation. Obviously, much non-occupational disease also reflects the harmful manifestations of inflammation. The purpose of this chapter is to identify and describe briefly examples of the harmful manifestations of inflammation in certain target organs commonly the site of occupational disease of the pattern being discussed here. Skin and lung are the two target organs selected for discussion.

In the outlines which follow, the modes of action of the harmful inputs are related, as far as possible, to the response. The overall purpose is to link terms commonly encountered in descriptions of occupational disease with the underlying pathological processes and to indicate the modes of action. It should be emphasized, however, that what follows is a highly selective review of a very large field.

SKIN

The boundary or barrier function of the skin has been referred to in previous chapters, as has its role in bio-dumping of heat. It is considered here as a target organ only in regard to effects which can be reasonably classified under inflammation and cell death.

Figure 13.1 displays the principal structures of the skin. The horny layer on the outer surface consists of dead cells covered with skin fats and kept moist with sweat. This layer provides a moderately effective barrier against chemical substances and ultra-violet energy. The skin is to some extent protected against trauma by the cushioning effect of subcutaneous fat. Inflammation of the skin follows the same general pattern of inflammation in any organ of the body, that is to say there are vascular changes, there is increased vessel permeability (with resulting exudate and swelling) and there is migration of cells. In addition, in certain inflammatory conditions of the skin the production of skin cells from the basal cells is speeded up. This results in exfoliation—the shedding of excessive quantities of surface cells and scales. The exudate may find its way to the surface. Inflamed skin is painful or itchy, red and fissured, sometimes with exudate and scales.

Chemical substances attack the skin by pervasion or implantation through the horny layer or by way of the ducts which lead to the surface of the skin. The resulting inflammation is called *contact dermatitis*. Four broad classes of contact dermatitis are recognized (Hjorth and Fregert, 1972):
  (1) irritant dermatitis,
      (1.1) acute irritant dermatitis,
      (1.2) cumulative insult dermatitis,

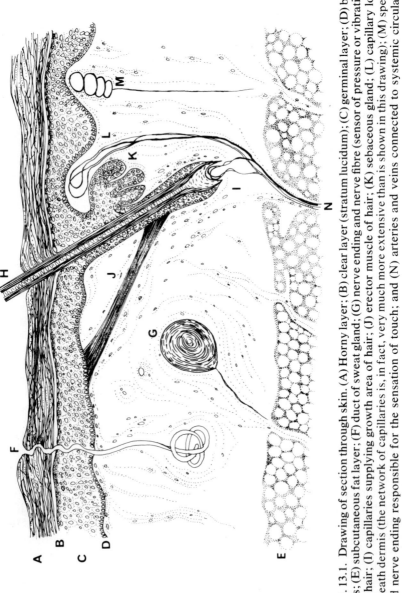

FIG. 13.1. Drawing of section through skin. (A) Horny layer; (B) clear layer (stratum lucidum); (C) germinal layer; (D) basal cells; (E) subcutaneous fat layer; (F) duct of sweat gland; (G) nerve ending and nerve fibre (sensor of pressure or vibration); (H) hair; (I) capillaries supplying growth area of hair; (J) erector muscle of hair; (K) sebaceous gland; (L) capillary loops beneath dermis (the network of capillaries is, in fact, very much more extensive than is shown in this drawing); (M) specialized nerve ending responsible for the sensation of touch; and (N) arteries and veins connected to systemic circulation.

(2) allergic contact dermatitis,
(3) phototoxic dermatitis, and
(4) photo-allergic dermatitis.

All four classes of contact dermatitis are characterized by inflammatory changes in the skin. In the brief discussion which follows, consideration is confined to the irritant types of contact dermatitis; the allergic types are considered under the general heading of 'Altered Sensitivity' in Chapter 14. The phototoxic and photo-allergic types of dermatitis are not discussed further.

Irritant dermatitis involves damage to the skin. When the concentration of the irritant is high, or it is particularly aggressive, irritant dermatitis or worse, destruction of skin, is inevitable. Irritants comprise acids, alkalis, caustics, detergents, emulsifiers, oils, metallic particles, solvents, oxidising and reducing agents, and biologically-active inputs generally.

Acute irritant dermatitis refers to the acute inflammatory changes resultant upon exposure of the skin to irritants sufficient in nature or extent to overcome the skin's natural resistance.

Cumulative insult dermatitis may develop after repeated insults by weak irritants over a long period. The resultant inflammation may very well be acute in pattern but the repeated insult takes place over a very long period of time. It would not be desirable to label this pattern of exposure 'chronic' because the word should be reserved to describe the temporal pattern of the inflammation and not the exposure.

The inputs, especially those causing cumulative insult, may act alone or in conjunction with micropredators and energies, which are themselves capable of causing skin inflammation. Modes of input of these agents include implantation, surface penetration, and irradiation. Climatic factors which cause extreme thermoregulatory response from the skin, particularly extreme sweating, may predispose the skin to the attack by irritants.

In general, damage results from abrasion, destruction of the horny layer, removal of protective fat, denaturing of skin proteins, or from specific chemical reactions. It is difficult to state precisely where irritant dermatitis stops and more destructive effects take over. For example, certain substances have a violently irritant effect on skin, causing blistering as well as inflammation. There may be a graded response to moderate exposure, but severe exposure leads to cell death and necrosis and possibly to secondary bacterial invasion. Examples of cell death and necrosis are seen in the chronic ulceration of the skin of the hands and nasal septum associated with chrome plating, and in the skin of the hands of people exposed to an excessive degree of brine or caustic agents.

The ducts with their openings on the skin provide a mode of input and they are also the site of inflammatory change. Thus *folliculitis* (oil acne) is associated with repeated exposure to mineral oil. Ultimately, this develops into a chronic condition called *multiple hyperkeratosis* (thickening of the horny layer). *Chlor-acne* is the name originally given to a skin disease observed in chlorine workers. These workers developed small gatherings of pus in the sebaceous gland ducts, gatherings known as sebaceous pustules, which are associated with the plugging of the sebaceous gland ducts. They then contracted secondary infection. However, it soon became apparent that chlorine was not the cause. Substances such as chloronaphthalenes are now regarded as the causative agents (Hunter, 1975).

Micropredators of various types are the cause of inflammatory lesions of the skin, some of which are diseases of occupation. *Anthrax*, for example, causes severe localized inflammatory lesions of the skin, which is probably the site of entry of *Bacillus anthracis*. However, this much-feared disease responds to antibiotics. Glanders, an infectious disease caused by the micropredator *Pfeiferella mallei*, is caught by man from contact with infected horses. It causes severe granulomatous lesions (chronic inflammation) of the skin.

RESPIRATORY PATHWAY

It will be recalled that the respiratory pathway is divided into conducting airways and respiratory units. Both provide a mode of input and both can be the target organ for a wide variety of harmful inputs capable of causing inflammation, degeneration, or cell death.

The respiratory pathway is vulnerable to attack by many irritants and by other inputs which damage skin. In order to reach the pathway the inputs need to be in the form of aerosols (defined in Chapter 3), gases, mists, vapours or fumes. The pathway is also vulnerable to thermal attack from air which is hot.

The terminology of inflammatory change in the lung follows the anatomy. Thus there is rhinitis (inflammation of the nasal passages), laryngitis, tracheitis, bronchitis and pneumonia (the latter being synonymous, for our purposes, with pneumonitis and bronchopneumonia). Each may be an acute or a chronic condition, although the chronic patterns of infection merge into disorders of repair (see Chapter 15) and the terminology changes. Hence 'chronic pneumonia' is a term not normally used.

The features of acute inflammation of the respiratory tract follow the general pattern of inflammation. However, the effects of exudation and swelling can in themselves be important if they result in narrowing, or even total obstruction, of the small conducting airways.

In the alveoli the exudation of fluid which follows increased permeability of the pulmonary capillaries can obstruct respiratory gas exchange, even to a fatal degree.

Chapter 22 contains case histories of gassing accidents: one exemplifies pulmonary oedema (exudation following irritation), which can occur several hours after exposure and occasionally can have disastrous consequences. The pattern of events associated with the inhalation of irritants falls under the general heading of gassing accidents.

The findings of Gross *et al.* (1967) suggest that in cases of inflammation caused by aggressive gases the point of most intensity is determined, at least in part, by the solubility of the gas. Chlorine and ammonia, for example, are soluble and highly aggressive. Even in quite low concentrations they dissolve in the moisture of the upper conducting airways and cause an acute tissue response of such extreme reaction that men run for their lives at the first whiff of an escape. Sulphur dioxide is also soluble, but according to Gross and his colleagues there is the complicating factor of its being absorbed by otherwise inactive (so-called nuisance) particles within the respirable size range. By this means sulphur dioxide may reach far along the conducting airways, perhaps as far as the respiratory units.

Oxides of nitrogen (apart from NO_2) and ozone are relatively insoluble. They are not, therefore, totally dissolved in the moisture of the upper conducting airways and consequently penetrate to the respiratory units. Here they set up acute inflammation, perhaps followed by delayed pulmonary oedema.

Chronic inflammatory effects are also identified with repeated low-grade exposure to irritants. Chronic bronchitis, an effect of particular concern in the discussion on respiratory disease in the coal industry (Chapter 21), is often referred to as 'English disease', an epithet which reflects the unevenness of its geographical distribution. Factors definitely associated with chronic bronchitis are tobacco smoking and atmospheric pollution. The role of micropredators is uncertain. The link with occupation has, in the past, been difficult to establish epidemiologically, because work people with chronic bronchitis often lived in geographical areas contaminated by smoke. Despite repeated attempts by the Trades Union Congress, chronic bronchitis has not so far been scheduled for the purposes of the National Insurance (Industrial Injuries) Act. The occupational link is, however, becoming more apparent.

In the past 20 or so years there has been a progressive drive towards cleaner air in industrial areas. Certainly, the particulate matter is now very much less in evidence. In 1970, Lloyd Davies and others from the Department of Employment published a survey of respiratory disease in 1997 foundrymen. Those surveyed were compared with a control population of 1777. The risk of foundrymen contracting chronic bronchitis was expressed as follows: 'A non-smoking foundryman, aged 65 years, with a lifetime of work in foundries, has the same expected prevalence of bronchitis as a man aged 65 years with heavy smoking habits ... who has spent his life in non-foundry and non-dusty work.'

Even though the risk of chronic bronchitis is clearly substantial, the confounding influence of atmospheric pollution in the general milieu could not be entirely excluded.

In cases of chronic bronchitis the exudation of fluid is manifest as sputum (phlegm), coughed up by the sufferer in amounts roughly proportional to the severity of the disease. There is progressive respiratory disablement, punctuated by 'flare-ups' of the disease, followed by permanent deterioration. In severe cases the small conducting airways become obstructed and narrowed, in which instance the condition may be termed chronic obstructive bronchitis. Evidence of this was found among the foundrymen studied by Lloyd Davies and his colleagues.

Chronic inflammatory changes are seen in the occupational lung disease which follows exposure to beryllium. This is discussed, along with certain other chronic occupational pulmonary diseases, in Chapter 15, under 'disorders of repair'.

BLOOD, PLASMA, BLOOD-FORMING ORGANS, CIRCULATORY SYSTEMS

The blood and its cellular elements are both subject to disease, but both also exhibit changes as a result of response to disease in most if not all organs, tissues and systems. For example, a rise in the number of circulating white cells is a generalized indicator of disease. The pattern of the change, in terms of the type of cell showing the increase, is also of diagnostic significance.

In occupationally significant contexts the target organs, blood, plasma, blood-forming organs and circulatory system, are subject to acute and chronic inflammation, degeneration, and cell destruction. Arbitrarily, the macrophage system has also been included with this collection of target organs, although it might just as readily have been included with the lymphoid system and lymphatics. The macrophage system outlined in Chapter 4 is the target organ of an occupationally significant disease caused by micropredators belonging to the genus *Brucella*, a disease known as brucellosis or undulant fever. The disease, risked by people working with cattle, is characterized by a fever which waxes and wanes and which is eradicated from the human body only with difficulty.

The destruction of red blood cells due to exposure to arsine gas is called haemolysis. The red cell normally contains no nucleus, so it is not capable of prolonged self-sustaining existence. We thus speak of the destruction rather than the death of red cells. In haemolysis, destruction of the red cells causes haemoglobin to be dispersed within the circulation. It is then concentrated, in changed form, in the kidney, where it causes haemoglobinuria (haemoglobin in the urine) and, if the exposure has been sufficiently severe, renal failure.

Atheroma (synonymous with arteriosclerosis and atherosclerosis) is a degeneration of the arterial walls associated with high concentrations of cholesterol and phospholipids in the lesions. The role of cholesterol in diet as a *cause* of such lesions has been the subject of debate. Of occupational significance is the observation by Wald *et al.* (1973) that atheroma is associated with moderate chronic exposure to carbon monoxide. For many years the view has been held that carbon monoxide poisoning was a strictly all-or-none (alive or dead) phenomenon, but the link between moderate low-grade exposure and atheroma has thrown doubt on that dogma.

KIDNEY AND URINARY TRACT

The kidney and the urinary tract are important organs of excretion of non-nutrients (see Chapter 11). Some non-nutrients result from the detoxification processes which take place in the liver and in other organs, and which are discussed in Chapter 8. The kidney, however, is also an organ vital in the maintenance of homeostasis (see Chapter 6). It maintains the constancy of the milieu interieur in respect of volume of body water, osmolarity of blood and tissue fluid, acid–base balance and arterial blood pressure (by means of hormones produced in kidney), while it also exercises some control on red cell formation. (The transport mechanisms of kidney were mentioned in Chapter 11.)

The kidney and the urinary tract are subject to acute and chronic inflammation, cell death and degeneration of occupational significance. Kidney inflammation is often known as Bright's disease, but the detailed classification is complex. The disease may primarily affect the glomerulus, or the tubules, or the interstitial tissues of the kidney, or be a combination of glomerular, tubular or interstitial disease.

Among the inputs capable of damaging one or other part of the kidney are metals (see Chapter 19) and ionizing radiations (see Chapter 17), although in the latter case damage is usually a sequel to

X-irradiation treatment for malignant tumours. Chronic interstitial nephritis is an important effect of a number of nephrotoxic agents and is discussed in Chapter 15, in connection with disorders of repair.

Inflammation follows the cytotoxic action of substances acting directly on the cell of the kidney; it may also be associated with altered sensitivity. But other pathogenetic processes may be involved. Certain substances severely reduce the blood flow in the kidney causing oxygen lack, damage, and subsequent inflammation. Homeostatic alterations in pH may increase the solubility of substances, thereby facilitating inflammation. Irrespective of origin, inflammation initiates repair processes which may effectively combat harmful effects.

The renal tract, especially the bladder, can be the subject of acute and chronic inflammation resultant upon the excretion of certain substances. This point tends to be overlooked in considerations of chemically induced bladder cancers which, understandably, are the centre of concern.

LIVER

The *Kuppffer cells*, which line the liver sinusoids, are part of the macrophage system and are probably the first site of attack in infective hepatitis and serum hepatitis. Both are inflammatory conditions of the liver caused by viruses, and both can be occupational in origin. The former is infectious and is spread in the course of everyday human contact. Serum hepatitis is transmitted by implantation of infected serum and is a disease to which laboratory workers are obviously at risk.

The liver (as we saw in Chapter 8) is the site of numerous detoxication processes, its cells dying and undergoing necrosis as a result of gross exposure to a number of halogenated hydrocarbons. Carbon tetrachloride is a well-documented example. The lipid solubility of these substances, as well as their detoxication in liver, is a factor of importance in their pathogenicity.

BONE

Acute and chronic inflammation of bone occur. In the occupational sphere, necrosis of the jaw bone by phosphorus (phossy jaw) is historically the most important (see Chapter 19). Necrosis of bone is observed in compressed air disease probably as a result of interference with the blood supply to bone caused by the formation of bubbles in the blood on decompression.

GUT

The terminology of acute and chronic inflammation of the gut reflects the anatomical structure: thus stomatitis is inflammation of the mucous membrane of the mouth (the mouth can be classified as the uppermost part of the gut for our purposes) resulting from excessive exposure to mercury vapour. Gastritis, inflammation of the stomach, is caused by ingestion of irritant substances. Gastro-enteritis, is a non-specific term often given to gut disturbance caused by micro-predators. Both gastritis and gastro-enteritis are characterized by vomiting and diarrhoea. Because they have numerous causes such as chemical substances, micropredators and the toxins produced in food by certain microorganisms they may present difficulties in diagnosis when there are outbreaks among workers.

SPECIAL SENSES

Photopthalmia is an acute inflammation of the conjunctiva and cornea of the eye, following exposure to intense light or ultraviolet light. There are a number of 'trade names' for the condition, including arc-eye and welders' eye.

Cataract is degeneration of the lens of the eye (see Chapter 17); according to Hunter (1975) it can be occupationally linked through mechanical trauma, electric shock, exposure to heat and infra-red energy, light, X-rays, radium and fast neutrons and possibly through exposure to certain chemical substances.

By a stretch of terminology it is also possible to regard the loss of sense of smell (anosmia), a condition associated with exposure to cadmium fumes, as a degeneration of the organ of olfaction.

REFERENCES

GROSS, P., RINEHART, W. E. and DE TREVILLE, P. P. (1967). The pulmonary reaction to toxic gases, *American Industrial Hygiene Association Journal*, **28**, 315–21.

HJORTH, N. and FREGERT, S. (1972). 'Contact dermatitis' in *Textbook of Dermatology*, 2nd Edition, A. Rook, D. S. Wilkinson and F. J. G. Ebling, eds, Oxford: Blackwell.

HUNTER, D. (1975). *The Diseases of Occupations*, 5th Edition, London: English Universities Press.

LLOYD DAVIES, T. A., EUINTON, L. E., RITCHIE, G. L., TROT, D. G., WATT, A., WEST, D. J. S. and WHITLAW, R. (1970). *Survey of Respiratory Disease in Foundrymen*, London: HMSO.

WALD, N., HOWARD, S., SMITH, P. G. and KJELDSEN, K. (1973). Association between atherosclerotic diseases and carboxyhaemoglobin levels in tobacco smokers, *British Medical Journal*, 1, 761–5.

Chapter 14

Altered Sensitivity

OBJECTIVES OF THE CHAPTER

This chapter provides:
  (1) definitions of commonly used concepts relevant to changes in the dose–effect relation of substances and energies: susceptibility, sensitivity, hypersensitivity, sensitization, altered sensitivity,
  (2) an outline of the classification of disadvantageous immune responses (immunopathological processes), and
  (3) a brief outline of the character of certain altered sensitivities in certain of the target organs listed in Table 12.1.

DEFINITIONS

Susceptibility, sensitivity, hypersensitivity, altered sensitivity, and sensitization represent whole concepts of wide importance in human safety. For example, in the description of Threshold Limit Values (TLVs), the American Conference of Governmental Industrial Hygienists (1976) make clear that TLVs represent conditions under which it is believed that nearly all workers may be repeatedly exposed without adverse effect. However, it is recognized that because of wide variation in individual *susceptibility*, a small percentage of workers may experience effects at the TLV which, for the majority of people, is associated with no effects. 'Susceptibility' as used by ACGIH, and others, implies that people differ in their susceptibilities to harm. *Sensitivity* can be used to describe a susceptibility of an individual to harm from a stipulated input *relative to* the average for a population of people exposed to a stipulated input under stipulated conditions.

The concept of sensitivity can be illustrated by reference to the dose–effect relation between occupational deafness and noise; see Figs. 12.1 and 12.2 and also the relevant definitions in Chapter 12. The ordinate of the graph represents the degree of deafness; the abscissa represents the dose of noise. The curves represent centiles of the population. The median (the centile 50) represents the 'average' members of the population. Thus, for any stipulated dose of noise the median degree of deafness can be read off from the centile 50 curve.

The centiles above 50 represent those people who are above average in their sensitivity; the centiles below 50 represent persons of below average sensitivity. The same concept can be put in another way: for any stipulated dose there will be a distribution of deafness within a population. Those whose deafness is above the median are regarded as displaying evidence of more-than-average sensitivity; those whose deafness is less than the average are said to be showing evidence of less-than-average sensitivity.

It is normal to assume that individuals maintain a constant degree of sensitivity to any particular agent/stimulus throughout their working lives. Thus, a population of people, all of whom are being exposed to the same noise, will accumulate noise dose at the same rate throughout their working lifetime (this, of course, assumes no non-occupational noise exposure). People may maintain their relative position within the population as the dose accumulates. Thus the sensitivity of an individual would not alter with time, even though there is progressive deafness with accumulating noise dose. (Matters such as the calculation of noise dose from noise energy are referred to in Chapters 12 and 17.)

However, if an individual were to change position relative to the population, yet his/her exposure was the same as that for the population as a whole, an alteration in sensitivity could be said to have taken place. The alteration in sensitivity might represent a shift in the direction of hypersensitivity. Alternatively, it might be a shift to a lower sensitivity, that is to say towards an increase in *resistance* or an increase in 'toughness' in respect of the input. Although alterations in sensitivity to noise are not frequently identified, alterations in sensitivity to other forms of energy and towards chemical substances and micropredators are a frequent occurrence. The purpose of this chapter is to explore, in outline, types of processes involved in alterations in sensitivity as described above. The alterations in sensitivity may take place rapidly within minutes, hours, or days; or progressively develop. A further category which can be recognized is those which are delayed in onset.

On the basis of considerations presented so far, the following definition can be put forward: alteration in sensitivity represents a shift in the dose–effect relation for an individual exposed to a stipulated input. The input may be a chemical substance, a micropredator, or form of energy.

Alteration in sensitivity in either direction can be brought about by the operation of the immune response; this is to be seen in operation in a number of injuries and diseases related to occupation and is discussed below; certain features of the immune response have been outlined already in Chapter 5. The chief concern of this chapter is with the immune response, although other types of altered sensitivity are remarked upon. For example, apart from the immune response, structural or functional changes in the body can occur, changes which alter the sensitivity of the body to a particular input. Simple examples of this include the toughening of the skin of the hands in response to rough work (which results in a decreased sensitivity to trauma) and hypertrophy (overgrowth) of voluntary musculature, a response to heavy manual work (which results in decreased sensitivity to muscular fatigue).

IMMUNE RESPONSE

It will be recalled (from Chapter 5) that the immune response has three principal functions:
  (a) surveillance,
  (b) disposal of antigens, and
  (c) self-disposal.
All three depend crucially upon the immune system being able to distinguish between self and non-self. The central ideas of *self/non-self recognition* and their implications are presented by Burnet (1970). In several important diseases (including, for example, auto-immune disease) this ability to make the distinction goes awry.

The following paragraphs investigate further the nature of the immune response, but it almost need hardly be said that there are terminological difficulties. Throughout this study 'immune response' is used to include the advantageous, neutral, and disadvantageous responses, though some authorities would prefer the use of 'allergic reaction' instead of 'immune response'. These authorities regard the term immune response as applicable only to those aspects of the process which 'exempt' the body from harmful inputs; inputs which have succeeded in reaching and penetrating the defensive boundaries.

As was indicated in Chapter 5, the effects of immune responses can be advantageous, neutral, or disadvantageous. The main concern of this chapter is with the disadvantageous effects. These are described by Coombs (1974) as 'immunopathological mechanisms' and a classification for these put forward by Gell *et al.* in 1973 and modified by Coombs in 1974, is in wide but not universal use.

On the classification, four types of response are recognized: types I, II, III, and IV. Figure 14.3 (adapted from Coombs, 1974) summarizes in diagrammatic form the four types of immunopathological mechanisms.

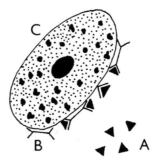

FIG. 14.1. Type I reaction between antigen, A, and antibody, B, taking place on surface of mast cell or basophil; histamine and substances with related effects liberated from mast cell or basophil. (After Coombs, 1974.)

Type I Reactions

Type I reactions (Fig. 14.1) are characterized by the antigen–antibody reaction taking place on the surfaces of mast cells and basophils (see Chapter 4). This reaction liberates substances including histamine which increase vascular permeability. Contraction of involuntary muscles such as those associated with conducting airways may also form part of the action. Asthma and hay fever are examples of type I reactions. IgE immunoglobulins are believed to be involved in a number of type I reactions, some of which can take place very rapidly.

Type II Reactions

Type II reactions (Fig. 14.2) involve antibody combining with an antigen which is on or part of a cell. Some chemical substances such

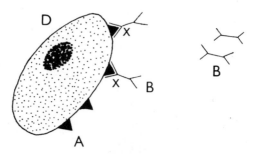

FIG. 14.2. Type II reaction between antigen, A, which is on or part of cell, D, and antibody, B; the combination of antigen and antibody links with either complement or lymphocytes at site X. (After Coombs, 1974.)

as metals or low molecular weight proteins can be adsorbed onto cellular membranes or tissue surfaces, subsequently to take part in type II reactions. The antigen–antibody combination links with either lymphocytes or complement. Complement is a plasma protein but distinct from antibodies. Complement is not specifically induced by antigens and, in some respects, complement has properties akin to those of enzymes.

Type III Reactions

Type III reactions (Fig. 14.3) take place between antigens and antibodies where the molecules of both are freely dispersed throughout the tissues and tissue fluids. The antigen and antibody combination fixes complement; fragments of the fixed complement break off and these display chemotaxis for polymorphs, and these stimulate phagocytosis. Indeed, phagocytosis may be stimulated to an excessive degree causing liberation of lysosomal factors into the surrounding tissue, thereby causing damage.

Type IV Reactions

Type IV reactions (Fig. 14.4) are at the root of delayed hypersensitivity. The antigen may be freely dispersed, attached to cells, arising from outside the body (exogenous) or arising within the body (endogenous). The reaction takes place between antigens and lymphocytes which appear to possess antigen-specific receptors. The type IV reaction differs from the other three in that the immunoglobulins are not involved; the role of antibody is taken by the lymphocytes and

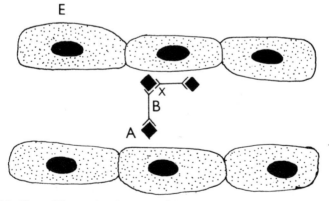

Fig. 14.3. Type III reaction between freely dispersed antigen, A, and antibody, B; complement links at site X; cells of sinusoid, E, where reaction could be taking place. (After Coombs, 1974.)

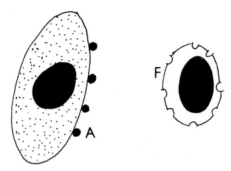

FIG. 14.4. Type IV reaction between antigen, A, in this case part of a 'foreign' cell, and F, the antigen-specific lymphocyte. (After Coombs, 1974.)

possibly monocytes. Once the lymphocytes are attached to the antigen-carrying cell they become cytotoxic to the antigen-bearing cell; the cytotoxic reaction attracts macrophages whose phagocytic activity is stimulated.

It should be emphasized that the classification relates specifically to immunopathological mechanisms and not to immune responses which are advantageous or neutral. Nor does the classification relate to disease in which the immune response is deficient in any way. It must also be noted that the classification relates to the initiating events of the immunopathological mechanisms. A further point to be added here is that Coombs's immunopathological mechanisms are also sometimes referred to as anaphylactic reactions or allergic reactions leading to tissue damage and disease.

A general point, which relates to many mechanisms underlying alterations in sensitivity and not just immune response, is that the alterations in sensitivity which result from various inputs can be influenced by factors within the individual, as well as by interactions between the inputs. Alteration in sensitivity can also take place in instances of substances lodged within one or other of the body's storage compartments, even after many years of storage.

Sensitization, used loosely, can refer to a state of immunological preparedness. Thus an individual who, as a result of initial exposure, possesses antibodies against a specific antigen can be said to be sensitized to that antigen. It must be noted, however, that the term sensitization is used with other meanings, not relevant here.

The following paragraphs provide, where appropriate, selected examples of altered sensitivity in those target organs listed in Table 12.1.

SKIN

A number of important categories of skin disease result from increased sensitivity, involving all types of immune response (Parish, 1972). The pattern is often that of a rapidly increasing sensitivity, although delayed increase is also seen. The picture is complicated, however, because factors 'internal' to the subject have an especially important influence in dermatoses (a non-specific term referring to skin diseases generally). Thus, in atopic dermatitis, a genetically determined disorder, there is increased liability to form IgE antibodies (mentioned in Chapter 5), coupled with a susceptibility to skin disease closely resembling contact sensitization dermatitis.

The role of emotional factors is far from being completely understood, but their importance is recognized to be of great significance in the cause and progress of skin disease generally (Rook and Wilkinson, 1972). This is hardly surprising, because there is daily evidence of the influence of emotion on skin. For example, blanching, which is caused by constriction of the blood vessels in the skin, is associated with fear and anxiety. Flushing, a condition caused by dilation of the skin blood vessels, is associated with embarrassment and anger. These effects are universal, although there may well be individual differences in their manifestations.

An unfavourable psycho–social milieu, as a matter of everyday experience, can provoke transient vascular changes in the skin. The prolongation of exposure to such milieux can provoke or perpetuate skin conditions which have manifestations of increased sensitivity. There is a close but puzzling link between contact sensitization dermatitis and emotional factors associated with severe psychological pressure (synonymous, in this instance, with psychological stress). A strong element of individual variability undoubtedly exists, in that not everyone's skin shows these effects. There is, however, evidence to suggest that few people are wholly resistant to psychological pressure. Pruritis (itchiness) sometimes occurs in response to psychological pressure, without apparent exposure to sensitizing inputs. In these instances the skin is presumably responding to internal (intrinsic) rather than to external (extrinsic) factors. The psycho–social milieu and contact sensitization dermatitis are examples of interacting inputs, to which reference is made in Chapter 2.

Contact sensitization dermatitis of occupational origin is associated with a wide range of inputs. Definition presents difficulty, as Hjorth and Fregert (1972) have indicated. They quoted the description of disease 42 in the schedule of diseases prescribed for the purposes of the National Insurance (Industrial Injuries) Act. Disease 42 is: 'a non-infective dermatitis of external origin, including chrome ulceration of the skin, but excluding dermatitis due to ionizing particles of electro-magnetic radiations other than radiant heat.' As they pointed out, legally the establishment of an occupational link depends upon whether the individual could reasonably have been expected to contract the disease if he/she has not been engaged in that particular occupation or type of work.

Hjorth and Fregert emphasized that acceptance of a stringent standard of diagnosis requires that the following criteria should be satisfied. There must exist:

(1) Work in contact with an agent known to have produced similar skin changes.

(2) Occurrence of similar eruptions in fellow-workers or within the same occupation.

(3) Correct time-relation between exposure and dermatitis.

(4) Type and site of lesions consistent with information of exposure, and similar to other cases.

(5) Attacks of dermatitis appearing after exposure, followed by improvement or clearing after cessation of exposure.

(6) History and examination corroborated by patch-test results.

Even where these stringent criteria cannot be satisfied, there will always be cases where an occupational link cannot be excluded. It should also be remembered that irritants produce effects on skin which may be difficult to distinguish from sensitization dermatitis.

A comprehensive list of the principal irritants and sensitizers is also presented by Hjorth and Fregert. Finally, attention should be drawn to the fact that the occupational dermatoses, especially those resulting from irritants and sensitizers, account for a very high proportion of occupational diseases.

RESPIRATORY TRACT

Occupationally significant diseases of the respiratory tract associated with altered sensitivity fall into the following four groups: (a) occupational asthma, (b) external allergic alveolitis, (c) specific dust reactions, and (d) unclassified.

(a) Occupational Asthma

'Asthma' describes a functional change resulting from narrowing of conducting airways and of varying duration, a change strongly associated with immune reactions of type I. In the non-occupational sphere, there are often intrinsic factors as well as extrinsic factors.

The symptoms of occupational asthma, and indeed asthma

generally, are wheezing and an often very alarming difficulty with breathing. However, asthmatic attacks are seldom fatal. Irritants provoke similar symptoms when exudation accumulates in the small conducting airways. Initially, differentiation between irritant effects and altered sensitivity may be difficult, but, in general, the distinction is reasonably obvious.

Parkes (1974) gave various examples of occupationally linked altered sensitivity. His first example was that of asthma in flour mill workers, a condition directly attributable to antigens in wheat grain or, possibly, to contamination of the grain by wheat weevil. The second example cited by Parkes was that of workers concerned in manufacture of polyurethane. In the course of their work contact with, or absorption of, isocyanates is unavoidable, since use of isocyanates, of which the commonest are tolylene di-isocyanate (TDI) and diphenyl methane di-isocyanate (MDI), the former more volatile than the latter, is an essential feature of polyurethane manufacture. Use of excessive amounts of isocyanate causes evolution of carbon dioxide, which provides the basis for foamed plastics. The isocyanate grouping is highly active biologically, behaving as an irritant and as a sensitizer to the respiratory tract, and it is from these effects that, after exposure, the workers suffered.

The TLV has been set at 0.02 ppm, but even at this low level sensitized people can apparently develop asthmatic symptoms. The consequences of TDI exposure were observed by Adams (1975): the asthma can be extremely disabling in sensitized subjects.

Parkes also referred to:

(1) Proteolytic enzymes derived from *Bacillus subtilis* which were used as one of the active constituents of 'biological' washing powders. The asthma observed in workpeople employed in the manufacture of the washing powders appears to have had an immunological basis in a dominantly type I reaction, although Flindt (1969) found evidence of a type III reaction. Some of Flindt's subjects showed severe symptoms, although according to Parkes long-term effects on lung have not been finally established.

(2) Gum acacia and the dusts from certain woods such as Canadian red cedar, which are associated with occupational asthma.

(3) Complex salts of platinum, such as chloroplatinates, which cause rhinopharingitis, conjunctivitis, dermatitis and asthma (see for example Pepys *et al.* (1972)).

(4) Piperazine derivatives used, for example, in the manufacture of sheep dips which are reported to have provoked asthmatic conditions.

(5) Aluminium soldering flux, used for jointing aluminium cable, which produces fumes containing ammonia and amine vapours, to which an asthmatic response has been prescribed.

(b) External Allergic Alveolitis

The term 'external allergic alveolitis' is used to describe a group of occupationally significant diseases characterized by an immune reaction to organic dusts. The reaction takes place in the walls of the respiratory units and in parts of the conducting airways. There are two reactions—type I, which has asthma-like symptoms, and type III, which produces a slower reaction in the respiratory units, particularly to the alveoli. The natural history of the disease is progressive disablement, because the type III reaction triggers a disorder of repair (described in Chapter 15).

The range of disease embraced by external allergic alveolitis is shown in Table 14.1, which is also taken from Parkes. Of the conditions listed therein, farmers' lung is the most important in

TABLE 14.1

EXAMPLES OF OCCUPATIONAL DISEASES MANIFESTING EXTRINSIC ALLERGIC ALVEOLITIS

Occupational disease	Source of dust	Source of antigen
Farmers' lung	Mouldy hay	*Micropolyspora faeni; Thermoactinomyces vulgaris*
Bagassosis	Mouldy sugar cane	*Thermoactinomyces sacchari*
Mushroom workers' lung	Mushroom compost	*Micropolyspora faeni; Thermoactinomyces vulgaris*; possibly other micropredators
Malt workers' lung	Mouldy barley and malt	*Aspergillus clavatus; Aspergillus fumigatus*
Maple bark strippers' disease	Infested maple bark	*Cryptostroma corticale*
Suberosis	Mouldy oak bark in cork manufacture	*Penicillium frequentans*
Bird breeders' lung	Faeces of pigeons, budgerigars, parrots and hens	Serum proteins in faeces
Wheat weevil disease	Wheat, grain and flour infected by weevils	*Sitophilus granarius*

Adapted, with acknowledgements, from Parkes, W. R. (1974). *Occupational Lung Disorders*, London: Butterworths.

Britain, whereas bagassosis is an occupationally significant disease in areas of the world in which cane sugar is grown. Both result from the handling of mouldy crops, which liberates spores in a cloud of aerosols.

The condition of farmers' lung is characterized, following exposure, by a latent period of a few hours. Then, often in the night, the farm worker is taken ill with an attack resembling acute influenza (fever, sweating, cough, headache, nausea and vomiting). The victim may recover completely or may progress to the chronic, disabling respiratory disease associated with pulmonary fibrosis (see Chapter 15).

Symptoms and signs typical of extrinsic alveolitis were observed by Friend et al. (1977) in a group of 24 workers employed in a stationery factory. Water circulating through vacuum pumps and air compressors was contaminated by a number of microorganisms and these became airborne in the workplace.

The workers' symptoms and signs usually developed on a Monday, towards the end of the afternoon, and persisted for 12 to 24 h. Among those described to the investigators were chest tightness, shivering, breathlessness and cough. These symptoms are typical of extrinsic allergic alveolitis but Friend et al. found no evidence to suggest the likelihood of progression of the illness to chronic inflammatory disease of the lungs.

(c) Specific Dust Reactions

The effect of quartz particles in producing silicosis in lung tissue is remarked upon in Chapter 15. Although the processes underlying the disease are not completely understood, the theories advanced to explain them have included immune reactions, of various types. According to Parkes, silicotic lesions are unlikely to be *initiated* by an immune reaction. There is a possibility, however, that the dust diseases of lung, which are characterized by slowly-increasing sensitivity or delayed sensitivity, such as progressive massive fibrosis in coal workers, may involve immune reaction.

(d) Unclassified

Byssinosis is a disease contracted as a result of exposure to cotton dust, flax and soft hemp and is most prevalent in certain stages of the manufacture of textiles. It is regarded by Parkes as a form of asthma. The conducting airways are the target organ, while the effect appears to be contraction of the bronchial musculature, producing symptoms of chest tightness. Over a period of years breathlessness increases and obstruction of the airways appears to develop, although drugs causing dilation of the bronchi can relieve the symptoms in the early stages. Chronic bronchitis appears to develop in a proportion of sufferers from byssinosis, although this seems to be related, at least partly, to cigarette smoking.

The pattern of the symptoms varies from occasional tightness of the chest throughout only the first day of the working week (the mildest form, often known as 'Monday tightness') to the severe condition, which is marked by tightness on every working day and is accompanied by reduced lung function, related to narrowing of the conducting airways.

Three theories about the mechanism of byssinosis have been advanced. According to these, byssinosis may involve: (a) a chemical action on the airways produced by substances in the cotton dust or by substances contaminating it, or (b) an immune response in the walls of the conducting airways, or (c) the action of bacteria and of toxins, known as endotoxins, liberated from bacteria.

A recent study by Cinkotai *et al.* (1977) showed an undoubted link between the concentration of airborne microorganisms (identified as Gram-negative microbes) and the prevalence of byssinosis. Their results also favoured the endotoxin explanation. If they are established, these findings would also explain another problem connected with byssinosis, namely, why the prevalence is related to the concentration of the fine particles of cotton dust rather than to the concentration of the total cotton dust.

BLOOD, PLASMA, BLOOD-FORMING ORGANS AND CIRCULATORY SYSTEM

Metal fume fever has features of an immune response. Within a few hours of first exposure, persons exposed to fumes from a number of molten metals (see Chapter 19) experience an illness similar to influenza, with which it may be confused. It lasts about 24 h. After recovery, and despite continued exposure, the risk of recurrence is decreased. This is because sensitivity is decreased. However, the decrease in sensitivity is observed to 'wear off', and, after a holiday from work, the illness may recur.

It is convenient to include metal fume fever here, in connection with blood and plasma as target organs, because these may well be the site of the immune response. This, however, is admittedly speculative.

When immune responses of various types take place one of the cells of the leucocyte family, the eosinophils, increases in number in the circulating blood. The blood count of eosinophils provides an indication that immune response is taking place somewhere in the body, though not necessarily in the blood.

By a stretch of imagination, vibration-induced 'white finger' can be regarded as an altered sensitivity to cold. This takes place in the arteries of the fingers and is the result of prolonged exposure to hand-held vibratory tools. The condition is trivial in the early phases but, according to Taylor and Pelmear (1975), it may become progressively worse, until there are changes of a permanent nature in the finger. These changes presumably result from prolonged hypoxia (lack of oxygen to the tissues). Unless the progression is halted necrosis will ultimately occur in the fingertips. Arterial narrowing was shown by James and Galloway (1975) to be present in workers with severe vibration-induced white finger.

Vibration-induced white finger may be regarded as an altered sensitivity to cold but there is no suggestion that immune response is in any way involved. The triggering factor, however, is cold, although other factors such as central body temperature, metabolic rate, vascular state generally and the emotional state, are also involved. Attacks of vibration-induced white finger last about 15 min, but can be prolonged for as much as 2 h in advanced cases.

KIDNEY AND URINARY TRACT

Certain types of glomerulonephritis (see Chapter 13) are associated with immune responses and with deposition of antigen–antibody complex in the glomerular capillary walls. However, it is not clear whether immune responses are central to occupationally significant diseases of the liver and urinary tract.

SPECIAL SENSES

The conjunctiva of the eye displays immune responses to sensitizers, often in parallel to the skin. It is debatable, however, whether it is appropriate to apply the concept of altered sensitivity to the changes in visual and auditory performance which follow certain types of stimuli. In general, the altered sensitivity is reflected in changes of performance in psycho–physical tests, rather than in decrease in sensitivity to harmful inputs. Even so, it could be argued that the ear becomes decreasingly sensitive to noise after continued exposure.

REFERENCES

ADAMS, W. G. F. (1975). Long-term effects on the health of men engaged in the manufacture of tolylene di-isocyanate, *British Journal of Industrial Medicine*, **32**, 72–8.

BURNET, M. (1970). *Self and Not-Self*, Melbourne: University Press.

CINKOTAI, F. F., LOCKWOOD, N. G. and RYLANDER, R. (1977). Airborne microorganisms and prevalence of byssinotic symptoms in cotton mills, *American Industrial Hygiene Association Journal*, **38**, 554–9.

COOMBS, R. R. A. (1974). Immunopathological mechanisms, *Proceedings of the Royal Society of Medicine*, **67**, 525–9.

FLINDT, M. L. H. (1969). Pulmonary disease due to inhalation of derivatives of *Bacillus subtilis* containing proteolytic enzyme, *Lancet*, **2**, 1177–81.

FRIEND, J. A. R., GADDIE, J., PALMER, K. N. V., PICKERING, C. A. C. and PEPYS, J. (1977). Extrinsic allergic alveolitis and contaminating cooling-water in a factory machine, *Lancet*, **2**, 297–300.

GELL, P. G. H., COOMBS, R. R. A. and LACHMANN, E. J. (1973). *Clinical Aspects of Immunology*, 3rd Edition, Oxford: Blackwell.

HJORTH, N. and FREGERT, S. (1972). In *Textbook of Dermatology*, 2nd Edition, Oxford: Blackwell.

JAMES, P. B. and GALLOWAY, R. W. (1975). In *Vibration White Finger in Industry*, W. Taylor and P. L. Pelmear eds, London: Academic Press.

PARISH, W. E. (1972). In *Textbook of Dermatology*, 2nd Edition, Oxford: Blackwell.

PARKES, W. R. (1974). *Occupational Lung Disorders*, London: Butterworth.

PEPYS, J., PICKERING, C. A. C. and HUGHES, E. G. (1972). Asthma due to inhaled chemical agents—complex salts of platinum, *Clinical Allergy*, **2**, 391–6.

ROOKE, A. and WILKINSON, D. S. (1972). In *Textbook of Dermatology*, 2nd Edition, Oxford: Blackwell.

ROOKE, A., WILKINSON, D. S. and EBLING, F. J. G. (eds) (1972). *Textbook of Dermatology*, 2nd Edition, Oxford: Blackwell.

TAYLOR, W. and PELMEAR, P. L. (eds) (1975). *Vibration White Finger in Industry*, London: Academic Press.

Further Reading

GRAY, D. F. (1970). *Immunology*, 2nd Edition, London: Arnold.

VAN LANCKER, J. L. (1976). *Molecular and Cellular Mechanisms in Disease*, Berlin: Springer-Verlag.

WEIR, D. M. (1973). *Immunology for Undergraduates*, 3rd Edition, Edinburgh: Churchill.

Chapter 15

Disorders of Repair

OBJECTIVES OF THE CHAPTER

This chapter presents in outline a description of:
   (1) how the processes of repair which ordinarily come into play
       when tissue has been damaged become disordered, in certain
       diseases, so that the repair processes themselves become the
       origin of disease,
   (2) the normal process of repair and the formation of scar tissue,
   (3) the role of collagen in repair and how collagen is produced from
       fibroblasts,
   (4) how tissue is 'made good' following damage and the nature of
       granulation tissue,
   (5) types and terminology of disorders of repair,
   (6) how the classification of pneumoconiosis is related to disorder
       of repair,
   (7) the terminology of disorders of repair in the lung, apart from
       pneumoconiosis,
   (8) specific categories of pneumoconiosis with special reference to
       silicosis,
   (9) theories relating to silicosis showing how these relate to dis-
       orders of repair,
  (10) fibrosis and emphysema, and
  (11) disorders of repair in skin.
The term 'disorders of repair' is used to embrace some of the
disease processes of some of the most notorious occupational diseases.

REPAIR

Tissues and organs are composed of supporting tissues and the
specialized functional cells. Damage to organs or tissues from a wide
variety of bio-destructive inputs may result in gaps or breaches in
structure, unless the bio-destruction has been of a trivial nature.
Gaps, involving loss of tissue, or breaches, involving incisions without
loss of tissue, are made good by processes of repair which involve, as
a first step, the removal of dead cells, exudate, or alien particles by
phagocytosis by microphages or macrophages. The 'making good' is

followed by regeneration of supporting tissues and, in some instances, the regeneration of the specialized cells. By this process the damaged tissue is restored; but the restoration will only be complete if the specialized cells have regenerated, and not only the supporting tissues.

Supporting tissues are generally capable of regeneration but specialized cells vary greatly in this capability: brain and muscle cells never regenerate; liver cells normally do so and bone cells always do.

If the damage has not involved much tissue damage then the repair will take place without the formation of *scar tissue*. Where the damage is at all extensive, or where repair is for any reason delayed, scar tissue will form. This is composed of a fibrous protein called collagen, which is formed by the polymerization of the natural amino acids glycine (Gly), proline (Pro) and hydroxyproline (Hypro). Polymerization is the formation of long-chain molecules by the linking together of numerous smaller ones.

Collagen has properties which are appropriate to a structural role: it is insoluble, and its molecular secondary structure is a three-strand superhelix the strands of which are closely bonded together. This structure makes it strong to tensile forces and it possesses only slight elasticity. In living tissues, collagen is found in white or colourless wavy bundles.

The Normal Collagen-Centred Repair Processes

The supporting tissues are made up of a network of fibrous proteins which resemble a three-dimensional fishnet. The 'catch', so to speak, contains the 'general duty' cells of the tissues: (i) fibroblasts, involved in the manufacture of the fibrous protein; (ii) macrophages, whose function, as has been seen, is phagocytosis and involvement in immune responses; and (iii) other cells, which do not concern us here.

There is some debate about how exactly fibroblasts produce collagen. The following outline provides a framework sufficient for this chapter's purposes.

Fibroblasts, see Fig. 4.6, like other cells, have ribosomes on the endoplasmic reticulum. The production of the Gly–Pro–Hypro units from which the collagen is ultimately polymerized takes place here. The units are packaged in the Golgi apparatus of the cell and find their way to the fibroblast's external membrane in secretory vacuoles, from which trocollagen is exuded. This material polymerizes outside the fibroblast, either to collagen or to another fibrous protein called reticulin, which differs from collagen in appearance and character. Some authorities look upon reticulin as young collagen, but others see it has having an independent role. However, as a matter of definition 'fibrosis' is universally recognized as meaning over-production of collagen fibres and not reticulin fibres (Parkes, 1974).

Gaps in Tissues

The nature of the repair process is dependent upon the nature of the damage and, in particular, whether there has been loss of tissue.

Gaps in tissues occur from wounds or ulcers, and involve the destruction of tissues. In broad categories gaps originate from: (i) traumatic tissue loss caused by, for example, cuts and lacerations; (ii) tissue destruction by aggressive chemical substances; (iii) tissue destruction due to excessive application of energies such as radiation and electricity; (iv) the actions of tissue-destroying micropredators; and (v) reaction to the removal of cells killed by oxygen deficiency or by other gross interference with tissue metabolism.

Supporting tissue regeneration is accompanied by scar tissue formation. If the gap is small the edges of the wound are held together by the adhesive effects of fibrin. Damaged tissue is phagocytosed and any particulate matter is dealt with by macrophages, perhaps in the form of giant cells. The capillaries in the adjacent intact tissues then grow into the fibrin-filled gap and form a vascular network. This capillary invasion is quickly followed by fibroblasts which lay down a framework upon which the scar tissue necessary to make good the gap subsequently develops. Once the scar tissue is established the extent of the capillary network declines, because oxygen requirements of scar tissue are relatively small.

Where large gaps occur and there is insufficient fibrin to provide temporary stopping, *granulation tissue* develops. This is a delicate tissue, brownish in colour and velvety in texture, which is composed of leucocytes, macrophages, capillaries and fibroblasts and is perfused with exuded tissue fluid. As repair progresses the granulation tissue is gradually taken over by scar tissue formed from the fibroblasts. Finally, the capillary network declines, leaving a pale, relatively inelastic scar tissue.

TYPES OF DISORDER OF REPAIR

The repair process can be disordered if tension is applied before the collagen has developed sufficient mechanical strength. This results in the reopening of gaps and the consequent formation of additional scar tissue.

Repair can be delayed by inadequate dietary intake of nutrients, vitamins, and, possibly, essential micronutrients. Excessive heat or cold retards repair.

A point of great importance is that the repair processes can be overstimulated so that the scar tissue formed is in extent disproportionate to the injury. In some diseases (mentioned later) scar formation does not stop, but instead continues until the scar reaches proportions which distort the structure or embarrass the function of the organ or tissue. Such overgrowth of scar can, for our purposes, be referred to as 'fibrosis'. On the surface of the body fibrosis may cause disfigurement but little functional embarrassment. In internal organs, however, the functional embarrassment may present severe problems.

Chemical substances and chronic inflammation can initiate the fibrosis which then progresses independently. Fibrosis of this type represents the most important category of disorder of repair in the occupational sphere.

Difficulties with Terminology

In some textbooks the process of scar formation is called 'fibrosis'. However, 'fibrosis' is used elsewhere to imply disorder of repair. For the sake of clarity it is preferable to use 'fibrous union' rather than 'fibrosis' to denote scar formation. 'Sclerosis' is also used as a synonym for 'fibrosis' and to denote the overproduction of collagen associated with the thickening and hardening of skin seen with exposure to vinyl chloride, mentioned in Chapter 23 and below.

Another possible source of confusion may be encountered in a group of diseases which used to be referred to as 'collagen disease' (Thomson and Cotton, 1976). Changes in collagen are common to these diseases, but the terminology 'collagen disease' has been discarded and the diseases are now referred to as 'connective tissue diseases', one of which is rheumatoid arthritis. 'Rheumatoid' coal pneumoconiosis (Caplan's syndrome), a disease described in coal workers, is characterized by large nodules (up to 3 cm in diameter) in the lung. These nodules exhibit necrosis as well as fibrosis, while, in addition, the joints display signs of rheumatoid arthritis. Apart from Caplan's syndrome the 'collagen diseases' appear to have no occupationally significant links.

DISORDERS OF REPAIR IN THE SKIN AND THE RESPIRATORY PATHWAY

In this chapter only two target organs are considered: skin and respiratory tract. Skin provides clear-cut examples of disorders of repair, but the respiratory tract is the commonest site of occupationally significant diseases which fit that category. Repair processes involving collagen scar tissue formation are, however, to be expected in *all* target organs following significant structural damage.

The liver, for example, is a site of necrosis following damage, *inter alia*, from chemical substances. Regeneration of the liver cells may take place, but so too may fibrosis. A distinction is drawn between fibrosis and cirrhosis of the liver, the latter involving more of the liver and greater alteration of its architecture consequent upon necrosis of cells. It may be plausible to regard cirrhosis as a disorder of repair affecting the liver, but it is not a condition comparable in *occupational* significance to pneumoconiosis.

Fibrosis, described in response to damage to kidney from chemical substances and irradiation, can be very extensive, although it is not clear whether this should properly be categorized as a disorder of repair.

SKIN

Disorders of repair in the skin are represented by hypertrophic* scars and cheloids (sometimes spelt 'keloid'); these are localized areas of excessive scar formation following a wound. These lesions are of little significance in occupational health and safety, but they are mentioned here because they exemplify disease caused by disorders of repair. In hypertrophic scars the connective-tissue response to skin injury exceeds the limit of physiological needs appropriate to the degree of injury and to the site (Rook, 1972). Hypertrophic scar tissue differs from ordinary scar tissue only in degree. One explanation of the overgrowth of scar tissue is that foreign material in the healed wound continues to provide a stimulus for the collagen fibre formation. Rook described the practice common to certain African peoples of deliberate and repeated inoculation of foreign substances into the skin, which produces so-called ornamental cheloids until the desired degree of hypertrophy has been achieved.

RESPIRATORY TRACT

Disorders of repair constitute an important group of occupational respiratory diseases loosely referred to as 'pneumoconiosis'. This is *not* a single disease; the question of which diseases should be embraced by the term has been the subject of considerable debate.

---

*There is a certain amount of terminological confusion about what constitutes 'hypertrophy', the word being used to denote overgrowth generally; in this instance the overgrowth refers to scar tissue, but in other contexts 'hypertrophy' is used to describe disorders of growth, which form the subject matter of the next chapter in which the terms are defined.

In 1971 at the International Labour Office's fourth International Pneumoconiosis Conference, a working group produced a definition of pneumoconiosis:

> In recent years a number of countries have included under pneumoconiosis, because of socio–economic reasons, conditions which are manifestly not pneumoconiosis, but are nevertheless occupational pulmonary diseases. Under the term 'disease' are included for preventive reasons the earliest manifestations which are not necessarily disabling or life shortening.
>
> Therefore the working group has undertaken to re-define pneumoconiosis as the accumulation of dust in the lungs and tissue reactions to its presence. For the purpose of this definition, 'dust' is meant to be an aerosol composed of solid inanimate particles.
>
> From a pathological point of view pneomoconiosis may be divided for the sake of convenience into collagenous or non-collagenous forms. A non-collagenous pneumoconiosis is caused by a non-fibrogenic dust and has the following characteristics: (i) the alveolar architecture remains intact, (ii) the stromal reaction (my note: this can be read as connective tissue reaction) is minimal and consists mainly of reticulin fibres; (iii) the dust reaction is potentially reversible.
>
> Examples of non-collagenous pneumoconiosis are those caused by pure dusts of tin oxide (stannosis), and barium sulphate (barytosis).
>
> Collagenous pneumoconiosis is characterised by: (i) permanent alteration or destruction of alveolar architecture; (ii) collagenous stromal reaction of moderate to maximal degree; (iii) permanent scarring of lung. Such collagenous pneumoconiosis may be caused by fibrogenic dusts or by altered tissue response to a non-fibrogenic dust.
>
> Examples of collagenous pneumoconiosis caused by fibrogenic dusts are silicosis and asbestosis, whereas complicated coal workers' pneumoconiosis or progressive massive fibrosis (PMF) is an altered tissue response to a relatively non-fibrogen dust.
>
> In practice, the distinction between collagenous and non-collagenous pneumoconiosis is difficult. Continued exposure to the same dust, such as coal dust, may cause transition from a non-collagenous to a collagenous form. Furthermore, exposure to a single dust is now becoming less common and exposures to mixed dusts having different degrees of fibrogenic potential may result in pneumoconiosis which can range from the non-collagenous to the collagenous forms.

Further Difficulties with Terminology

(i) There is a tendency to confuse the meaning of the term 'fibrogenic dusts'. Under the ILO definition 'fibrogenic dust' is dust capable of causing collagenous pneumoconiosis. Unfortunately, 'fibrogenic' is easily confused with 'fibrous', the latter referring specifically to dusts composed of fibres, whether fibrogenic or not.

(ii) *Scar tissue, fibrosis, collagenous pneumoconiosis*: scar formation results from any tissue destruction of more than trivial proportions. Collagen will dominate the scar. Inflammation following acute and massive exposure to aggressive irritants will, if not fatal, lead to destruction of the lining cells of the conducting airways or respiratory units and, therefore, to scarring. This scarring, however, will normally be consistent with the degree of tissue destruction and not progressive as is collagenous pneumoconiosis.

The ILO definition of penumoconiosis does not attempt to deal with one difficult issue, the extent to which fibrosis should be considered a progressive, self-propagating condition. It seems reasonable to reserve the term 'collagenous' pneumoconiosis for the pattern of disease in which fibrosis is both excessive and progressive. Such a condition provides a basis for distinguishing collagenous pneumoconiosis from 'post-damage' scarring, i.e. that which follows the inhalation of highly aggressive irritants. Such scarring within the lungs is likely to differ in distribution from that observed in collagenous pneumoconiosis, while it will also differ in being neither excessive (in relation to the original injury) nor progressive.

(iii) Non-occupational fibrosis: fibrosis can occur in the lung in cases of non-occupational disease associated with severe, chronic inflammation of the chest. 'Fibrosis of the lung' should not necessarily be regarded as an occupational-linked disease, but rather as the end product of damage or chronic inflammation of the lung tissues.

COMPLICATIONS OF PULMONARY FIBROSIS

It should be recognized that collagenous pneumoconiosis and, in all probability, post-injury scarring both predispose lung tissues to the development of tuberculosis. The role of tuberculosis as agent in coal-workers' pneumoconiosis is not fully understood. Tuberculosis possesses strong socio–economic determinants, its incidence having shown a marked decline, in many countries, over the last generation or so. Rising standards of nutrition, better housing, and the provision of more comprehensive social services as well as improved medical treatment have contributed to this. However, in countries where socio–economic development is still incomplete, and/or where medical facilities are inadequate, tuberculosis endures as a sequel to occupation-linked collagenous lung disorder.

However, the disappearance of tuberculosis in prosperous countries has not removed the problem of complications, for example, from asbestos exposure. As will be discussed in Chapter 20, asbestos fibres give rise to a collagenous pneumoconiosis, known as asbestosis, and there is a proven link between the contraction of asbestosis and the contraction of lung cancer in asbestos workers. Cigarette smoking increases the risk of contraction of lung cancer. Formerly, asbestos workers with asbestosis were prey to tuberculosis; latterly, however, the virtual disappearance of tuberculosis has unmasked, so to speak, the increased risk of lung cancer.

SILICOSIS

Silicosis is a collagenous pneumoconiosis caused by free silica. The disease is generally declining in incidence, although, as Parkes (1974) pointed out, it has by no means disappeared from all countries. Silica is ubiquitous, can be encountered in a wide range of occupations and makes unexpected appearances from time to time as, for example, in certain domestic scouring powders of which the manufacture ceased only in recent years.

Terminology relating to silicosis is confusing, because, according to context, the words 'pneumoconiosis', 'fibrosis' and 'silicosis' have acquired different shades of meaning. Four forms of the disease silicosis are recognizable:

(a) *Nodular silicosis*, which is associated with exposure to dusts dominated by quartz or flint; the disease progresses over a period of years.

(b) *Acute silicosis*, which is a disease involving fibrosis and accumulation of lipoprotein material in the alveoli; the disease develops over a period of weeks following massive but short-term exposure to quartz, and is fatal within a year or so.

(c) *Mixed dust fibrosis*, a disease which occurs when silica is inhaled together with certain non-silica dusts, for example iron oxide. The presence of non-silica dust (often referred to as 'inert' dust) appears to affect the pattern, though not the fundamental nature of the collagenous pneumoconiosis. The progress of the disease is similar to that of nodular silicosis. Life shortening is uncommon.

(d) *Diatomite pneumoconiosis*, a collagenous pneumoconiosis associated with exposure to the amorphous form of free silica known as diatomite. The disease normally progresses over a period of years, but occasionally it takes a more acute course.

The pathogenesis of silicosis has been the subject of intense study. The following outline is consistent with present knowledge, even though it represents a considerable simplification of what is a complex field. In Chapters 3 and 4 the link between respiratory clearance of aerosols and macrophage activity was highlighted. Now attention is turned to the effect on the macrophages of the substance which they phagocytose. As previously indicated, 'aerosols' can conveniently be renamed 'particles' if usage refers to the stage at which they are phagocytosed by macrophages. However, this terminological gloss should not be allowed to obscure the fact that particles may be droplets as well as solids.

Particles are phagocytosed and incorporated into phagosomes (see Fig. 4.1), to which lysosomes containing enzymes become attached. The enzymes are thereafter discharged into the phagosomes, the resultant enzymatic action producing a residue. With biologically-inactive substances many may remain harmless in the form of residual bodies within the macrophage's cytoplasm or, alternatively, be excreted by the macrophage as debris. Biologically active substances give rise to a process drastically different: the phagosome membrane is breached, allowing the contents to spread within the cytoplasm of the macrophage. As a result of this spread the macrophage is killed, the lethal factor in the phagosome contents being either the biologically active substance, or, possibly, both substance and enzyme. After it has been killed the macrophage is scavenged by other macrophages, but the lethal factors which killed the 'first line' macrophages are lethal also to this 'second wave'. They, too, die and there is a recurring cycle of macrophage mortality. Substances which initiate this cycle are said to be 'cytotoxic'. How, precisely, specific substances are cytotoxic to macrophages is still a matter for research in many instances.

The recurring cycle of macrophage mortality takes place in the interstitial spaces (see Chapter 3) following the phagocytosis of cytotoxic particles within the alveoli. As previously mentioned, the final details of the transport system for particles out of the alveoli remain to be unravelled. Possibly macrophages migrate from the alveoli through the alveolar membrane. Alternatively, there may be a more complicated system of 'hand-to-hand' transport involving more than one type of phagocytic cell. In any event, the result is that the interstitial spaces, especially those surrounding the bronchioles (see Fig. 3.5), become a metaphorical graveyard for the poisoned macrophages. These 'graveyards' provide the focal point for the overproduction of collagen by fibroblasts. In all likelihood, the phagosome contents (enzymes and cytotoxic substances) disturb events at the fibroblasts' ribosomes, a disturbance which causes the excessive production of units from which collagen is polymerized.

Theories to Explain Silicosis

Parkes (1974) has reviewed the most notable theories relating to silicosis. At one time it was believed that free silica particles directly injured the lung tissue, provoking fibrosis. That theory was abandoned when animal experiments showed that fibrosis did not result from sharp particles of corundum or diamond. Subsequent experiments confirmed that these substances are not cytotoxic. Another theory to explain silicosis centred on protein adsorption taking place on the surface of the silica particles. Substances other than silica adsorb proteins but do not cause fibrosis, so this theory, as it was first proposed, could not be sustained. However, the 'protein adsorption theory' did provide the basis for theories centred on immune responses—whether to the silica itself or to protein adsorbed onto the silica surface. It is now believed that fibrosis is not initiated by immune responses, although these may be involved in the progression of rapidly developing forms of collagenous pneumoconiosis.

For several years research was focused on the observation that silica slowly dissolved in tissues to form silicic acid. Theories based on this observation failed in the face of evidence that various forms of silica, equal in solubility, were not equal in their fibrogenic potential.

At present, opinion favours a theory based on macrophage lethality. This was advanced by Allison *et al.* in 1966, developed by Nadler and Goldfisher in 1970, and strongly supported by the work of Beck and colleagues (1971 and 1972).

OTHER FORMS OF COLLAGENOUS PNEUMOCONIOSIS

Apart from the four forms of silicosis mentioned above, collagenous pneumoconiosis is associated with (a) 'mixed dust fibrosis'; (b) certain of the later stages of coal-workers' pneumoconiosis, progressive massive fibrosis (see Chapter 21); (c) asbestosis (see Chapter 20); (d) 'talc' pneumoconiosis (see below); (e) certain metals identified in Chapter 19, notably aluminium, beryllium and cobalt; and (f) farmers' lung (see Chapter 14). All the dusts associated with conditions (a) to (f) are fibrogenic; perhaps the only other common characteristic (apart from being, or being able to become dusts) is their potential for causing excessive production of collagen in lung.

Mention has been made of progressive massive fibrosis (PMF), and further reference will be made to it in this chapter.

Other forms of carbon besides coal are generally regarded as non-fibrogenic. However, a 1972 report on the Sri Lankan graphite industry by Ranasinha and Uragoda found that in many respects graphite pneumoconiosis closely resembled coal-workers' pneumoconiosis. In their view it appeared to present characteristics somewhere between pure silicosis and carbon pneumoconiosis, in that X-ray changes observed in the group of workers may have been suggestive of an element of fibrosis. The possibility of the Sri Lankan graphite-workers' pneumoconiosis being a mixed dust fibrosis cannot be ruled out.

This chapter was titled 'Disorders of Repair' in order that emphasis be placed upon the pathogenesis of collagenous pneumoconiosis, a group of diseases unpleasant, disabling and sometimes life shortening. The chapter title excludes, however, all non-fibrogenic dusts which accumulate in lung but have only minimal tissue reaction. Their sometime description as 'inert' dusts is, in the author's opinion, misleading, because: (i) inert dusts may carry on their surfaces irritant gases, (ii) far too little is known about the interaction of dusts and other factors, and (iii) questions involving alterations in sensitivity remain to be resolved.

Lack of evidence of a dust's fibrogenicity should not, therefore, be taken as affirmative evidence of the harmlessness of that dust.

CLASSIFICATION OF SEVERITY OF COLLAGENOUS PNEUMOCONIOSIS

The classification of pneumoconiosis used in this chapter is based upon knowledge of the lung tissues' reactions to dust. Normally lung tissues can only be viewed under the microscope during a post-mortem examination. In the living subject the nature of, extent of, and prognosis for occupational respiratory disease has to be understood by the physician on the basis of (i) the subject's history, (ii) the physician's examination of the subject's chest, including measurements of the subject's respiratory performance, and (iii) the X-ray picture.

Although relevant questions are primarily medical, the administration of lung function tests is not the prerogative of the medical practitioner; perhaps more surprisingly, neither is the reading of X-rays. Surveys of industrial disease among large populations (e.g. coal workers) show that non-medically qualified personnel are just as competent in reading X-ray films as the medically qualified.

A typical history of occupational respiratory disease involves (i) exposure to dust, (usually in the absence of respiratory protection), (ii) breathlessness (dyspnoea) on exertion, (iii) a cough (often associated with sputum), and (iv) sometimes, pain in the chest. Chest examination in the early stages may reveal very little collagenous pneumoconiosis, but in the later stages signs may be present. For example, fibrosis eventually involves the pleura and produces characteristic sounds in the physician's stethoscope. Functional tests of the respiratory system may show a loss of elasticity of lung tissue owing to fibrosis, while the X-ray picture will show shadows due to both dust and fibrosis.

In the early stages it may not be possible to distinguish between collagenous and non-collagenous pneumoconiosis. Both appear as nodular opacites in the X-ray film, and, in the early stages, there is no way of distinguishing between radio-opaque dust particles and fibrosis. In the later stages of diseases such as asbestosis or coal workers' pneumoconiosis the pattern of the opacites lends itself to interpretation, and various systems for classifying the pattern have been proposed. One is the ILO/UICC classification of radiographic appearances of pneumoconioses produced by the ILO, in 1971.* Further reference to the classification of radiographs is made in Chapter 21.

SITE OF FIBROSIS

A detailed discussion of the distribution of fibrosis associated with various collagenous pneumoconioses in the lung is beyond the scope of this book. However, the following points should be emphasized:

(a) fibrosis often starts in the interstitial spaces surrounding the small bronchioles, the site of origin of the silicotic nodules,

(b) the fibrosis associated with asbestosis is diffused throughout the lung, involving the alveolar walls, the interstitial spaces and the pleural linings of the lung, and

(c) in the reaction to coal dust, known as progressive massive fibrosis (PMF), large patches of fibrosis develop in various parts of the lung.

EMPHYSEMA

Because emphysema is often associated with lung fibrosis of occupational origin it is convenient to mention it here. Emphysema has been defined (Reid, 1967) as a condition of the lung in which the air spaces of the respiratory units are increased beyond their normal size.

*This is reproduced in many standard medical texts including Price's *Textbook of the Practice of Medicine*, edited by Sir Ronald Bodley Scott (1973).

Emphysema is not regarded as a disease in its own right. Rather it is a change observed in a number of diseases, including certain occupational lung diseases. Severe fibrosis distorts lung architecture, producing emphysema in adjacent parts of the lung tissue. However, emphysema occurs even where fibrosis is not found, for example, in coal-workers' pneumoconiosis. This emphysema is normally found only at post-mortem examination because, in most instances, it is too slight to be detectable in life.

In recent years research into emphysema has focused on the possibility that proteolytic enzymes are released in the lung by, for example, macrophages. It is believed that these are normally held in check by an inhibitor, α-1-antitrypsin (Robin and Simon, 1976). A possible explanation for emphysema is that the inhibitors are absent or are themselves inhibited. Perhaps some such mechanism explains the severe emphysema associated with cadmium exposure (see Chapter 19).

The key concept for this chapter is that the ordinary process of repair can become the origin of important occupational diseases.

REFERENCES

ALLISON, A. C., HARINGTON, J. S. and BIRBECK, M. (1966). An examination of the cytotoxic effects of silica on macrophages, *Journal of Experimental Medicine*, **124**, 141–54.

BECK, E. G., HOLT, P. F. and NASRALLAH, E. T. (1971). Effects of chrysotile and acid-treated chrysotile on macrophage cultures, *British Journal of Industrial Medicine*, **28**, 179–85.

BECK, E. G. and MANOJLOVIĆ, N. (1972). Comparison of effects on macrophage cultures of glass fibre, glass powder and chrysotile asbestos, *British Journal of Industrial Medicine*, **29**, 280–6.

BODLEY SCOTT, R. Sir (1973). *Price's Textbook of the Practice of Medicine*, 11th Edition, London: Oxford University Press.

NADLER, S. and GOLDFISHER, S. (1970). The intracellular release of lysosomal contents in macrophages that have ingested silica, *Journal of Histochemistry and Cytochemistry*, **18**, 368–71.

PARKES, W. R. (1974). *Occupational Lung Disorders*, London: Butterworth.

RANASINHA, K. W. and URAGODA, C. G. (1972). Graphite pneumoconiosis, *British Journal of Industrial Medicine*, **29**, 178–83.

REID, L. (1967). *The Pathology of Emphysema*, London: Lloyd Luke.

ROBIN, E. D. and SIMON, L. M. (1976). In *Pathophysiology*, 2nd Edition, E. D. Frohlich, ed. Philadelphia: Lippincott.

ROOK, A. (1972). In *Textbook of Dermatology*, 2nd Edition, A. Rook, E. S. Wilkinson and F. J. G. Ebling, eds, Oxford: Blackwell.

THOMSON, A. D. and COTTON, R. E. (1976). *Lecture Notes on Pathology*, 2nd Edition, Oxford: Blackwell.

Chapter 16

Disorders of Growth

OBJECTIVES OF THE CHAPTER

This chapter:
(1) shows which pathological processes can be regarded as disorders of growth, and why,
(2) shows how cancer may be classified as a disorder of growth,
(3) presents terminology, meanings of terms, and general considerations of cancer,
(4) outlines the histogenetic classification of tumours, and explains its basis,
(5) outlines the nature of occupational cancer, and
(6) in relation to occupational cancer: summarizes statements made with reasonable certainty by authorities; summarizes qualified statements by authorities; highlights certain problems with occupational cancer related specifically to prediction; and shows the way forward as seen by one authority.

INTRODUCTION

'Disorders of growth' include some of the most important occupation-linked diseases, such as cancer induced by chemicals. Some of the most challenging problems for research in occupational health and safety can be embraced under the heading 'disorders of growth'. The entire subject thus deserves to be studied in detail. There is, however, uncertainty and even ignorance about the underlying processes, and it is not, therefore, surprising that the terminology is sometimes confusing. Pathologists, toxicologists, epidemiologists and occupational physicians cannot be blamed for using different terminologies, but this does cause difficulties for those new to the subject. In recognition of these difficulties this chapter has been divided into two parts:
Part (1) terminology, meanings, and general consideration of neoplasia, and
Part (2) occupation-linked neoplasia in certain target organs.
As far as is possible, the main consideration of this chapter has been relevance to occupational health and safety. Inevitably, the decisions to include or exclude topics reflect the way in which the writer perceives the subject matter. The reader should recognize that

the information presented here is but a selection from a very complex assortment of information, the chapter's purpose being simply to provide a framework for further understanding, an understanding obtainable from more advanced writings.

Part (1):   TERMINOLOGY AND MEANINGS

CLASSIFICATION OF DISORDERS OF GROWTH

The following classification is taken from Thomson and Cotton's (1976) classification of abnormalities of cell growth. Suffixes such as '-plasm' and '-plasia' are widely encountered. 'Plasm' means 'image' and 'plasia' means the processes by which the image is 'moulded'. As was pointed out in Chapter 9, most growth in tissues occurs by cell division, but there are occasions when tissues increase in dimension by other means. Decrease in tissue dimension will normally represent a diminution in cell population, although there are other causes; but, overall, the most important elements in disorder of growth are abnormalities of cell division and the structural and functional consequences of those abnormalities.

(a) *Aplasia* is the failure of development of cells, tissues or organs. The thalidomide tragedy, which resulted in approximately 10 000 deformed children being born throughout the world (Lenz, 1962; Taussig, 1962), is the most widely known example of aplasia. Many 'thalidomide' children were born with phocomelia, a condition in which the hands and feet develop reasonably normally, but the arms and legs suffer from aplasia (synonym agenesis) or even from complete non-development.

Aplastic anaemia, which is associated with exposure to benzene is something of a misnomer. The condition involves a severe reduction in the number of circulating red and white blood cells and results from atrophy (see below) of the bone-marrow (the site of origin of the cells and one of the target organs for benzene); 'atrophy of the bone-marrow' would be a better description than aplastic anaemia.

(b) *Hypoplasia*, the failure of organs to develop to full size, is a condition regarded as a less severe abnormality than aplasia. Some of the thalidomide children have incomplete development rather than complete non-development of limbs.

(c) *Atrophy* is a condition wherein an organ previously of normal dimensions subsequently diminishes in size, owing to a reduction in the number of cells or the size of cells, or to simultaneous reduction in both. There are numerous causes of atrophy, including undernutrition of the individual or undernutrition of an organ or tissue following interference with blood supply. Atrophy occurs with disuse, whether from strictly physiological causes (such as the decline in lymphoid tissue after adolescence) or because of enforced disuse through, for example, paralysis caused by traumatic severance of the nerve supply to a muscle, or decrease in blood supply. Atrophy can result from prolonged pressure on tissues. An example of pressure atrophy may be the atrophy of the muscles of the hand mentioned by Legge (1934) in his discussion of the effects of compressed air tools. Cachexia is a general wasting or atrophy of tissues associated with malignant neoplasms (see below), chronic infections such as tuberculosis and disturbances of biosynthesis.

(d) *Hypertrophy* is an increase in the size of tissues resulting from an increase in cell size without an increase in cell number; however, the size of individual cells may increase. It is frequently accompanied by hyperplasia.

(e) *Hyperplasia* describes the increase in size resulting from an increase of cell numbers, although the cells themselves may retain their original size. Hypertrophy and hyperplasia are associated with compensation and adaptation: the loss of one kidney causes hypertrophic changes in the other; the voluntary musculature of manual workers is hypertrophic compared with that of sedentary workers; demands on the defensive cell systems accompanying infection or disease cause hyperplasia of lymphoid tissue. Hyperplasia is normally reversible although the dividing line between hyperplasia and neoplasia (see below) is not always easily definable.

(f) *Metaplasia* is a change from one cell type to another. It usually involves a change from a specialized type of cell to a less specialized type, or to a different type of cell, and is associated with prolonged chronic irritation from chemical substances or infections. Metaplasia is observed in the epithelial lining of the conducting airways of heavy smokers (see Fig. 16.1), and in the epithelial lining of urinary bladder. In this latter instance it is associated with urinary excretion of chemicals and their conjugates and transformates (see Chapter 8). Metaplasia may be reversible if the cause is removed, but metaplastic cells may undergo neoplastic changes (see later). Metaplasia occurs in connective tissues in association with repair processes, the ability of fibroblasts (see Chapter 15) to change into other connective tissue cells being a good example. Since this change of nature may be a necessary part of the repair process (for example, fibroblasts may change into bone-forming cells) this connective tissue metaplasia should not be classified as a disorder of repair.

(g) *Dysplasia* is alteration in the orderliness of epithelial cells and

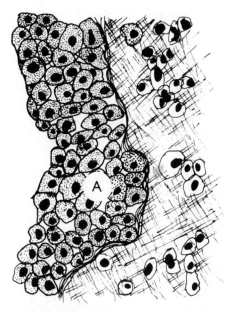

FIG. 16.1. Metaplasia of the lining of the conducting airways; compare with Fig. 3.7. Specialized cells of bronchial epithelium have been replaced by less specialized cells (A); characteristic features of bronchial epithelium have disappeared.

is associated with chronic irritation or inflammation. In cases of dysplasia, the normal orderly progression of tissue development is upset and there may be frequent mitosis (see Chapter 9). The condition is reversible if the causal factor is removed, but it can progress to neoplasia or be displaced by neoplasia.

(h) *Mutagenesis* describes the processes in which abnormalities are passed from one generation of cells to another. It was mentioned in previous chapters and is discussed again below.

(i) *Teratogenesis* describes certain of the processes by which abnormalities occur in offspring—termed congenital abnormalities. Becker (1975) pointed out the need for a careful distinction between teratogenic and toxicological causes of congenital abnormalities. 'Congenital' means present from (or at) birth. Congenital abnormalities (*i*) can have hereditary origins, (*ii*) can derive from deviation in embryological development, and (*iii*) can derive from pathogenic substances affecting the unborn offspring after the 16th week of development. The embryological differentiation* between cells and tissues is generally a short-lived phenomenon which follows the union of the gametes (sperm and ovum). After embryological differentiation, development and growth of the formed tissues continues until, and after, birth. 'Teratogenic change' refers specifically to (i) and (ii), and not to (iii), which falls under the heading of 'toxicological change'. Teratogens, i.e. agents able to induce (i) and (ii), include chemicals and ionising radiations. Toxicological changes, i.e. (iii), can be brought about by any substance able to pass the placental barrier, or any other harms which can reach the foetus.

It should be borne in mind that there is no reason why mutagenesis in the non-reproductive cells of the foetus should not occur in response to pathogenic factors. These, however, would be mutatenic effects falling under Becker's heading of toxicological change if they took place after differentiation had ceased.

Mutagenesis, teratogenesis and neoplasia are believed to have at least one factor in common, namely that the abnormalities are manifested in the sequence of events embracing the replication of DNA and the transcription of RNA (see Chapter 10).

(j) *Neoplasia* describes the condition of new, abnormal growth. Some neoplasms are not very harmful, but others are lethal. Words such as 'cancer' are, in non-specific terminology, used to denote the more harmful neoplasms. However, 'cancer' also serves as a synonym for 'carcinoma', which is a specific category of harmful neoplasia, namely malignant tumours of epithelium.

Neoplasia is the term used to describe new growth in tissue. This consists of a mass of cells which have undergone a fundamental, inheritable and irreversible change in their physiology and structure, a change which leads to a continuous and unrestrained proliferation. A neoplasm composed of a solid mass of cells is called a tumour.

Berenblum (1974) identified the following defining characteristics of tumour cells:

(a) tumour cells are irreversibly altered cells,
(b) they are endowed with new properties not shared by normal cells, and
(c) in their malignant form they are harmful to the rest of the body.

Berenblum sees tumours as being characterized by:

(a) progressive growth in size which is not subject to growth equilibrium as in normal tissues, and

---

*The term 'differentiation' is used here in its broadest sense to imply the development of two or more types of cells, tissues or organs from the multicellular mass which develops after fertilization. Differentiation is complete for most tissues by the 16th week of development of the human foetus.

(b) a disturbed growth equilibrium and not an increased mitotic rate (although an increase in mitotic rate can exacerbate the effect by accentuating growth disequilibrium).

Any cell type can give rise to a neoplasm, although some do so more frequently than others. As a general rule, cells which do not undergo regeneration or replacement—such as nerve and muscle cells—are the ones least likely to give rise to tumours. Cells undergoing active division are more susceptible to carcinogenic (= cancerogenic) stimulus.

The change from normal to neoplastic cells often occurs abruptly. It is also permanent, so that the neoplastic cell transmits its new characteristics to its descendants. The neoplasm continues to develop independently of the cells of origin, from which cells it differs, often to a considerable degree.

Over the past 60 years cancer research has sought the key to the transformation from normal to neoplastic cell. A lot of evidence suggests that changes in the nuclear DNA are the cause. As was seen in Chapter 10, the DNA carries the coding sequence which determines characteristics inherited by the daughter cells after cell division. Changes in the DNA cause abnormal characteristics in succeeding generations of cells. In outline, this is the somatic mutation theory, a theory supported by experimental evidence, which shows that many chemical carcinogens can directly interact with DNA. One example is the alkylating agents, such as ethyl methanesulphanate.

Most neoplasms start at a single site in the body. Those developing in lymphoid and certain other tissues may, however, appear in several parts of the body simultaneously. These neoplasms are said to be multicentric. It is difficult to say whether multicentricity represents the simultaneous appearance of cancer at several sites or whether it reflects rapid communication between lymph nodes by way of lymphatic ducts.

The rate of neoplasm growth varies greatly. Some take years to expand to any noticeable degree, whereas others increase in size and extend beyond their site of origin over a matter of days. Behavioural differences in growth rate are used as the basis of a broad classification of neoplasm, under the headings *benign* and *malignant*. This distinction is of the utmost practical importance in treatment and prognosis.

In general, benign neoplasms grow slowly and have little or no tendency to spread. They are often encapsulated by fibrous tissue. Berenblum (1974) described the capsule formation as resulting from the benign tumour's growth pushing aside the normal tissue. The pushing results in condensation of the neighbouring tissue and the disappearance of ordinary cells from that tissue; within the pressurized area, fibrous tissue develops—forming the capsule. No capsule develops with malignant tumours because the encroachment is too rapid.

Benign cells, commonly, are not grossly different from those in the original tissue. Because of this, the neoplastic cells are said to be *well differentiated*, that is to say they display the specialized features of the cells of the original tissues. Where neoplastic cells differ considerably from those of the tissue of origin they are said to be *poorly differentiated*, that is they display almost none of the specialized features of the cells of the original tissue.

As just stated, in benign neoplasms the cells tend to be similar in appearance to those in the tissues of origin, whereas in malignant tumours the opposite tends to be the case. However, certain benign neoplasms display no marked deviation in cell morphology, even though biochemical function of the cells is grossly distorted. This is seen in certain liver neoplasms, known as minimal-deviation hepatomata. The neoplastic cells are morphologically similar to normal liver cells, but biochemically their function is very different.

The abnormal functioning of neoplastic cells—whether benign or malignant—is sometimes the cause of severe functional disorder as, for example, in endocrine glands, which may be caused to oversecrete hormones. Functional disorders apart, benign tumours do not affect well-being or shorten life unless they occur in, or interfere with, some vital organ.

Malignant neoplasms range from well differentiated to poorly differentiated cells. *Anaplasia* describes the appearance of cells which bear little or no resemblance to the cells of the tissue of origin. It is a condition which, as indicated above, tends to be found in malignant rather than in benign neoplasms. Anaplastic cells take on a very primitive appearance, coming more and more to resemble the simplest type of cell.

Mitotic figures (see Chapter 9) are seen in cells which are undergoing division. Because of the generally low rate of growth in benign neoplasms, mitotic figures are just as rare as they are in normal tissue. Malignant neoplasms, on the other hand, grow rapidly and consequently display mitotic figures in abundance. These commonly display morphological abnormalities, owing to faults in the mitotic process. Often the neoplastic cells have increased numbers of chromosomes. There may be two, three, four or more times the correct number; an abnormality known as polyploidy. Under the microscope this has the appearance, in some cells, of nuclei stained in varying degrees of density—heterochromatic nuclei. In general, malignant neoplasms are characterized by a wide variation in the size of the nucleus and cytoplasm—known as pleomorphism. Figure 16.2

shows comparisons between normal, benign, and malignant tumour cells from the urinary bladder, and other sites.

The rapid and disorganized pattern of growth characterizing malignant neoplasms is the key to the harm they cause. The invasion of surrounding tissues is a highly destructive process. Apart from this, the malignant neoplasms may not develop a blood supply of their own sufficient to nourish the rapidly growing neoplastic tissue, even though they can induce some growth in neighbouring blood vessels. The neoplasm then parasitizes on the host tissues. The blood supply may not be adequate to support both tumour and host tissue, and death of cells occurs in both. The dead cells are subject to necrosis, to bacterial invasion and to further breakdown, which produces harmful substances with grossly debilitating effects on the body as a whole (cachexia).

Malignant neoplasms infiltrate surrounding tissues, causing severe damage to the invaded tissues. As well as infiltration, malignant neoplasms spread by a process of seeding. Small clumps of malignant cells are transported by the blood stream or the lymphatic system to distant organs. There, these secondary deposits (known as metastases) set off a fresh outbreak of the neoplastic process. These in turn may seed further metastases. Commonly, the metastases do more harm than the primary tumours.

The secondary neoplasms formed by metastasis have the morphological characteristics of the primary neoplasm. Primary neoplasms in lung commonly metastasize to kidney. The secondary neoplasms in the kidney display the morphological characteristics of the primary lung neoplasm. Certain neoplasms appear to have a selective tendency for producing metastases in certain tissues. Primary neoplasm of the lung, for example, metastasizes in liver and brain as well as kidney.

There are three principal ways in which neoplasms metastasize: by the blood stream, by the lymphatic system and by transcoelomic spread. Spread by the blood takes place when the neoplasm has invaded the capillaries. Arteries and veins in general are too thick-walled for ready penetration by invading malignant cells. Once the neoplasm has penetrated the capillary, small fragments break off, forming malignant emboli. These circulate in the blood stream and become trapped in the capillary networks, such as those in lung and liver. Once trapped, the malignant emboli divide and develop into secondary neoplasms. Lymphatic spread takes place by a comparable process, in that lymphatic vessels are infiltrated by a process of permeation. It should be recalled that the lymphatic system functions as a drainage channel for tissues generally. When permeation by malignant cells has taken place, small clumps break off and are

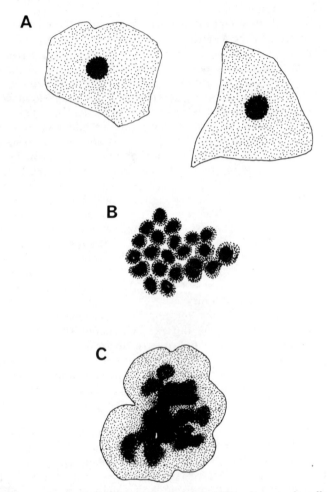

FIG. 16.2. (A) Two normal epithelial cells observed in urine sample collected for exfoliative cytology showing regularly shaped nucleus; (B) cluster of benign tumour cells observed in urine sample collected for exfoliative cytology showing large nuclei which are somewhat irregular in outline; (C) a malignant cell observed in urine sample collected for exfoliative cytology showing large, deeply stained, highly irregular nucleus. Magnification differs between A, B and C. (All drawn from photographs kindly supplied by Dr G. Parkes.)

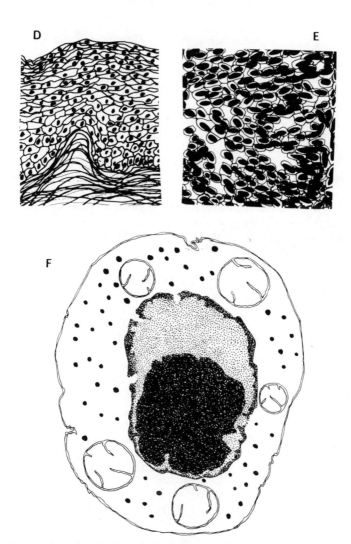

FIG. 16.2—*contd.* (D) *Normal epithelium*: note the regular appearance of the cells with the tendency of flattening towards the exterior surface, and the nuclei small in relation to the cell which they occupy. (E) *Malignant cells*: the malignant cells sketched here can be compared with the normal cells in (D). Note that the cells have no tendency to differentiation; that they appear not to be closely adhering to each other; and that the nuclei are large in relation to the cells. (F) *Malignant cell*: this impression shows four characteristics of a malignant cell: (1) a large irregular nucleus containing a nucleolus, itself large in relation to the nucleus; normally, the nucleolus is small in relation to the nucleus; (2) ribosomes scattered throughout the cell cytoplasm without attachment to endoplasmic reticulum of which there is little or no sign (in this drawing the ribosomes are exaggerated in size); (3) the mitochondria appear swollen and abnormal. A more normal-looking mitochondrion is shown in Fig. 9.1; and (4) overall, no sign of differentiation—the whole appearance is that of a highly primitive cell.

carried in the lymphatic channels to adjacent lymph nodes. The malignant cells then give rise to a secondary malignant growth which in due course takes over one or more of the lymph nodes. The lymph nodes at first act as a barrier to neoplastic spread and the malignant growth may be retained for an appreciable time within the structure of the lymph node. Eventually, the lymph node is overwhelmed and further spread takes place. In surgical procedures the aim is to eradicate the primary neoplasms and any secondary neoplasms which may be developing in adjacent lymph nodes. When the spread has extended beyond the regional lymph nodes, and there are blood-borne secondary deposits, the prospects of successful surgical intervention are poor although chemotherapy and radiotherapy have roles in stopping the spread of metastases.

As indicated in Chapter 3, the lymphatic system joins the blood stream in the thoracic duct. When once the lymphatic system is overwhelmed, neoplastic cells may spill over into the blood circulatory system, thereby causing further and more widespread secondary neoplasms.

Transcoelomic spread—spread across a space within the abdomen or chest—takes place when a malignant tumour has infiltrated all the way through an endothelial surface of one of the main cavities. Subsequent growth may take place within the cavity in the form of isolated secondary neoplasms or as a continuous sheet of malignant cells.

The frequent occurrence of secondary neoplasms in certain organs can be partly explained in terms of the direction of blood supply and lymphatic drainage. However, there may well be other factors influencing the process, such as the reaction of the host tissue.

CLASSIFICATION OF NEOPLASIA

Classification of neoplasia presents considerable difficulty. Present trends favour systems of classification based largely on the name of the tissue in which the neoplasia occurs, a system known as histogenetic classification. Therefore, in order to explain the classification of neoplasia it is first necessary to classify tissues.

Classification of Tissues

The classification of tissues used here follows that found in *Gray's Anatomy* (Warwick and Williams, 1973) although minor modifications have been made in order that it may better coincide with the histogenetic classification of tumours. Four types of tissue can be recognized: (a) epithelium, (b) connective tissue, (c) nerve tissue, and (d) muscle tissue. Each type is composed of cells which determine its structure and function.

It must be emphasized that organs always contain more than one type of tissue, a fact which needs to be kept in mind in any consideration of target organs. As discussed in Chapter 12, it would be preferable, although not at present feasible, to discuss effects in terms of target cells or target tissues. This is so because of the multi-tissue nature of organs and because the various cells and tissues of a target organ can be expected to differ in their dose–effect relation to a specified harmful input.

(a) *Epithelium* is the tissue which lines body surfaces such as the external surface, the internal cavities, the gut and the urinary tract. The epithelial tissues in the fully developed human are mixtures of the embryological cell layers. The names given to epithelial tissues in certain parts of the body reflect the embryological background. For example, tissues lining the internal cavities are known as *mesothelia*, those lining the blood vessels are termed *endothelia*, and those lining the external surfaces, including gut and part of the urinary tract, are known as *epithelia*. It would thus be possible to speak of mesothelial epithelium, endothelial epithelium and epithelial epithelium but in practice these terms are not used.

Epithelium can be classified into three types: (i) simple epithelium, in which there is a single layer of cells on a basement membrane, (ii) stratified epithelium, where there is more than one layer and where cell replacement occurs from the basal layer (i.e. that nearest the basement membrane), and (iii) transitional epithelium in which replacement occurs in any layer (Fig. 16.3).

Simple and stratified epithelium can be further subdivided according to the shape of the cells: squamous (flat, like a paving-stone), cuboidal, and columnar.

Glands, including the endocrine glands which produce the hormones, are all derivatives of epithelial tissues.

(b) *Connective tissues* (Fig. 16.4) are characterized by cells distributed widely in an intercellular material, which is often secreted by the cells themselves. Connective tissues have important structural and defensive functions, many of which have been outlined in previous chapters. Connective tissues are divided into (i) ordinary types, and (ii) special types. The ordinary types of connective tissue, comprising supporting tissues and blood, are widely distributed throughout the body. The special types include cartilage and bone. All connective tissues comprise cells such as fibroblasts and macrophages, as well as extracellular matrix (which consists of fibres and an amorphous, viscous substance).

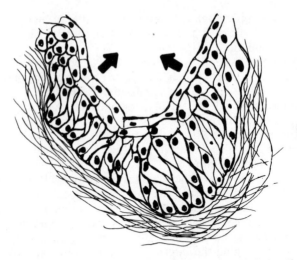

FIG. 16.3. Transitional epithelium of bladder. Arrows show direction of exfoliation; cells from the outer layer become detached as part of the ordinary replacement processes of the epithelium. The exfoliated cells are voided in the urine, as seen in Fig. 16.1.

Blood is categorized as a connective tissue, although it has specialized cellular elements and there are normally no fibres.

(c) *Nerve tissue* contains a functional tissue of a highly specialized nature, known as 'excitable tissue', the general properties of which are discussed in Chapter 17 and illustrated in Fig. 17.2. A remarkable degree of specialization is observed in excitable tissue in the special senses (eye, ear, or nose). For example, the olfactory threshold for many substances is all too often beyond the reach of measuring instruments.

(d) *Muscle tissue* is specialized tissue which produces an effector response of movement. There are three types of muscle: (i) striated muscle associated with the skeletal system, (ii) striated muscle peculiar to the heart, and (iii) non-striated muscle found in many organs, but not under voluntary control.

The differences between benign and malignant neoplasms have already been outlined. These differences are summarized in Table 16.1 (taken from Tighe, 1976), which provides a guide based primarily on behavioural differences of the neoplastic cells and, to a certain extent, on their microscopic appearances. The distinction between the two kinds of neoplasm is crucial, even though a high degree of pathologist's skill may be required to make the distinction.

FIG. 16.4. Diagram of connective tissue showing cells and fibres with spaces between cells filled with a ground substance in the form of a featureless gell. (A) collagen fibres; (B) eosinophil; (C) neutrophil; (D, E, H) mast cells; (F) lymphocyte; (G) plasma cell; (I) ground substance; (J) nerve fibre; (K) fat-bearing cells; (L) elastin fibres; (M) pericyte surrounding capillary; (N) capillary with neutrophil passing from lumen into surrounding tissues; (O) fibroblast; and (P) macrophage. Redrawn, with acknowledgements, from *Gray's Anatomy*.

Various taxonomies have been used in the classification of neoplasia. Even now, there is no one single system in universal use. As previously stated, the scheme which has gained the widest acceptance is based both on tissue of origin (histogenetic classification) and on degree of malignancy. In general, names of neoplasms end in the suffix '-oma'; this implies tumour, which simply means swelling. Taxonomic classes are based on embryological terminology, each class being divided into benign and malignant according to the tumour's growth rate.

TABLE 16.1
CHARACTERISTICS OF BENIGN AND MALIGNANT NEOPLASMS

Benign	Malignant
Remain localized	Can form metastases
Slow growth	Rapid growth
Often circumscribed with a fibrous capsule	Capsules incomplete, show infiltration of surrounding tissues
Mitotic figures rare	Mitotic figures common
Similar to cells of origin	Dissimilar to cells of origin— often totally undifferentiated
Cells of uniform size and appearance	Cells and nuclei variable in size and structure
Degenerative changes uncommon	Degenerative changes common including necrosis

(a) *Neoplasms in Epithelium*
Descriptive terms are commonly used: *papilloma* is an epithelial neoplasm with a craggy surface; *polyp* has a smooth surface (but not all polyps are neoplasms).

(i) *Benign tumours in epithelium.* *Squamous cell papillomas* resemble warts; the former is neoplastic but the latter is hyperplastic. *Transitional cell papillomas* occur in the urinary bladder or any transitional epithelium. *Adenomas* grow in glands sometimes protruding from the epithelial surface as papillomas or polyps.

(ii) *Malignant tumours in epithelium.* These are termed carcinomas, and their classification parallels that of benign epithelial tumours. *Squamous cell carcinomas* often arise in stratified squamous epithelium (these particular tumours are sometimes termed epitheliomas). Sometimes, however, squamous cell carcinoma arises in other epithelial tissues. For this to happen metaplasia first occurs (from one type of epithelium to another) followed by neoplasia. 'Cigarette smoker's cancer' commonly displays this pattern in the lining of the conducting airways. *Basal cell carcinoma* arises from the basal-cell layer of the stratified squamous epithelium of the skin. One form typically affects the side of the face and is often called 'rodent ulcer'. *Transitional cell carcinomas* occur in transitional epithelium of, for example, the urinary tract. They commonly recur even after surgical removal, and they are highly invasive. Consequently any evidence of hyperplasia, or of papillomas of bladder, is always suspected for future malignancy.* *Adenocarcinomas* arise in glan-

*The cells illustrated in Fig. 16.2 are used as the basis for screening tests aimed at identifying the presence of tumour cells in urine of workers exposed to certain chemicals associated with bladder cancer, as discussed in Chapter 18.

dular elements of epithelium. Some adenocarcinomas develop fibrous tissue which makes them firm to the touch.

Anaplastic tumours occur in all epithelial tissues but their degree of undifferentiation makes identification difficult or impossible.

(b) Tumours in Connective Tissues

Connective tissues tumours are classified according to the predominant tumour cell. The distinction drawn between ordinary and special types of connective tissue is only partly relevant because connective tissue tumours are especially prone to metaplasia. Terminology generally derives from the tissue of origin combined with the suffix '-oma' for benign tumours (as with benign epithelial tumours) and '-sarcoma' for malignant tumours. Table 16.2 lists the principal examples.

The terminology does not apply to the blood, plasma, blood-forming organs or macrophage system: *myeloproliferative disorder* denotes diseases in blood-forming organs and *malignant lymphoma* refers to lymphoid tissues. It should be noted that myeloproliferative disorders and malignant lymphomas are almost invariably malignant; benign forms are either rare or non-existent (opinions vary on this point). Principal examples are listed in Table 16.3.

(c) Neoplasms in the Nervous System

Neoplasms of neurones which are the nerve cells proper are extremely rare because these cells almost never divide. Most primary tumours of the nervous system arise, in fact, in the connective tissues of the nervous system. Table 16.3 lists the principal types of tumour. Most are malignant, although the degree of malignancy varies considerably. But benign or malignant tumours in the nervous system,

TABLE 16.2
CONNECTIVE TISSUE TUMOURS*

Tissue	Benign	Malignant
Fibrous tissue	Fibroma (Myxoma)	Fibrosarcoma (Myxosarcoma)
Fatty tissue	Lipoma	Liposarcoma
Bone	Osteoma	Osteosarcoma
Cartilage	Chondroma	Chondrosarcoma
Blood vessels	Haemangioma	Haemangiosarcoma
Lymphatics	Lymphangioma	Lymphangiosarcoma
Synovium	Benign synovioma	Synovial sarcoma
Mesothelium		Mesothelioma

*Modified from Tighe (1976).

TABLE 16.3
NEOPLASMS OF THE LYMPHO-RETICULAR SYSTEM*

(i) Myeloproliferative disorders	(ii) Malignant lymphoma
Chronic myeloid leukaemia	Lymphocytic lymphoma
Chronic lymphatic leukaemia	Lymphoblastic lymphosarcoma
Myelomonocytic leukaemia	Histiocytic lymphoma ⎫ Reticulum
Acute leukaemia	Stem cell lymphoma ⎭ cell sarcoma
Polycythaemia vera	
Thrombocythaemia	
Erythraemia	Hodgkin's disease
Myelofibrosis	
Myelomatosis	Thymoma
Neoplasms of the nervous system	
Brain and spinal cord	Glioma
Meninges	Meningioma
Adrenal medulla and	
sympathetic chain	Ganglioneuroma
	Neurobostoma
	Phaeochromocytoma
Nervous system connective tissue	Neurofibroma
	Neurofibrosarcoma
	Neurilemmoma

*Modified from Tighe (1976).

especially brain, radically disrupt function by their very presence. Metastases, commonly from lung, frequently form in the brain.

(d) *Tumours of Muscle Tissue*
Muscle tissue has a limited capacity for division and tumours of muscle tissue itself are uncommon. Muscle, however, contains connective tissue which can be the originating tissue of 'muscle' tumours. Benign tumours of smooth muscles are called *leiomyomas* and those of striated muscle are *rhabdomyomas*. The malignant types are *leiomyosarcomas* and *rhabdomyoscarcomas*, respectively.

(e) *Mixed Tumours*
Tumours occur involving both epithelial and connective tissues. The tumours included under this heading, with one exception, do not appear often in reports relating to toxicology, so they are not mentioned here. The exception is the group known as teratomas. These are thought to be caused by cells becoming misplaced during foetal development, resulting in a bizarre mixture of tissues such as teeth,

hair and brain cells developing as a tumour—almost as a person-within-a-person. Teratomas may become malignant; links between teratogenesis and teratomas are at present speculative.

PART (2): OCCUPATION-LINKED NEOPLASIA IN CERTAIN TARGET ORGANS

Carcinogenesis* has stimulated a vast amount of research effort resulting in a body of literature which is, frankly, daunting but within it there are data relative to occupation-linked neoplasia. The modern study of cancer (for the remaining purposes of this chapter 'cancer' rather than 'neoplasia' will be used as the general title for the subject under discussion) is a multidisciplinary matter. But until recently, according to Doll (1977, see below), studies were often conducted with little contact between epidemiologists and laboratory researchers drawn, in any case, from a variety of major disciplines. Doll is, however, optimistic for future co-operation and communication through agencies such as the World Health Organization's International Agency for Research on Cancer (IARC), some of whose publications are listed in this chapter's bibliography. There is, nevertheless, a legacy of diverse scientific literature and problems with differences of terminology which cause comprehension problems for people who are specialists in neither epidemiology nor the relevant laboratory research but who wish to understand the nature of problems of cancer.

In the next few pages is given an outline intended to convey information about cancer under four headings: (A) statements made with reasonable certainty by authoritative people and which non-specialists can take as established knowledge (with, of course, all the necessary reservations generally applicable to knowledge in science regarded as established); (B) statements made with qualifications by those authorities, that is to say the uncertainty is greater, but the statements are stronger than speculation; (C) problems with occupational cancer which are of concern to students of occupational health and safety as well as epidemiologists and laboratory workers; and (D) the way forward indicated by one authority.

*The reader will meet 'oncogenesis' in other texts, though not in this one. An explanation of the term may be helpful. Berenblum (1974) regards 'carcinogenesis' and 'oncogenesis' as being essentially the same process. He finds it convenient to refer to 'carcinogenic' chemicals and 'oncogenic' viruses in order to emphasize the difference between them as causal factors.

(A) REASONABLY CERTAIN STATEMENTS FROM AUTHORITATIVE SOURCES

Sir Richard Doll gave the Sir Ernest Kennaway lecture at the Royal Institution in November 1976 and published a review article based on the lecture in 1977 in which he stated: 'It is now clear ... that most, if not all cancers have environmental causes and can in principle be prevented.'

This crucial, and hopeful, statement depends on evidence drawn primarily from the variation in incidence of different types of cancer in different communities in different parts of the world, together with evidence of the shifting temporal pattern of certain types of cancer.

Although some of the geographical variations may be linked with heredity this has been ruled out by Doll, on epidemiological grounds, as a major factor. Although heredity appears comparatively unimportant in man, the converse is partly true for in-bred animals (Dinman, 1974); indeed, strains of rodents are bred specifically for cancer-proneness.

Cancers in certain organs are increasing in both men and women and the increase cannot be dismissed as artefact produced by better diagnosis (Doll, 1977). Concomitantly, there is a decrease in some cancers which cannot be attributed wholly to better medical treatment. Of the total prevalence of cancer in a population at any one time only a small proportion can be linked directly with an occupational origin (that is the cancer occurs in work people and is caused by their work). As shown by Doll, an unknown proportion is industry-linked, however, because of the contamination of the general external milieu by known carcinogens such as polycyclic hydrocarbons or asbestos (see Table 16.4). Another unquantifiable source of cancers is iatrogenic (caused by treatment or diagnostic procedures such as X-ray photography); however, Doll believed that Illich (1975) has overestimated iatrogenic cancer.

Carcinogens, whether occupational or not, are very diverse in nature; see Table 16.5 which shows information from many sources. Indeed, the only factor which they all possess in common is their carcinogenicity. There are numerous difficulties in attaching the label 'carcinogen' to any factor. Of these the most important ones appear to be:

(1) The interactions between factors which increase, perhaps multiplicatively, the incidence of cancer associated with one or other of the factors. Interactions may have the effect of rendering carcinogenic one factor which would be non-carcinogenic—or appear to be so—in the absence of the other factor (see below). Mode of input and context of mode of input complicate the interactions; for example, cigarette

TABLE 16.4

OCCUPATIONAL CAUSES OF CANCER CONTRIBUTING TO GENERAL
ENVIRONMENTAL POLLUTION*

Agent	Site of cancer
Ionizing radiations	Bronchus
	Skin
	Bone
	Marrow (leukaemia)
Polycyclic hydrocarbons	Skin, scrotum
in soot, tar and oil	Bronchus
Arsenic	Skin
	Bronchus
Asbestos	Bronchus
	Pleura, peritoneum
Vinyl chloride	Liver (angiosarcoma)
Ultra-violet light	Skin
Aromatic amines	Bladder
2-naphthylamine	
1-naphthylamine	
benzidine	
4-aminobiphenyl	
bis-chloromethyl ether	Bronchus
Benzene	Marrow (leukaemia)
Mustard gas	Bronchus
	Larynx
	Nasal sinuses
Nickel ore	Bronchus
	Nasal sinuses
Chrome ore	Bronchus
Cadmium (?)	Prostate
Agents in isopropyl oil	
hardwood furniture manufacture	Nasal sinuses
leather goods manufacture	

*Taken from Doll (1977).

smokers may be exposed at work to substances absorbable through skin.

(2) Alterations of substances, brought about in the metabolic pathways by transformation and conjugation, resulting in intoxication (= activation), (see Chapters 8 and 18) are of widely acknowledged importance in carcinogenesis (see for example Süss, Kinzel and Scribner, *passim*, 1973; Berenblum, 1974; Dinman, *passim*, 1974; Weisburger, 1975; and Curry, 1977).

(3) Acknowledged difficulties are encountered in transposing data

TABLE 16.5
OCCUPATIONAL CARCINOGENESIS

Occupation	Agent	Site
Chimney sweeps	Combustion products	Scrotum
Distillers of brown coal	of coal	Other parts of skin
Makers of 'Patent Fuel'	Shale oil	Skin
Makers of coal gas		
Road workers	Polycyclic	Skin
Boat builders	hydrocarbons	
Others exposed to pitch and tar products		
Cotton mule spinners	Mineral oils (polycyclic hydrocarbons)	Scrotum and skin
Miners in Schneeberg	Ionizing radiation from radon	Lung
Farmers and sailors	Sunlight (UV light)	Skin
Arsenic ore smelters	Arsenic	Scrotum
Dye manufacturers	Benzidine (1- and 2- naphthylamine)	Bladder
Radiologists, radiographers	Ionizing radiation and X-rays	Skin
Makers of coal gas	Polycyclic hydrocarbons	Lung
Makers of mustard gas	Mustard gas	Lung, larynx, nasal sinuses
Chemical workers	4-aminodiphenyl	Bladder
Manufacturers of PVC	VC monomer	Liver
Sheepdip manufacturers		
Vineyard workers	Arsenic	Lung
Cobalt smelters		
Rhodesian goldminers		
Haematite miners	Radon	Lung
Asbestos workers	Asbestos	Lung
Insulation workers		
Asbestos workers	Asbestos	Pleural and peritoneal mesothelium
Chromate manufacturers	Chrome ore and chrome pigments	Lung
Makers of ion-exchange resins	Bis (2-chloromethyl) ether	Lung

TABLE 16.5—*contd.*

Occupation	Agent	Site
Nickel refiners	Nickel ore	Lung and nasal sinuses
Furniture makers	Hardwood dusts	Nasal sinuses
Makers of isopropanol	Isopropyl oil	Nasal sinuses
Workers with glues and varnishes, etc.	Benzene	Bone marrow (Myeloid leukaemia)
Luminizers	Ionizing radiation from radium	Bones and blood
Makers of coal gas	1- and 2-naphthylamine	Bladder
Rubber workers	1- and 2-naphthylamine	Bladder
Fluorspar miners	Ionizing radiation from radon	Lung
Leather workers	Not known	Nasal sinuses
Newspaper printers	Printing inks	Lung

obtained in animal experimentation to man (Dinman, 1974). This is not to say, however, that animal experimentation is useless. Doll instanced those substances for which human carcinogenicity was heralded by animal experimentation: 4-aminobiphenyl (syn: 4-amino-diphenyl), mustard gas, vinyl chloride, and a powerful group of carcinogens, the nitrosamines 'that are still looking for human cancers to induce'.

(4) Epidemiological difficulties arise from uncertainties about the real extent of exposure to putative occupational carcinogens (Dinman, *passim*, 1974) and there are problems with estimation of non-occupational exposures to carcinogens. Doll strongly advocated improvements in record linkage, so that patterns in disease could be linked with changes in social, dietary, occupational and general environmental factors. Numerous observations confirm the point that occupationally induced neoplasms may not differ at all from 'naturally occurring' neoplasms. Cancers of the renal tract associated with occupational factors are not, for example, generally distinguishable from cancers in the same site without an apparent occupational link.

Authorities appear to agree that the concept of a 'safe threshold' for carcinogens needs to be approached with great caution; at the present state of knowledge the concept can neither be confirmed as generally valid nor rejected as wholly invalid. In regard to chemical carcinogenesis there is acceptance of the following observations:

(a) many chemical carcinogens react with DNA and thereby induce mutations (whether by affecting the DNA directly, referred to as *genetic change*, or by alteration of non-genetic material which in turn affects the DNA, referred to as *epigenetic change*);

(b) chemical substances can be mutagenic without necessarily being carcinogenic (not all mutations, that is to say, result in the loss of control implied by carcinogenesis), and the possibility cannot be excluded of substances being carcinogens without being mutagens (such an observation would mean that mutation was not, of itself, sufficient explanation of carcinogenesis);

(c) there is abundant evidence that chemical carcinogenesis is dosage dependent for a whole range of substances (Weisburger, 1975), but the concept of dosage dependency introduces important difficulties; and

(d) in relation to cancer generally, and not just chemical carcinogenesis, it is accepted that factors in the host, such as endocrine activity, affect the incidence and the growth of cancer.

(B) STATEMENTS MADE WITH QUALIFICATIONS BY THE AUTHORITIES

(i) Two-stage or multi-stage hypotheses are needed to explain the long latent periods observed with chemical carcinogens and the interactive effects labelled 'initiation' and 'promotion' (see below) but a two-stage effect has been established with certainty only for mouse skin and not with other experimental animals (Süss, Kinzel and Scribner, *passim*, 1973). Initiation induces the so-called latent tumour cells; these proceed to tumour formation only after the application of a promotion factor. One example of an initiator is the carcinogenic hydrocarbon 3,4-benzpyrene for which the promotion agent is croton oil. However, it should be noted that the 'active' ingredient in croton oil is itself a weak carcinogen.

(ii) Initiation and related phenomena are irreversible. Hence, chemical carcinogens are cumulative because each dosage given results in an irreversible change in some cells until the irreversibly changed cells are present in sufficient numbers for a tumour to result (Weisburger, 1975).

(iii) Among the host factors discussed in relation to carcinogenesis generally are what Berenblum called 'immunological influences'. These include immunological surveillance (mentioned in Chapter 5) in which the body reacts immunologically against the cells transformed into neoplastic cells. There is also the possibility that the body could react immunologically against the induction process—a chemical, say. Immumological responses in either direction would be examples of advantageous immune responses, although there is nothing to indicate that immunopathological processes, as described in Chapter 14, might not also occur.

(iv) Viruses are definitely established as agents in certain instances of animal carcinogenesis but the evidence supporting a viral aetiology for human cancer is equivocal, although there is extensive evidence to *associate* viruses with Burkitt's lymphoma and carcinoma of the nasopharynx (Doll, 1977). That interest should be drawn to viruses is obvious from the facts that viruses are largely composed of nucleoprotein (there are RNA and DNA viruses) and that they are normally incapable of independent existence because they lack the apparatus necessary for replication. Instead, they instruct cells upon which they are parasitizing to produce viral rather than cellular components. Viruses are often described as particles on the boundary between living and non-living matter; their involvement with, and distortion of, the cellular replication mechanisms make them, as has been said, obvious targets for suspicion, but at present their role in human carcinogenesis appears surprisingly slight.

A viral explanation for human carcinogenesis would not preclude, necessarily, chemicals as activators of the virus.

(v) No doubt is now expressed that carcinogenesis is bound up with the genome (= the whole genetic 'kit' carrying the cell's hereditary instructions which appears, during mitosis, as a diploid set of chromosomes). What is not clear is how the genome misinterpretation occurs; it might be, for example by *transformation*, in which a carcinogen transforms a normal cell into a neoplastic cell from which a new cell population develops not subject to the normal processes of control of growth equilibrium, or it might be *selection*, in which abnormal cells already present are allowed to develop instead of being suppressed, or it might be by *isolation*, in which normal cells are removed from the ordinary growth-regulating influences (Süss *et al.*, 1973), perhaps involving enzymes concerned with repair (Weisburger, 1975). *Transformation* is also used to denote cells which have undergone morphological or functional changes in tissue culture but this observation *per se* provides no proof that transformation is the central process of tumour formation in living tissue.

(vi) The nineteenth century irritation theory of carcinogenesis was long ago discarded as being an oversimplification. However, it is now recognized that there is a link with irritation as, for example, that seen between chronic irritation (or stimulation) and metaplasia and the subsequent development of neoplasia at the site of the metaplasia.

(vii) Atmospheric contamination's role as a carcinogenic factor is a matter for debate. The evidence of Doll is that non-smokers have a low incidence of lung cancer irrespective of where they live. Heimann

(1961), on the other hand, categorized lung cancer as a disease linked with general atmospheric pollution.

(viii) Auerbach (1976) discussed the somatic mutation theory of carcinogenesis. This theory was put forward, at about the turn of the century, by the German biologist Boveri. Briefly, the theory holds that cancer cells derive from normal ones by somatic mutation (that is not in a reproductive cell but in a cell, or cells, of the body). Support for the theory came from observations that most potent mutagens, for example X-rays, ultra-violet light, and alkylating agents are also carcinogens. However, the opposite did not appear to be true—until recently when it was found that certain powerful carcinogens, for example 2-aminofluorine, methylcholanthrene, and dibenzanthracene are, in fact, converted by metabolic processes (such as those described in Chapter 8) to mutagens. Consequently the theory was revived.

(C) PROBLEMS OF WIDE CONCERN RELATING TO OCCUPATIONAL CARCINOGENESIS

(1) Can carcinogens be predicted anterior to human exposure? This all-important question is at present the subject of intensive research. These are indications of progress, but the problem is far from solved.

(2) The place of animal experimentation is not yet fully defined. Nevertheless it seems reasonable to conclude from the literature relating to carcinogenicity generally that human exposure to chemical carcinogens has been avoided because data were available from experiments involving animals (and microbes).

(3) Terminology presents a knotty problem which needs attention. As mentioned previously, dosage dependency is a reality for many carcinogens. Very often the term 'dose–response relation' is used as a synonym for dosage dependency. The difference between dose–response relation, as defined in Chapter 12, and dosage dependency is a subtle but important one. 'Dosage' means the quantity of carcinogen actually injected, implanted or instilled into the tissues. For example, Süss, Kinzel and Scribner (1973) examined Bryan and Shimkin's work, about which they wrote:

> Increasing quantities of the carcinogenic hydrocarbons ... were injected subcutaneously in mice, and the sarcomas arising at the site of the single injection were counted. It is seen that 0.1 mg of methylcholanthrene suffices to produce a 100 percent sarcoma yield, but that about ten times as much benzpyrene is required for the same effect.

Contrast this statement with part of Nordberg's (1976) statement that dose may have to be estimated from (*inter alia*):

> ... occupational exposure: this usually involves the concentration in air, rate of inhalation, time, and the appropriate deposition, retention and absorption factors. In addition, dermal exposure and ingestion during work time should be considered, noting that the worker is also exposed to ambient air, drinking water and food, as a member of the general population.

Now contrast these previous statements with that of Hatch (1972):

> A basic principle of occupational disease prevention rests upon the reality of threshold levels of exposure for the hazardous agents of industry, below which man can cope successfully.... The concept derives from the quantitative characteristic of the dose–response relation according to which there is a systematic downward change in the magnitude of man's response as the dose of the offending agent is reduced.

It should be clear from the three quotations that the term 'dose', as used in 'dose–response relation', carries more than one interpretation, and that the occupational hygienists' concept differs fundamentally from that of the toxicologists. The hygienists are making estimations of dose on the basis of exposure to contamination and with all the body defence processes fully operational, whereas the toxicologists, by their injecting, implanting or instilling of doses, are bypassing many defences and overwhelming others. *'Dosage dependency', as studied in toxicology, does not correspond to 'dose–response relation' as used in occupational hygiene* but unfortunately 'dose–response relation' is frequently used in both contexts without the necessary clarification of 'dose', sometimes with misleading results.

(D) THE WAY FORWARD

Doll (1977) pointed the way forward when he wrote:

> ... we can no longer assume that thresholds exist for chemical or physical agents [which cause cancer] and we ought neither to ignore nor to condemn them until we have derived quantitative relationships between the dose to which the individuals are exposed and the resultant incidence of the disease. At present we can do this only very crudely. Nevertheless, any quantitative evidence is better than none.

Some of the questions relating to quantitative evidence are discussed in later chapters.

REFERENCES

AUERBACH, C. (1976). *Mutation Research*, London: Chapman & Hall.

BECKER, B. A. (1975). In *Toxicology*, L. J. Casarett and J. Doull, eds, New York: MacMillan.

BERENBLUM, I. (1974). *Carcinogenesis as a Biological Problem*, Amsterdam: North-Holland.

CURRY, S. H. (1977). In *Current Approaches in Toxicology*, B. Ballantyne, ed., Bristol: John Wright.

DINMAN, D. D. (1974). *The Nature of Occupational Cancer*, Springfield: Thomas.

DOLL, R. (1977). Strategy for detection of cancer hazards to man, *Nature*, **265**, 589–96.

HATCH, T. F. (1972). Permissible levels of exposure to hazardous agents in industry, *Journal of Occupational Medicine*, **14**, 134–7.

HEIMANN, H. (1961). Effects of air pollution on human health, in *Air Pollution*, Monograph Series 46, Geneva: World Health Organization.

ILLICH, I. (1975). *Medical Nemesis*, London: Caulder and Boyars.

LEGGE, T. (1934). *Industrial Maladies*, London: Oxford.

LENZ, W. (1962). Thalidomide and congenital abnormalities, *Lancet*, **2**, 1332–3.

NORDBERG, J. F. (ed.) (1976). *Effects and Dose–Response Relationships to Toxic Metals*, Amsterdam: Elsevier.

SÜSS, R., KINZEL, V. and SCRIBNER, J. D. (1973). *Cancer: Experiments and Concepts*, New York: Springer-Verlag.

TAUSSIG, H. (1962). A study of the German outbreak of phocomelia, *Journal of the American Medical Association*, **180**, 1106–14.

THOMSON, A. D. and COTTON, R. E. (1976). *Lecture Notes on Pathology*, 2nd Edition, Oxford: Blackwell.

TIGHE, J. R. (1976). *Pathology*, 3rd Edition, London: Baillière.

WARWICK, R. and WILLIAMS, P. L. (eds) (1973). *Gray's Anatomy*, 35th Edition, Edinburgh: Longman.

WEISBURGER, J. H. (1975). In *Toxicology*, L. J. Casarett and J. Doull, eds, New York: MacMillan.

BIBLIOGRAPHY: PUBLICATIONS OF THE INTERNATIONAL AGENCY FOR RESEARCH ON CANCER

Liver Cancer, IARC Scientific Publications, No. 1, 1971.

Oncogenesis and Herpesviruses, IARC Scientific Publications, No. 2, 1972.

N-Nitroso Compounds Analysis and Formation, IARC Scientific Publications, No. 3, 1972.

Transplacental Carcinogenesis, IARC Scientific Publications, No. 4, 1973.

Pathology of Tumours in Laboratory Animals—Volume I—Tumours of the Rat, Part 1, IARC Scientific Publications, No. 5, 1973.

Host environment Interactions in the Etiology of Cancer in Man, IARC Scientific Publications, No. 7, 1973.

Biological effects of Asbestos, IARC Scientific Publications, No. 8, 1973.

N-Nitroso Compounds in the Environment, IARC Scientific Publications, No. 9, 1974.

Chemical Carcinogenesis Essays, IARC Scientific Publications, No. 10, 1974.

Oncogenesis and Herpesviruses II, IARC Scientific Publications, No. 11, 1975, In two parts.

Screening Test in Chemical Carcinogenesis, IARC Scientific Publications, No. 12, 1976.

Evaluation of Carcinogenic Risk of Chemicals to Man, IARC Monographs: Volume 1 (1972)–Volume 14 (1977).

Section 5:
CASE HISTORIES AND EXAMPLES

Section 5 comprises Chapters 17 to 23 covering energies, aromatic amines, certain metals and substances resembling metals, asbestos, respiratory disease in the coal industry, gassing accidents, and vinyl chloride.

The case histories and examples are intended, in general, to provide support for the various points put forward, on a theoretical basis, in previous chapters. Sufficient referencing and annotation is given in order to provide authority for the factual statements being made. In the selection of topics for the case histories the deciding factor was whether a topic appeared to contain relevant points. For some of the topics certain conclusions are presented but in general the explicit drawing of conclusions has been avoided because the facts as presented may justify more than one point of view in their interpretation.

'Case histories' differ from 'examples' in that the case histories are generally more detailed and better documented. Chapter 17 presents both 'case history' and 'example' material concerned with the energies. Chapter 19 is example material presented as an outline guide intended to emphasize those features possessed by pathogenic metals which unify them with other aspects of toxicology.

The chapter of gassing accidents deals with specific examples of accidents and substances, and it also provides two approaches to a framework for these accidents.

In each of the case histories there is information about the strategy adopted in connection into the specific dangers. These data are provided in summary form but in sufficient detail to allow the reader to compare strategies from one substance to another, and from one era to another. This material, as well as providing an outline review of current approaches to control, also provides background for the final chapter of the book.

Chapter 17

Radiating Energies

OBJECTIVES OF THE CHAPTER

This Chapter:
(1) provides, by means of case histories, examples of ionizing radiations acting as inputs, as described in Chapter 2,
(2) emphasizes the similarity between the effects of ionizing radiation and those of chemicals at the cellular level,
(3) provides an outline of the mode of action of ionizing radiation which links with outlines given in earlier chapters on growth and disorders of growth,
(4) outlines the modes of action of radiating energies generally on normal cells, and
(5) outlines the modes of action of excessive application to living tissues of certain types of energy other than ionizing radiations.
The chapter is divided into four parts:
(A) case histories of accidental exposures to ionizing radiations,
(B) modes of action of ionizing radiations,
(C) normal cellular response to non-ionizing energies, and
(D) modes of action of certain non-ionizing radiations, and certain other forms of radiating energy.
'Radiating energies' is a convenient label for most of the energies discussed in this chapter—ionizing radiations, non-ionizing radiations, radiant energy, and certain other forms of energy capable of passing through air. Mechanical vibration has been included because of its similarities to infrasound and sound, not because it should be categorized as a radiating energy.

PART (A): CASE HISTORIES OF ACCIDENTAL EXPOSURES TO IONIZING RADIATIONS

The case histories presented here are selected from among the numerous reports, dealing with accidental exposures, which have been published over the past 75 years. The case histories are classified according to the six modes of input identified in Chapter 2. As will be seen, accidental exposures to ionizing radiations have involved five out of the six possible modes of input; no report has been located for the one exception but it is, nevertheless, recognized as a possible mode of input in connection with ionizing radiations. Before, however, the case histories are embarked upon, one difficulty requires mention: 'ionizing radiations' presents a problem with definition.

The Ionizing Radiations (Sealed Sources) Regulations, 1969 (SI 1969 no. 808 amended by SI 1973 no. 36), for example, give the following:

> Ionizing radiation means electromagnetic radiation (that is to say, X-rays and gamma rays) or corpuscular radiation (that is to say, alpha particles, beta particles, electrons, positrons, protons, neutrons, or heavy particles) being electromagnetic radiation or corpuscular radiation capable of producing ions and emitted from a *radio-active substance* or from a machine or apparatus that is intended to produce ionizing radiations or from a machine or apparatus in which charge particles are accelerated by voltage of not less than 5 kilovolts . . . .

It should be noted that ultra-violet radiation is omitted from the definition of ionizing radiations yet it causes ionization. It is, moreover, linked with skin cancer (the evidence for this is touched upon briefly in Chapter 16). The principal source of human exposure to ultra-violet light is, of course, sun-light. But occupational exposure to 'non-natural' sources of ultra-violet light also takes place and there would be justification for including these sources in any administrative definition of ionizing radiations for the purposes of occupational health and safety. Nevertheless, in the case histories which follow, ultra-violet light has been excluded, in accordance with the legal definition of ionizing radiation.

INHALATION

Inhalation is an important mode of input for radioactive substances which, once inside the body, emit whatever forms of ionizing radiation accompany their radioactive decay. Clearance time is an important consideration for radioactive particles; the relatively rapid clearance time of the ciliary escalator may be short enough to rid the conducting airways of, say, silica particles but it may be long enough to permit radioactive substances to deliver a damaging dose of radiation to the conducting airways and adjacent tissues.

(1) One case of historical as well as scientific interest is that of Dr

S. A. von Sochocky ('Dr S'), a research chemist and technical director of a company in New Jersey in which luminous paint was prepared and used for luminous dials on clocks and watches (the case is reported by Martland, 1929 and 1931). He was involved with work on radium, and between 1913 and 1921 he personally extracted about 30 g of radium from ore. During this time he was exposed continually to a high level of radioactivity, and also to the inhalation of radioactive dust in the laboratory. Additionally, on four occasions he had been exposed to explosions of tubes containing high concentrations of radium and mesothorium.

In 1925 he devised a method for measuring the radioactivity of bones at post-mortem examination (the reasons for his doing this are explained below). At the same time he tested the radioactivity in his own exhaled air and realized, from the results, that he was beyond medical care. In 1928 he died of 'aregenerative anaemia' (aplastic anaemia, as defined in Chapter 16). Post-mortem examination revealed radioactive deposits in the lungs, ascribed to inhalation.

(2) A high incidence of lung neoplasms has been observed in miners in the uranium mines of Schneeberg and Joachimsthal, 30 km to the south of Schneeberg on the southern slope of the Erzgebirge mountains of Eastern Europe, and in the mines of the Colorado plateau region in the USA. The high incidence of the disease in both areas is not disputed. Hueper (1942 and 1966) estimated that between 1869 and 1939 at least 400 miners from the Schneeberg mine had died of lung cancer. According to Weber in 1926 (cited by Hueper), three out of every four miners dying during the period up to 1926 had been killed by neoplasia of the respiratory system.* It is believed that the lung cancer originates from radon. This is a radioactive gas which is completely chemically inert (other inert gases, though not radioactive, are helium, neon and argon).

Radon and its short-lived radioactive daughters (polonium-218, lead-214, bismuth-214, polonium-214) are alpha emitters which attach to dust particles, which are then inhaled; this is believed to be the cause of the lung cancer. The data for the Colorado uranium mines display the same general picture but the excess incidence, in epidemiological terms (see Chapter 24), is not great. Archer and Wagoner (1973), for example, have shown that the miners were heavy smokers, see Fig. 17.1, and consequently the attribution to radon of the lung cancer in the miners is not entirely straightforward. The findings in an earlier report by Wagoner et al. (1964) were questioned by, for example Hunter (1975). Despite the problems of interpretation,

---

*It was from the pitchblende found in these mines that Madame Curie obtained radium.

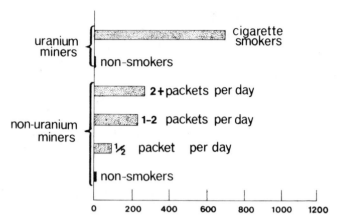

FIG. 17.1. Comparison of incidence of lung cancer in uranium miners, smokers, and non-smokers. Abscissa: number of lung cancer cases per hundred thousand population per year. Redrawn from Bair, W. J. (1970), in *Inhalation Carcinogenesis*, M. G. Hanna, P. Nettesheim and J. R. Gilbert, eds, Springfield: US Atomic Energy Commission.

however, Hobbs and McClellan (1975) accepted that the aetiology of the lung cancer is basically the alpha radiation from the radon and radioactive daughters attached to inhaled dust particles.

INGESTION

Ingestion, that is the entry of substances to the human body by means of the gut, is not normally regarded as a mode of input of great significance in human safety. However, that this mode of input must not be ignored is clearly demonstrated by the following case history.

The New Jersey factory at which Dr Sochocky was exposed to radioactivity provided, as did another factory doing similar work in Connecticut, well-documented evidence of the risk of ingestion of radioactive materials (for example Martland, 1929, 1931; Martland and Humphries, 1929; and Flinn, 1927). Martland (1931) reported that during the years 1917 to 1924 a total of about 800 girls were employed in the New Jersey factory, although there were never more than 250 girls working at any one time. Many of the girls were said to have worked for only a short period, although Martland (1931) emphasized the point that all the known cases of radium poisoning affected girls who had worked in the plant for more than one or two years.

Commenting on the number of known deaths (18 by May, 1931), he pointed out that the number of girls likely to have died as a result of this work was unknown owing to their being dispersed throughout the USA and to difficulty on the part of medical practitioners in making the exact diagnosis.

The deaths were ascribed to the use of luminizing paint which came into use in about 1908.

In the two factories the girls involved in the painting had the habit of pointing the brush between the lips. This was confidently identified as the reason why ingestion was the mode of input for the radioactive paint. Hunter (1975) cited an inquiry by the International Labour Office into watch factories in Switzerland, France, Germany, Austria, Britain, and Belgium in which it was found that brushes were not in use (and therefore not pointed with the lips), instead stylets were used; in all of these countries luminizing had been carried out without, apparently, ill-effects.

The luminizing paint contained radium-226 which is a naturally-occurring radionuclide. Radium emits alpha particles and also some weak gamma radiation; the alpha particles are likely to have been principally involved in the development of the diseases in the luminizing workers. Radium is a metabolic analogue of calcium. Its transport, deposition, and storage is thus likely to correspond closely with bone metabolism. Hobbs and McClellan (1975) reported that a portion of radium taken into the body is diffused throughout the skeleton while another portion is deposited in parts of the skeleton where the biological commerce of calcium is in an active phase at the time when the exposure to radium takes place. These localized areas of relatively high concentration become sources of alpha emission, irradiating the neighbouring tissues. The osteosarcomas, and carcinomas of mastoid and paranasal sinuses which occurred in the luminizing workers are explicable on the basis of the localized high concentration and consequent irradiation of adjacent tissues.

PERVASION OF THE SKIN

Here pervasion is interpreted to mean the passage of substances, rather than radiations, through the intact skin. That this mode of input is feasible was recognized by the International Commission on Radiological Protection (ICRP, 1959) and clearly it must be anticipated within the planning of decontamination procedures. However, no accidents involving this mode of input have been identified among the published reports consulted.

IMPLANTATION

Forceful breaches of the skin by, for example, punctures with sharp-edged hand tools or fragments of swarf, or broken glass, may carry radioactive particles into the tissues. This mode of input could, if not treated effectively and promptly, 'internalize' radiation sources. Hence, a source of insufficient strength to cause much damage whilst confined to the body surface might, on internalization, be capable of producing a material degee of harm.

Hemplemann *et al.* (1973) published a detailed account of the 25 men who worked with plutonium on the 'Manhattan project' during World War II. This project was concerned, as is well known, with the development of the atomic bombs used in World War II. According to Hemplemann and his colleagues, the men worked under 'extra-ordinarily crude working conditions'. Amongst a host of extremely important observations, these authors noted that in several subjects, known to have had hand wounds while working with plutonium in 1944 and 1945, traces of plutonium could be detected by radioactivity measurement. At the time of observation (1973) the subjects' skins overlying the active sites appeared to be normal apart from minimal scaling. In one subject the site of the implantation was located and the amount of radioactivity could be measured quite accurately. They reported cases of four subjects who cut their hands with contaminated objects. In these particular cases, the tissues excised surgically from around the cuts contained little radioactivity. This was an unexpected finding because implantation of radioactive materials is generally regarded as an inevitable complication of such accidents. It is normally recommended that even quite trivial puncture wounds should be treated with relatively extensive surgical dissection in order to secure adequate decontamination. In the instances just mentioned this precaution was not necessary, seemingly.

SURFACE PENETRATION

Surface penetration is here defined to mean penetration of the epidermis (see Fig. 13.1) by ionizing radiations which reach as far as the basal cells but do not pass it in any significant quantity. The basal layer and the layers of cells immediately exterior to it are the germinal layer of the skin in which the epidermal cells undergo mitosis as part of the normal process whereby the skin replaces cells lost, from the outer surface, by normal wear and tear. Hence, penetration by ionizing radiations as far as the germinal cells may initiate disorders of growth, including neoplasia. From the historical point of view, surface penetration is associated with the first observations of the adverse effects of radiations on people.

According to Hunter (1975) within three months of Roentgen discovering X-rays, in November 1895, it became known that T. A. Edison and W. J. Molton had suffered from conjunctivitis after some hours exposure to X-rays. Stevens and Daniel appear to have made the first reports, in 1896, of the injurious effects of X-rays on the skin (cited by van Lancker, 1976).

In 1898, Gassmann (cited by McLean, 1973) reported acute radiodermatitis from radium gamma rays. In the years following there were numerous reports of skin effects resultant upon exposure to X-rays. The acute effect was quickly recognized to be the reddening of the skin (erythema) which, according to McLean, was used in the earliest days as a guide to dose, no accurate methods of physical dosimetry being available.

In 1902, Frieben (cited by Hunter, 1975) recorded cancer of the skin as a sequel to ulceration caused by X-rays. Hesse (1911) reported 94 cases of cancer of the skin of which 54 cases occurred in people associated with medical uses of radiation. By 1909 Paton had reported a case of cataract resulting from exposure to X-rays. Apart from chronic ulceration and skin cancer, it was also recognized, by Cassidy (1900), that intense exposures to X-rays could cause deep burns which were painful and which healed only with difficulty.

Early medical pioneers of X-rays failed to recognize the chronic sequels to the acute condition. In Birmingham, one such pioneer, Dr Hall-Edwards (1904), reported in detail his own experiences of chronic X-ray dermatitis describing it as 'one of the most persistent, painful and disfiguring maladies it has been my misfortune to meet, and although in the light of our present knowledge, it is avoidable, there appears little or no hope for those who have contracted it'.

He continued to describe how shortly after 'Professor Roentgen's discovery' and long before it was recognized that X-rays had any physiological or therapeutic effects it was a common practice for experimenters to expose their hands freely to the rays. 'With no knowledge of the fearful results which were to follow' they would continue with this practice until an acute dermatitis appeared. Dr Hall-Edwards was one of those who started his experiments within a few weeks of Roentgen's report. During his experimentations he continuously exposed his hands to the X-rays. Early in 1896 he gave a series of demonstrations lasting four evenings. On each of these occasions, for several hours, he exposed his hands to the X-rays. Two or three weeks later he noted that the skin round the roots of the nails was red and painful but he thought little of these symptoms, attributing them to the constant use of a photographic developer containing metol. He ignored the symptoms; the redness and soreness gradually disappeared only to be replaced by transverse lines upon the nails which became so brittle that they broke under the slightest pressure. Then longitudinal cracks made their appearance and the quicks became dry, showing a tendency to separate from the nails.

The nails of the index finger were most affected and thickened; the substance of the nails degenerated until they became shapeless masses. Both nails came off on several occasions. Then warty growths made their appearance and the increased epithelial growth became progressively more marked. At this stage he suffered no inconvenience; two to three years later, however, he began to experience pain. This he described as being of a dull, grinding character and it appeared to come chiefly from the bones. Then he noticed a small sore over the knuckle of his right middle finger which resisted all his attempts to heal it. Extremely painful, sensitive to the slightest touch; it was soon followed by other sores.

The warts continued to increase in size; he kept them under control with the aid of fine sandpaper or by cutting them with a razor. Without this crude treatment he found that the hard skin over the knuckles entirely lost its elasticity and cracked with the slightest exertion. The cracks, too, were very painful and resistant to healing. 'Lately,' he continued, 'I have experienced a marked increase in pain, especially when the hands are held in a dependent position. I have also noted a loss of power and some pain in the arms . . . .'

*   *   *

Surface penetration is of relatively small importance in the consideration of alpha particles. These are not normally considered a risk from external exposure (that is the source of the radiation lies outside the body) because alpha particles at the energies normally met with penetrate only short distances in soft tissue. However, intensive sources, such as accelerators, might produce beams of alpha particles with greater penetration. Beta particles have relatively greater surface penetration, although they have a limited range in living and other matter.

IRRADIATION

For the purposes of this chapter irradiation implies that the body is exposed to radiation and that effects of this exposure as well as those attributable to surface penetration need to be considered. *External radiation exposure* and *external emission* have the same meanings as irradiation.

Recognition of irradiation effects came as early as 1897 when Walsh reported, in the *British Medical Journal*, symptoms of abdominal pain and diarrhoea in an X-ray worker; the symptoms disappeared following protection with a lead shield.

Irradiation from exposure to external radiation, whether from natural or man-made sources, normally involves X-rays or gamma radiation. But other types of radiation may be encountered in accidents with industrial or other radiation sources. Radiotherapy and space flight are situations where external radiation of various types may be encountered. As previously stated, the attention of scientists was first called to injuries from X-rays in 1896 (four years earlier than the first report of harmful effects of radium) but it was not until 1944, however, with the atomic bombing of Hiroshima and Nagasaki that the effects of whole body radiation came to be fully appreciated (Hunter, 1975).

The history of harm from X-rays reads like a history of physics because so many of the physicists associated with the early research on X-rays themselves became the victims of X-ray injuries. Sixty names of distinguished scientists appeared on a death roll from X-ray disease, compiled by Meyer in 1937 (cited by Hunter, 1975). According to Faber (1923) the first death to be attributed to X-rays occurred in 1914. The victim was Dr Tiraboschi of Bergamo, a radiologist. He had a history of 14 years work with X-rays without protection; his illness began with a 'radio-dermatitis' of the left hand and right side of the face. He later developed aplastic anaemia and atrophy of the testes. By 1922, according to Ledoux-Lebard (1922), as many as 100 radiologists had died from malignant disease from X-rays.

An accidental irradiation in relatively recent times at the Oak Ridge Laboratories in Tennessee was described by Vodopick and Andrews (1974). Accidental exposure to gamma rays from an unshielded cobalt-60 source of 7700 curies caused irradiation of a 32-year-old male technologist; the exact date of the exposure is not given.

The exposure occurred when the technologist, unaware that the source was outside its shield, walked close by the source. After he had left the room, the operator of the irradiation facility realized that the source might have been unshielded and that serious exposure could have occurred. The technologist was admitted to hospital one hour after exposure. His dosimeter was sent to the supplier for processing and, when the results became available *28 hours later*, the exposure was established at 260 roentgens. From a reconstruction of the accident it was shown that the technologist had been irradiated for about 40 sec. Simulation showed that he had probably suffered a mid-line dose to the chest of about 127 rads and that dose for the marrow was 118 rads. It was believed that the right hand might have received between 800 and 1200 rads, and the left hand between 500 and 600 rads.

Vodopick and Andrews detailed the clinical course of the technologist's subsequent illness. He was first examined 2 h after the accident and he appeared apprehensive but not ill. At that stage a general medical examination revealed no abnormality. At $2\frac{1}{4}$ h post exposure he began vomiting, without preceding nausea, and vomiting recurred ten times during the next 24 h.

At 24 h post exposure he complained of itching and burning of his eyes and reddening was noted. These signs disappeared after a further 24 h. On the seventh day of hospitalization he was allowed to go home (to a bacteriological milieu more normal for him than the hospital's). He returned, daily, to the hospital for blood counts. On the 25th day post exposure he was re-admitted to hospital and placed in a special unit aimed at reducing bacteriological contamination. 'Haematological depression' was observed from the 25th to the 34th post exposure day (this corresponds to interference with blood cell formation, as mentioned in Chapter 4). On the 36th post-exposure day he developed a mouth infection which was successfully treated with oral penicillin. By post-exposure day 48 his blood picture had fully recovered and he was sent home.

There were no visible signs of radiation damage in his right hand which, as stated previously, was thought to have received more radiation than any other part of his body. However, he described a dull aching sensation in the finger and the palm, especially after he had used his hand for long periods. These symptoms started 3 days post exposure and continued intermittently for some time. The pain sometimes woke him at night. At four months post exposure the pain subsided but recurred several months later. Finally the pain in the hand disappeared altogether. For about four months after the accident he felt easily fatigued but by the eleventh week following the accident he was fit to return to full-time employment (in an area believed free of any abnormal radiation exposure). There is a clear implication that he made a full recovery from his accidental exposure.

The conclusions that Vodopick and Andrews drew from the technologist's case is that medical support for him through the period of maximum haematological depression resulted in full clinical recovery.

PART (B):   MODES OF ACTION OF IONIZING RADIATIONS

GENERAL COMMENTS

In the previous part of this chapter outlines were given of some of the accidental exposures to ionizing radiation which have been reported. When the nature of the risks from this form of energy is considered against the historical background of the use to which it has been put we should not find it surprising that it has been treated as a 'special case' for legal and administrative purposes. Considered purely as an input detached from the social and political implications of its use, however, the justification for seeing it as a special case becomes less obvious. Indeed, this form of energy has, not surprisingly, much in common with other forms of energy and, moreover, a number of factors underlying the modes of action of ionizing radiation also underlies the harmful effects of pathogenic chemicals. The following discussion on modes of action of energies emphasizes the unifying concepts.

Because living tissue contains so much water, the modes of action of ionizing radiations need to be viewed as a multi-stage process:

(1) the first stage is represented by the effects of ionizing energies on water, these include the formation of free radicals,

(2) the second stage can be represented by the combination of the free radicals with biological molecules, and

(3) the third stage comprises all the alterations in biological structure and function consequent upon the modification of the biological molecules by the free radicals.

A distinction needs to be drawn between ions and free radicals. It is useful to recall one of the introductory experiments in electrochemistry in which a pair of electrodes dip into an ionic solution. If the two electrodes are connected together through a current-measuring device, an electrical current will be observed to flow. If, now, an electrical potential difference is applied to the two electrodes a current will also flow. The former set-up is called a galvanic cell; the latter is called an electrolytic cell. In either case the electrode which releases electrons is called the cathode; that receiving electrons is the anode. Because of the progress of electrons from one electrode to another, the ionic solution separates into its components: positively and negatively charged ions. As we have already seen (Chapter 11) these phenomena are important in biology as, for example, with the Na^+ and K^+ ions associated with the electrochemical potential at biological membranes.

The free radicals produced by ionizing energies on water can be regarded as fragments of molecules without an electric charge, that is to say they are electrically neutral. Free radicals are intensely chemically reactive owing to abnormal electron configuration. Their intense reactivity leads them to react almost instantly with any molecules in the vicinity, including other molecules of water, and especially biological molecules. A large number of free radical reactions are possible; those from water alone represent numerous possibilities for chemical reactions. Free radicals may well also be formed from the effect of ionizing radiation on other molecules present in cells apart from water; water-derived radicals predominate, however, simply because cells contain so much water.

MODES OF ACTION OF IONIZING RADIATION AT THE CELLULAR LEVEL

Van Lancker has summarized views as far as 1976. Changes in the DNA molecule (see Chapter 10) appear to be one factor, if not the key factor, responsible for the manifestations of cellular injury caused by the formation of the free radicals which are the direct product of the ionizing radiation. The changes to the DNA molecule caused by the radicals may or may not be repaired. If no repair takes place then the result is alteration in the transcription of DNA to RNA. Alteration in transcription may result in:

(a) cellular death,
(b) block in DNA synthesis,
(c) delay in mitosis, and
(d) distortion of gene expression.

DNA synthesis and mitosis (see Fig. 9.2) seem particularly sensitive to ionizing radiation. The radiation-induced blocks in DNA synthesis and mitosis lead to cellular death. Thus populations of frequently dividing cells are depleted after exposure to X-radiation. Neoplastic cells divide frequently; hence these, too, are depleted by X-radiation. Herein lies the principle of therapeutic radiation; it also explains the apparent paradox how the one form of energy can both initiate and impede tumours.

Van Lancker's summary provides serviceable concepts for the understanding of the interaction between ionizing radiation and biological tissue at the cellular level. Certainly, many of the observable effects of ionizing radiation on living tissue can be reconciled with the summary. Other explanations have, of course, been considered in the course of the development of knowledge. In particular, mention should be made of the theories which held that these energies interfere with enzyme production. From the central nature of enzymes in the regulation of cell metabolism this is not at all an implausible theory. However, various experiments have suggested that interference with enzymes is not the central effect of ionizing radiations at the cellular level.

As stated, informed views up to 1976 emphasized damage to DNA as the central phenomenon. Enzyme mediated disturbances were considered to be a reflection of the damage to the DNA rather than the primary mode of action.

MODES OF ACTION IN RELATION TO TRANSPORT AND STORAGE

'Internal emitters' is a term used to describe all radioactive materials which irradiate the body from within. Internal emitters gain access to the body by means of any of the modes of input available for any chemical substance.

There are numerous radionuclides which are potential internal emitters; for example, radium, strontium, uranium, and plutonium. With these, and other radioactive substances, the first considerations are the chemical and morphological nature of the internal emitters. The mode of input of an internal emitter will be determined by its morphology; what happens to it inside the body will depend on its chemical properties because these determine which of the metabolic pathways it joins.

Substances such as those just mentioned evidence strong bone-seeking properties because they behave as analogues for calcium. Radon, as we have seen, may initiate disease in the respiratory system which is therefore the target organ as well as the organ of input for radon.

If the lung neoplasia in the Erzgebirge and Colorado miners were, in fact, due to alpha emission from radon contaminated dust particles, then this would indicate that the residence time of the particles in some part of the respiratory pathway is sufficiently long for neoplastic changes to be initiated. (Factors affecting residence time in the respiratory pathway have been outlined in Chapter 3.)

The distribution of radioactive substances within the body, their transport to the target organs, and their residence time in those organs can be understood in terms of transport systems and storage—all as outlined in connection with chemical substances. Thus, consideration has to be given to radionuclides' metabolic transformation and conjugation; and their anabolism and catabolism. Radiation toxicology obviously has a great deal in common with non-radiation toxicology;

indeed, a number of advances in knowledge relevant to toxicology have resulted from studies directed at radiotoxicological problems.

With regard to the quantitative aspects, the relation outlined in Chapter 11 is relevant. This, it will be recalled, relates the concentration at various times, the elimination constant and biological half-time. As shown below, the relation can be extended to include the radioactive half-life of radioactive substances. For the following outline, the example given by Hobbs and McClellan (1975) has been used.

Because decay of radioactivity from radioactive materials is exponential in character, half-life is used routinely in physics and the field of radiology to characterize the duration of radioactivity from a radioactive material. As we saw in Chapter 11, it is possible to characterize in exponential form the amount of substance disappearing from or accumulating in an organ. It was necessary, however, to make certain assumptions.

A radioactive material in the body is subject to two processes of decline, both exponential: radioactive decay and biological elimination. The resultant of the two processes can be termed 'effective decay'. The indices used are radioactive half-life (T_p), biological half-time (T_b) and effective half-life (T_e). The three are related:

$$T_e = \frac{T_b T_p}{T_b + T_p}$$

The following expression can be used to calculate the cumulative dose from an alpha emitter:

$$\text{Cumulative dose} = 51.2 \bar{E} f \int_0^t C(t)\, dt$$

where

$$C(t) = C_0\, e^{-0.693t/T_E}$$

and C_0 = initial concentration in $\mu Ci/g$, $\bar{E}$ = average alpha energy, f = fraction of emitted energy deposited in tissue of interest, T_E = effective half-life in days for the radionuclide in the tissue of interest, and t = time in days.

The validity of calculations made with this equation depends on the strength of the assumptions made. We saw in Chapter 11 that biological elimination may not always be exponential and that further research is required in order to establish a range of mathematical models needed to explain biological elimination in all its realities. The weakness in present knowledge is evidently centred in the field of non-radiation toxicology, yet the problem is an important one for radiation toxicology.

MODES OF ACTION: IRRADIATION BY EXTERNAL EMITTERS

'External emitters' refers to all sources of radiation which can irradiate the body by external irradiation.

Irradiation is conventionally divided into whole-body and part-body exposure. The effects of irradiation are divided broadly into somatic and genetic effects. Somatic effects refer to the non-reproductive cells of the body; the genetic effects to the reproductive cells of the body. Of the somatic effects, acute and chronic conditions are recognized; a further division is made according to the length of time the effects take to appear following the exposure, that is a distinction is drawn between chronic effects and late effects.

Langham *et al.* (1965) have described the manifestation of acute whole-body exposure to radiation, pointing out that this is likely to happen only as a result of accidents or nuclear war. The acute, early effects appear to be dose-rate dependent so that their incidence and severity increase non-linearly with increasing dose.

Acute radiation sickness (acute radiation syndrome) has three forms of manifestation (see for example Langham *et al.*, 1965; Hempelmann *et al.*, 1952):

(a) effects on the blood forming organs (haemopoietic effects),
(b) effects on the gastro-intestinal system, and
(c) effects on the central nervous system.

The three are related to radiation dose (Fig. 17.2). At the highest dosage, death occurs within hours or minutes from degeneration of the central nervous system and probably the cardio-vascular system as well.

The gastro-intestinal phase is caused by destruction of the cells of the small intestine causing it to ulcerate. The severe disruption of the gastro-intestinal function causes gross interference with metabolism because of loss of vital bodily constituents such as electrolytes. In addition, the ulceration provides an opportunity for micropredators to enter the body where previously they would have been held harmlessly within the gut. Unless treatment is successful, death takes place in the gastro-intestinal phase at about 5 to 10 days following exposure. Obviously, there will also be interference with the absorption of nutrients from the irradiated gastro-intestinal tract but this effect is not, of itself, lethal in the short term.

At lower doses the bone marrow, the organ of haempoiesis, is injured resulting in changes in the circulating blood cells. Lymphocytes (see Chapter 4, and Figs. 4.2–4.6) are the cells which are the most sensitive to radiation, their numbers in the circulating blood being swiftly lowered, and taking weeks or months to return to normal proportion. The effects on the lymphocytes may be discernible within 2 days following exposure to 50 rads or so.

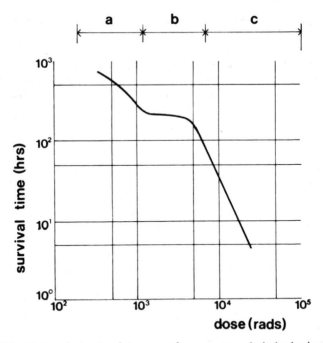

FIG. 17.2. Mode of death of humans from acute whole-body irradiation. Ordinate: means survival time in hours; abscissa: dose in rads. (a) range of haemopoietic effect, (b) gastro-intestinal effect, and (c) central nervous system effect. Modified, with acknowledgements, from Langham, W. H., Brooks, P. M. and Grahn, D. (1965), Biological effects of ionizing radiation, *Aerospace Medicine*, **36**, 1–55.

PART (C):   NORMAL CELLULAR RESPONSE TO NON-IONIZING ENERGIES

The nervous system contains excitable tissue cells. Some of these respond specifically to energies; for example, the specialized cells of the retina respond to electro-magnetic radiation in the visual range, the skin contains cells which respond specifically to temperature, pressure, and vibration. Specialized cells in the inner ear respond to sound, a form of mechanical energy, conducted through the inner ear from the external milieu. Electrical energy, whether developed as part of the normal biological electrochemical processes which take place constantly throughout the body, or externally applied, stimulates excitable cells.

A common response by an excitable tissue cell is for the applied energy to cause a change in the cellular membrane which spreads, wave-like, over the entire surface of the cell. The change may be manifested by alterations in permeability or polarization of the cell membrane. Other responses possible from an excitable tissue cell include secretion or release of a secretory product, and contraction in cells specialized for movement.

Harm may result when energy is applied to excess (whether in terms of rate or total quantity delivered) or where the energy applied is not in a form for which the cell is specialized. The harm may be reversible or irreversible dependent on the quantitative and qualitative factors of the energy occasioning the insult. For example, occupational noise-induced hearing loss has both temporary and persistent forms both of which are known to have quantitative relations with sound incident upon the ear.

In general, injury from excessive stimulation or inappropriate stimulation will normally produce specific effects, that is to say the effects are confined to the specialized cell and are caused by a reasonably restricted range of inputs.

Application of energy to any living cell will cause damage and ultimately death of the cell if (a) the rate of energy delivered, and/or the total quantity delivered are beyond the cell's capability for withstanding that form of energy, and (b) if the defensive or homeostatic processes fail to provide the necessary safeguard or adaptation. Generally speaking, the result of this kind of excessive exposure will often be a non-specific pathological response, such as inflammation, degeneration or cell death; altered sensitivity; disorder of repair; or disorder of growth. It is likely, too, that the end-result of an excessive and severely damaging application of the *appropriate* energy to a specialized cell will also be one or other of the non-specific pathological changes.

PART (D): MODES OF ACTION OF CERTAIN NON-IONIZING RADIATIONS AND CERTAIN OTHER FORMS OF ENERGY

MICROWAVES

Michaelson (1975) summarized knowledge relating to microwaves up to about 1974. Microwave absorption in biological tissues leads to thermal effects which are well documented and well understood. The absorbed energy induces physiological responses resultant from increased body temperature or alterations in thermal gradients locally within the body. Most observations are consistent with physiological adjustment to thermal input whether locally or generally. However, some authorities suggest non-thermal or specific effects of microwave exposure.

Schwan (1975) discussed the dielectric properties of biological materials and showed that as frequency increases from a few hertz to the giga-hertz range the dielectric constant decreases from several million to only a few units: concurrently conductivity increases considerably. He commented that both dielectric data and conductivities at microwave frequencies are essentially those of electrolytes containing macromolecules.

Schwan showed that above a characteristic frequency all biological membranes have a capacitance which is reasonably independent of frequency. He pointed out that this capacitance corresponded to an impedance of about $50 \mu \Omega$ at $3 \mathrm{GHz}$. Hence, a microwave flux of $10 \mathrm{mW/cm^2}$, in tissues of a typical resistivity value of 100 to 200 Ω-cm, would produce a current density of about $3 \mathrm{mA/cm^2}$. Such a current density would induce a membrane potential of about $0.15 \mu \mathrm{V}$ which is about 1000-fold smaller than potentials reportedly necessary to excite excitable biological membranes. Moreover, the potential varies beyond the 'frequency response' associated with biological excitations.

Schwan noted that tissues' electrical properties were independent of field strength. He concluded that energies imparted by thermal molecular collisions are greater than those which can be applied with microwave field strengths of 'practical' magnitudes. Hence, field strengths encountered in practical situations are not likely to cause the denaturing of protein.

Schwan also considered forces which can be generated by the application of alternating fields (such as those associated with radar) and came to the conclusion that such 'electro-mechanical' effects may be responsible for the phenomenon of hearing pulsed radar waves.

The sensation of hearing (associated with microwaves) which, according to Schwan, can occur at energies as low as $1-10 \mathrm{mW/cm^2}$ may also be responsible for some of the so-called behavioural effects reported in connection with type of exposure. Table 17.1 is a summary of data presented by Schwan (1975) and by Gordon and Schwan (1973, cited by Schwan, 1975). In discussing the possible processes whereby microwaves might be 'heard' by the human ear, Michaelson (1975) commented on 'thermo-acoustic' conversion in which mechanical stress generated at a surface by thermal expansion during transient heating results from microwave or laser beams incident on some surface. To be effective in biological systems it

TABLE 17.1

MICROWAVE RADIATION LIMITS RECOMMENDED TO WHO (MODIFIED FROM SCHWAN, 1975)

Power range mW/cm²	Application/Standard	Effect	Recommendation to WHO
1 000	Therapeutic application for medical purposes	Thermal	Above 10 to 100 mW/cm² distinct thermal effects which may be hazardous at high intensity.
100	Threshold of observed hazardous effects		1 to 20 mW/cm² thermal effect, direct field effects such as the hearing of pulsed microwaves and other possible but as yet undefined phenomena.
10	US Standard audible effects / 'Behavioural' responses threshold of thermal sensation	'Subtle effects'	
1	US Department of Health, Safety and Welfare Standard for Microwave Patterns		Below 1 mW/cm² thermal effects improbable.
0.1		Non-thermal effects may be involved	
0.01	Russian Standard TV, radio		

would presuppose an interface where one material could expand differentially in relation to the other in order to produce the surface forces. The phenomenon cannot be ruled out for the inner ear because there is, in fact, a tissue/fluid interface in the cochlea.

Odland (1975) reviewed the problem of cataract induction by microwaves. He pointed out that the most likely mode of action is heat coagulation (denaturing) of the protein within the lens of the eye. As he pointed out, the lens does not have an adequate vascular system for heat exchange so that even slight temperature elevations threaten denaturing of lens protein. Such damage may progress, perhaps from cumulative insult, and ultimately form an opacity which obstructs the passage of light to a sufficient degree to interfere with normal vision. This is that stage at which the opacity becomes, in clinical terms, a cataract. According to Odland, experimental evidence indicated that all 'cataractogenesis' levels for 1-hour exposures are above $100 \, mW/cm^2$ in the frequency range 19 MHz to 24.5 GHz. In experiments on rabbits, single and multiple exposures below $120 \, mW/cm^2$ do not result in cataract production. He quoted the experience of the US armed forces in which the observed number of cataracts does not appear to have changed despite increasing use of microwaves. He concluded that the limit of $10 \, mW/cm^2$ is justified as a standard to safeguard people against the risk of cataract from microwaves.

LASERS

Lasers generate straight beams of intense light which cause heating of surfaces on which they impinge. High energy laser beams cause burns of tissues. Beams of relatively low energy may cause damage to the retina at the point of impingement. For further discussion on ocular hazards from lasers see Clarke (1970) and Hayes and Wolbarsht (1968).

ACOUSTICAL ENERGY

Ultrasound

Reid and Sikov (1972) summarized knowledge relating to ultrasound in 1971 and Michaelson (1975) summarized knowledge up to late 1974 concerning the interactions of ultrasound and biological tissues. Heat is produced when ultrasound is absorbed by tissue. Many, but apparently not all, of the reported effects of ultrasound can be explained on the basis of heat production. Some investigators have reported biological effects from exposure to ultrasound which cannot, apparently, be explained solely on the basis of heating.

Injury to tissue is produced when ultrasound is focused on localized areas of tissues so as to deliver intense exposures of ultrasonic energy. This principle underlies those surgical techniques which utilize ultrasonic energy. Thermal effects are recognized to be the principal agent of injury though whether these account for all the effects is not entirely clear.

Ultrasound is used in physiotherapy but at very much lower intensities than those employed in surgery. At 'physiotherapeutic' levels, temperature rise in tissues results only in mild and intended biological effects such as increased blood flow and, possibly, increased capillary permeability. Ultrasound is used for diagnostic purposes at even lower intensities than those employed in physiotherapy. It is believed that the diagnostic procedures involve energies well below any threshold for any change in tissues. No side-effects have been reported from the diagnostic use of ultrasound.

As Reid and Sikov (1972) pointed out, the question of genetic changes was responsible for most of the concern centred on ultrasound. Some of this arose because effects known to be associated with ionizing radiations were projected without justification onto ultrasound. It is true that certain experiments have resulted in alterations of chromosomes in insects and plant cells, but the conditions under which these experiments were done, according to Reid and Sikov, would have favoured mechanical damage to chromosomes. Moreover, as stated (Chapter 10), alteration in the morphology of chromosomes is not a good guide to the degree of disturbance of cellular genetic processes.

Reid and Sikov pointed out that mechanical rupture of cellular and sub-cellular components would be unlikely in fields of plane waves, at 'diagnostic' intensities in living tissue. Where there are discontinuities or boundaries of cellular dimensions it is possible that forces could be developed which would be sufficient to cause damage. Such boundaries could be provided by small bubbles of gas such as those said to occur in cavitation.

The evidence on cavitation is difficult to interpret. Cavitation effects have been observed in numerous studies. Lele (1975) has, however, questioned these reports on the basis that many of the phenomena allegedly associated with cavitation can, in fact, be explained on the basis of thermal mechanisms. Hill (1975), on the other hand, recognized without question the phenomenon of cavitation. Lele defined two aspects of thermal mechanisms: (a) the threshold temperature hypothesis under which it is held that specific changes are initiated in biological tissues once the local temperature in those tissues exceeds a certain threshold temperature, and (b) the thermal hypothesis under which it is held that a wide class of reversible and irreversible effects caused by the action of ultrasound on tissue can be reproduced by non-acoustic localized heating of the tissue, provided that the temperature history during heating and cooling duplicate the temperature history during irradiation.

The acoustical properties of biological materials were commented upon by Dunn (1975). The pattern of response is very complex, involving frequency as well as intensity and pattern of irradiation. Another variable factor is the nature of the tissue being irradiated. There are difficulties in defining and measuring the response of the tissue or cellular response. Moreover, it is not always possible to specify exactly the intensity of irradiation at localized areas of tissues.

The absorption of acoustical energy in air is proportional to the square of the frequency; hence airborne ultrasonic energy is not transmitted very far. Thus, in practice, fluid coupling is needed to effect energy transfer between the transducer of an ultrasonic generator and the tissues being irradiated. This relation also means that ultrasound is unlikely to be damaging to human tissues where the irradiation is purely airborne.

Sound in the Audible Range

The audible range extends by common convention from about 20 Hz to about 20 000 Hz. Sound energy in this range causes loss of hearing acuity. *Hearing acuity, hearing level* and *threshold of hearing* can be regarded as synonyms. All are measurable by means of an audiometer. This instrument is used to present to a subject's ear a pure tone of specified frequency. By means of ear-phones the subject is required to listen to the tone which the experimenter then adjusts until it is only just audible to the subject. That point is the subject's threshold of hearing which can then be specified in terms of the frequency and level of the *test tone* (see also Appendix 12.1).

Hearing loss may refer to changes (in the direction of diminished acuity) in auditory thresholds at stipulated frequencies which have taken place between succeeding measurements during the course of, say, a few years' exposure to noise. However, decline in hearing acuity occurs not only with noise exposure but also with advancing age.

For some purposes it may be necessary to estimate the degree of loss of hearing acuity attributable to noise separately from that attributable to age. By convention, it is assumed that the effects of noise and age are additive. Loss of hearing acuity attributable to noise is estimated, therefore, by subtraction of a value of 'normal hearing acuity' corresponding to the age of the individual for whom the inference is being drawn. By such means estimates are arrived at of *noise-induced hearing loss.*

Noise is measured on a frequency-weighted decibel scale, denoted dB(A). The frequency weighting mimics, to some extent, the frequency response of the human ear which, in its normal functional state, does not possess equal acuity for all frequencies. Hearing is most acute in the frequency range from about 250 Hz to about 4000 Hz. This range corresponds roughly to the frequency spectrum of the human voice. Hearing acuity tails off progressively above and below the speech range (see Fig. 12.A1).

The empirical basis for the use of dB(A) in connection with noise-induced hearing loss is from the studies reported by Burns and Robinson (1970). These studies demonstrated a mathematical relation between the extent and degree of noise-induced hearing loss and the A-weighted sound energy incident upon the ears of the exposed population. Burns and Robinson's results centred on steady noises. In 1971, Atherley and Martin showed that A-weighted sound energy of drop-forging noise predicted, from the Burns and Robinson equation, the extent and magnitude of hearing loss for drop-forgers. Atherley, in 1973, showed that recurrent impact noise from pneumatic hammers also fitted the Burns and Robinson equation. Other evidence in support of the A-weighted energy concept has been reviewed by Burns (1976). The mathematical aspects of the relation are shown in Appendix 12.1.

Although the A-weighted energy concept is widely accepted in Britain and throughout Europe it is not accepted in the USA. In the USA there is, instead, adherence to the principles accepted under the Walsh–Healey Act of 1967 and incorporated in the Occupational Safety and Health Act of 1970. These principles accept dB(A) but not the A-weighted energy principle. Under the USA principles it is held that the doubling of the exposure period needs to be offset by reduction in the noise level of 5 dB in order to maintain constant dose. The corresponding figure is 3 dB for the A-weighted energy concept. Hence, the two systems are sometimes designated the '5 dB rule' and the '3 dB rule', respectively. Table 17.2 shows a comparison between the two systems.

It is unfortunately true that there exist in parallel two systems based on differing mathematical models. Results of calculations made with the two systems differ substantially especially at high noise levels. Measuring instruments are available for both systems and publications reflect one or other depending on their country of origin.

TABLE 17.2

PERMISSIBLE NOISE EXPOSURE UNDER US OC-
CUPATIONAL SAFETY AND HEALTH ACT 1970
(OSHA) COMPARED TO DATA FROM UK
DEPARTMENT OF EMPLOYMENT CODE OF PRAC-
TICE FOR REDUCING THE EXPOSURES OF
EMPLOYED PERSONS TO NOISE. (1972, LONDON:
HMSO).

Duration per day (hours)	OSHA dB(A) (5 dB rule)	DE Code dB(A) (3 dB rule)
8	90	90
6	92	91.2
4	95	93
3	97	94.3
2	100	96
$1\frac{1}{2}$	102	97.3
1	105	99
$\frac{1}{2}$	110	102
$\frac{1}{4}$ or less	115	105

Because both systems utilize dB(A) it is not always clear which rule is being assumed.

There are advantages for the 3 dB rule in terms of simplification of instrumentation, standard writing, and practical measurements in the field. However, no side-by-side comparisons of the two rules have been located, consequently it is not possible to state whether one is better than the other in correlating noise data and hearing loss.

The mathematical relations between noise and noise-induced hearing loss are empirical, that is to say they correspond to *supporting statistics* as discussed in Chapter 2. Knowledge is incomplete about the modes of action of noise on the organ of hearing (corresponding to *generative mechanisms* as defined in Chapter 2). As yet no complete model of the mode of action can be drawn. Investigations such as those by Spoendlin (1975) and Spoendlin and Brun (1976) lead to the possibility that there is more than one pathological process involved. Their investigations indicate that the mode of action may be determined by the dose rate as well as the total quantity of A-weighted sound energy delivered. The absence of a complete damage model for the ear is not simply an academic point: it undermines the confidence which can be placed in noise standards which imply extrapolations beyond empirical data.

In summary, the position at present in regard to occupational deafness is that two comprehensive systems of measurement co-exist and one of these (the 3 dB rule) is supported by empirical evidence but not by a complete scientific explanation of the mode of action.

Infrasound

Infrasound is defined as sound energy, the frequency of which is less than 16 Hz (Westin, 1975, has reviewed effects of infrasound on man and the outline which follows is based upon his review). Infrasound has man-made and natural sources but because of the frequency response on the human ear much of it is inaudible. The lower frequency limit for audibility is not easily determined because below 20 Hz the threshold of audibility rises steeply with fall in frequency. With frequencies down to about 8 Hz there may be a sensation if the intensity is great enough. However, it is not certain whether the sensation is really audition proper, that is to say sensation arising from stimulation of the specialized sensory cells of the inner ear. Infrasound may instead (or as well) stimulate the sensory nerve endings in the skin lining the external ear canal, giving rise to a sensation which cannot properly be labelled audition.

Natural infrasound is often of frequencies less than 0.1 Hz although, according to Westin, there is little doubt that it also exists at frequencies above 1 Hz. Man-made infrasound, more common than generally believed, occurs mainly in the 1–15 Hz range.

Data on infrasound's effects on humans are sparse in comparison with audible sound, vibration, and ultrasound. The measurement of infrasound requires specialized techniques because even high grade acoustical apparatus is not adequate for measurement of infrasound. Exposures to infrasound seldom occur in the absence of exposure to audible sound and, sometimes, vibration as well. Westin observed that up to 1975 there were no data about the absorption of infrasound by living tissues. At 100 Hz, according to von Gierke (cited by Mohr, 1965) only about 2 per cent of acoustic energy incident on living tissues is absorbed. The coefficient of absorption of sound waves in gases and liquids is proportional to the square of frequency and hence, on theoretical grounds, high absorption would not be expected from infrasound. A heating effect would be unlikely, therefore, even with very high levels of infrasonic exposure. There may be sufficient surface penetration of the skin in order to stimulate the specialized receptors. There are specialized receptors for pain in most regions of the skin. According to Westin, pain from infrasound is experienced only in the region of the ear.

The effects which Westin felt able to attribute reliably to infrasound were (a) pain and discomfort in the ear and (b) disturbances of balance suggesting effects on the vestibular organ at intensities in the range 110 to 130 dB (sound pressure level at single frequencies).

Vibration

Mention has been made elsewhere of hand-arm vibration resulting in vibration-induced white finger. The mode of action of the vibration is not completely understood but James and Galloway (1975), on the basis of investigations of small arteries of hands and fingers of work people using pedestal grinders, came to the conclusion that exposure to the vibration led to damage to arteries of the fingers resulting in diminution of blood flow to the fingers. They quoted a previous study by Ashe and colleague, in 1962, in which it was shown that arteries of men exposed to hand-arm vibration were subject to fibrosis (the meaning of this term is discussed in Chapter 15).

REFERENCES

ARCHER, V. E. and WAGONER, J. K. (1973). Lung cancer among uranium miners in the United States, *Health Physics*, **25**, 351–71.

ATHERLEY, G. R. C. and MARTIN, A. M. (1971). Equivalent-continuous noise level as a measure of injury from impact and impulse noise, *Annals of Occupational Hygiene*, **14**, 1–23.

ATHERLEY, G. R. C. (1973). Noise-induced hearing loss: the energy principle for recurrent impact noise under noise exposure close to the recommended limit, *Annals of Occupational Hygiene*, **16**, 183–92.

BURNS, W. (1976). In *Disorders of Auditory Function*, 2nd Edition, S. D. G. Stephens, Ed. London: Academic Press.

BURNS, W. and ROBINSON, D. W. (1970). *Hearing and Noise in Industry*, London: HMSO.

CASSIDY, P. (1900). Report of a severe X-ray injury, *Medical Record* (New York), **57**, 180–1.

CLARKE, A. M. (1970). Ocular hazards from lasers and other optical sources, *Critical Reviews in Environmental Control*, **1**, 307–39.

DUNN, F. (1975). Acoustic properties of biological materials, in *Fundamental and Applied Aspects of Non-Ionizing Radiation*, S. M. Michaelson, M. W. Miller, R. Magin and E. L. Carstensen, eds, New York: Plenum Press.

FABER, K. (1923). Anemie pernicieuse aplastique mortelle chez un specialiste des rayons Roentgen, *Acta Radiologica*, **2**, 110–5.

FLINN, F. B. (1927). A case of antral sinusitis complicated by radium poisoning, *Laryngoscope*, **37**, 341–9.

HALL-EDWARDS, J. (1904). On chronic X-ray dermatitis, *British Medical Journal*, **2**, 993–6.

HAYES, J. R. and WOLBARSHT, M. L. (1968). Thermal model for retinal damage induced by pulse lasers, *Aerospace Medicine*, **39**, 474–80.

HEMPELMANN, L. H., LANGHAM, W. H., RICHMOND, C. R. and VOELZ, J. L. (1973). Manhattan project plutonium workers: a 27 year follow-up study of selected cases, *Health Physics*, **25**, 461–79.

HESSE, O. (1911). Das Röntgenkarzinom Fortschritte auf dem Gebiete, *Der Röntgenstrahlen und der Nuklermedizin*, **7**, 82–92.

HILL, C. R. (1975). Action of ultrasound on isolated cells and cell cultures, in *Fundamental and Applied Aspects of Non-Ionizing Radiation*, S. M. Michaelson, M. W. Miller, R. Magin and E. L. Carstensen, eds, New York: Plenum Press.

HOBBS, C. H. and McCLELLAN, R. O. (1975). In *Toxicology*, L. J. Casarett and J. Doull, eds, New York: Macmillan.

HUEPER, W. C. (1942). *Occupational Tumors and Allied Diseases*, Springfield: Thomas.

HUEPER, W. C. (1966). *Occupational and Environmental Cancers of the Respiratory System*, New York: Springer-Verlag.

HUNTER, D. (1975). *The Diseases of Occupation*, London: English Universities Press.

INTERNATIONAL COMMISSION ON RADIOLOGICAL PROTECTION (1959). Recommendation of the International Commission on Radiological Protection adopted 9 September 1958, London: Pergamon.

JAMES, P. B. and GALLOWAY, R. W. (1975). In *Vibration White Finger in Industry*, W. Taylor and P. L. Pelmear, eds, London: Academic Press.

LANGHAM, W. H., BROOKS, P. M. and GRAHN, D. (1965). Biological effects of ionizing radiations, *Aerospace Medicine*, **36**, 1–55.

LEDOUX-LEBARD, R. (1922). Le cancer des radiologistes, *Paris Medical*, **12**, 299–303.

LELE, P. P. (1975). Thermal factors in ultrasonic focal destruction in organized tissues, in *Fundamental and Applied Aspects of Non-Ionizing Radiation*, S. M. Michaelson, M. W. Miller, R. Magin and E. L. Carstensen, eds, New York: Plenum Press.

MARTLAND, H. S. (1929). Occupational poisoning in manufacture of luminous watch dials, *Journal of the American Medical Association*, **92**, 466–552.

MARTLAND, H. S. (1931). The occurrence of malignancy in radioactive persons, *The American Journal of Cancer*, **15**, 2435–516.

MARTLAND, H. S. and HUMPHRIES, R. E. (1929). Osteogenic sarcoma in dial painters using luminous paint, *Archives of Pathology*, **7**, 406–17.

McLEAN, A. S. (1973). Early adverse effects of radiation, *British Medical Bulletin*, **29**, 69–73.

MICHAELSON, S. M. (1975). Sensation and perception of microwave energy, in *Fundamental and Applied Aspects of Non-Ionizing Radiation*, S. M. Michaelson, M. W. Miller, R. Magin and E. L. Carstensen, eds, New York: Plenum Press.

ODLAND, L. T. (1975). Military role in safe use of microwaves, in *Fundamental and Applied Aspects of Non-Ionizing Radiation*, S. M. Michaelson, M. W. Miller, R. Magin and E. L. Carstensen, eds, New York: Plenum Press.

PATON, L. (1909). Case of posterior cataract commencing subsequent to prolonged exposure to X-rays, *Transactions of the Ophthalmological Society*, **29**, 37–9.

REID, J. M. and SIKOV, M. R. (1972) *Interaction of Ultrasound and Biological Tissues*, Rockville: US Department of Health, Education and Welfare.

SCHWAN, H. P. (1975). Dielectric properties of biological materials and interaction of microwave fields of the cellular and molecular level, in *Fundamental and Applied Aspects of Non-Ionizing Radiation*, S. M. Michaelson, M. W. Miller, R. Magin and E. L. Carstensen, eds, New York: Plenum Press.

SPOENDLIN, H. (1975). In *The Effects of Noise on Hearing*, D. Henderson, R. P. Hamernik, D. S. Dosanjh and J. H. Mills, eds, New York: Raven Press.

SPOENDLIN, H. and BRUN, J. P. (1976). In *Disorders of Auditory Function*, 2nd Edition, S. D. G. Stephens, ed., London: Academic Press.

VAN LANCKER, J. L. (1976). *Molecular and Cellular Mechanisms in Disease*, New York: Springer-Verlag.

VODOPICK, H. and ANDREWS, G. A. (1974). Accidental radiation exposure, *Archives of Environmental Health*, **28**, 53–6.

WAGONER, J. K., ARCHER, V. E., CARROLL, B. E., HOLADAY, D. A. and LAWRENCE, P. A. (1964). Cancer mortality patterns among US uranium miners and millers 1950–1962, *Journal of the National Cancer Institute*, **32**, 787–801.

WAGONER, J. K., ARCHER, V. E., LUNDIN, F. E., HOLADAY, D. A. and LLOYD, J. W. (1965). Radiation as the cause of lung cancer among uranium miners, *New England Journal of Medicine*, **273**, 181–8.

WALSH, D. (1897). Deep tissue traumatism from Roentgen ray exposure, *British Medical Journal*, **2**, 272–3.

WESTIN, J. B. (1975). Infrasound: a short review of effects on man, *Aviation, Space, and Environmental Medicine*, **46**, 1135–43.

Chapter 18

Aromatic Amines and Occupational Cancer of the Renal Tract

OBJECTIVES OF THE CHAPTER

This chapter provides:
  (1) an outline of the industrial and chemical background of certain aromatic amines,
  (2) a summary of the early and recent history of knowledge about occupational cancer of the renal tract caused by aromatic amines,
  (3) an explanation of why the site of the tumours is predominantly the bladder,
  (4) an explanation of the cyanosis associated with aromatic amines,
  (5) an outline of the history of action against occupational cancer of the renal tract from aromatic amines,
  (6) a commentary on the influence of compensation on the action taken in connection with occupational cancer of the renal tract,
  (7) an outline of the provisions for statutory control as they stand at present, and
  (8) certain epidemiological data illustrating aspects of the problem of occupational cancer of the renal tract.

INDUSTRIAL AND CHEMICAL BACKGROUND

Benzene and compounds which resemble benzene in their chemical behaviour are known as aromatic compounds, while an amine can be defined as having the general formula RNH_2, R_2NH or R_3N, where R is any alkyl or aryl group. Aromatic amines, as opposed to aliphatic amines, are those in which the nitrogen is attached directly to an aromatic ring, being derivatives of the simplest aromatic amine, aniline. In general, aromatic amines are soluble in water, although their solubility is relatively low in non-polar solvents such as ether and ethyl alcohol. However, many aromatic amines are readily absorbed through the skin, aniline being one of the best-known examples. The non-lipid solubility of aromatic amines requires us to postulate active transport systems for aromatic amines into cells. A further ground for this claim is the highly biologically active nature of the amino groupings.

EARLY HISTORY OF THE OCCUPATIONAL CANCER
PROBLEM*

Aromatic amines are of wide industrial importance, their historical prominence arising from their being the basis of the first synthetic alternatives to vegetable dyes. In the late 1850s, in attempting to synthesize quinine, the young British chemist Perkin instead discovered a mauve dye, the commercial importance of which was soon recognized, especially in Europe. Within the next half-century an important chemical industry for the manufacture of synthetic dyes was to develop in Germany, where coal was used as the raw material to feed the aniline industry. By the turn of the century the six largest German chemical works employed 650 chemists and engineers, while the entire British coal industry commanded no more than 30–40. In Germany were thus created the circumstances for the development of accidental chemistry, which accompanies the progress of industrial chemistry, circumstances which have been set out by Scott (1962) upon whose account the following summary is based.

In 1895 the German Surgical Society held a congress in Frankfurt, an important centre of the chemical industry. To the congress, Rehn, a surgeon, reported on his patients. He had observed the frequent occurrence of a blue tinge to the lips, ears and finger nails of chemical workers whom he had treated, and noted the more serious fact that some were passing blood in their urine. Among these men three were found to have tumours of the bladder, while a tumour was suspected in a fourth. All four men had worked in the same dye-manufacturing factory, the total workforce of which numbered only 45. In ordinary surgical experience, tumours of the bladder were uncommon, and four cases out of a total of 45 men was a matter for remark.

Rehn concluded that: (1) the fumes developed during the manufacture of the dye led to disturbances in the urinary system, that (2) after years of work in the industry, tumours occurred as a result of continuous irritation (three of his cases having been employed in the factory for 15, 29 and 20 years, respectively, and having passed blood in the urine for 4 years, $1\frac{1}{2}$ years and 6 months, respectively), and that (3) the injurious effect was due to the inhalation of aniline vapour.

The disease identified by Rehn became known as aniline cancer, but his observations were questioned by Grandhomme in 1896. Grandhomme, an official of a firm which had been manufacturing dyes on a large scale for the previous 30 years, expressed strong doubts about the connection between the occupation of the men and their disease. He acknowledged that the substances involved in dye

*With acknowledgements to C. Clutterbuck.

manufacture might irritate the bladder, producing pain and blood, but denied the link with the tumours. The argument continued, until in 1898 Leichtenstern attributed a bladder tumour to 'naphthylamine', this, as Scott pointed out, being the first mention of a naphthylamine as a cause of tumours. Significantly, the case was reported 18 years after the start of manufacture of 2-naphthylamine in Germany.

In 1904, Rehn reported a further 20 cases to the Surgical Congress and was asked to carry out a survey of aniline factories. In 18 factories he found a total of 38 cases, the chemicals identified as significant contacts being aniline, fuchsin, benzidine and naphthylamine. Those, like Grandhomme, who doubted the occupational link preferred instead a geographical hypothesis but in 1912, when Lueunberger reported 18 cases of bladder tumours among aniline dye workers in Basel, this explanation was effectively destroyed. Lueunberger used epidemiological methods to show that tumours of the bladder were 33 times more common among aniline dye workers than in the rest of the male population of Basel. By the outbreak of World War I the link between the occupation of aniline dye workers and tumours of the bladder was therefore firmly established and accepted on the Continent of Europe. At that time, however, Britain's chemical industry was tiny and no cases had been reported, although the annual report of the Senior Medical Inspector of Factories for the years 1912 and 1913 had drawn attention to the danger.

In 1921 the International Labour Office (the one still-enduring remnant of the ill-fated League of Nations) produced a report drawing attention to the bladder tumour problem in the German and Swiss chemical industries. It recommended control procedures and emphasized the importance of keeping detailed records, while also analyzing the many compounds inculpated in the various researches. The review stated that although aniline was frequently mentioned as a cause of cancer the amino compounds generally were suspect, in particular, benzidine and 2-naphthylamine.

World War I had deprived many countries of German dyestuffs, and chemical industries had rapidly developed in countries previously supplied by Germany. By the end of the war the German industry was thus crippled and substantial industries had been established elsewhere, principally in the countries of her enemies.

As Scott said: 'the problem of bladder cancer in these countries outside Germany and Switzerland was forgotten, if it had ever been seriously believed or heard of, and a complete repetition of the German and Swiss experience was set in motion in Great Britain, Italy, France and the United States of America. The aftermath endures to this day.'

In 1926, HM Senior Medical Inspector of Factories reported 12 deaths from bladder tumours in chemical workers during the preceding 4 years. In 1929, Wignall reported 14 cases of bladder tumours in chemical workers handling aromatic amines, including 1-naphthylamine, benzidine and colour manufacture intermediates. In the USA the first case was reported in 1931, although in 1934 a further 27 cases were reported. By 1937, when 83 cases were confirmed, the disease was firmly endemic. One feature common to all the reports dating as far back as Rehn's original observations was the long interval between the start of exposure and the onset of disease—15–20 years.

American studies, conducted by Gehrmann *et al.* (1949) finally demonstrated that aniline was not a tumour-producing agent. Rehn's one false assumption had finally been overturned, but he had given a title to a disease which endured long after aniline's reputation had been cleared—so far as tumours were concerned.

In 1938 came laboratory confirmation of the tumour potential of 2-naphthylamine, Hueper and his colleagues having induced bladder tumours in dogs by oral administration of 2-naphthylamine. Two questions remain for us to answer: what was the origin of the cynosis which Rehn observed, and why should the bladder be the site of the tumours? The bladder is scarcely accessible to substances in the external milieu, and a direct mode of input to the bladder is hardly feasible. Since the tumour-producing aromatic amines would have entered the men's bodies by skin, gut or lung, why were tumours not observed in these tissues? (In fact, tumours do occur in lung, as discussed later.)

THE SITE OF THE TUMOURS

According to Scott no case of a person suffering from chronic liver damage had been reported in connection with occupational exposure to aromatic amines, even though aromatic nitro-compounds such as trinitrotoluene and dinitrophenyl definitely affect the liver. By contrast, liver damage as well as tumours in the liver can be induced in animals by many aromatic amines. The tumours may be benign or malignant, while in some cases they are preceded by cirrhosis of the liver. In other cases, for example when 4-aminodiphenyl is used, no pre-tumour damage can be found in the liver. Long-term experiments on rabbits and dogs with benzidine have shown a disturbance of the liver which led to cirrhosis. It has been found that increasing liver damage is accompanied by a rise in the amount of free benzidine excreted in the urine, indicating that the liver metabolizes benzidine.

The liver is the principal site for metabolic transformation and conjugation of non-nutrients, as discussed in Chapter 8. The detoxification pathways taken by a number of aromatic amines have been traced in some detail, and it appears, at least in the case of 2-naphthylamine, that a conjugate is produced and extracted from the circulating blood by the kidneys. In fact, as regards 2-naphthylamine several possible conjugates and more than one form of transformation are involved (Parke, 1968), while there are several pathways, one of which leads to the formation of 2-amino-1-naphthol, a substance (Bonser, 1947; Clayson, 1950; Bonser *et al.*, 1951) which will produce bladder tumours in mice after exposure for 30 weeks or so.

That the bladder is the site of aromatic amine-induced tumours is accounted for by the fact that the amines are subject to metabolic transformation and conjugation in the liver. The transformates and metabolites subsequently extracted from the blood by the kidneys are then acted upon by enzymes capable of inducing hydroxylation, with the result that 2-amino-1-naphthol, as well as other substances, is created. This is a clear example of intoxication as described in Chapter 8.

RECENT HISTORY OF THE OCCUPATIONAL CANCER PROBLEM

The aromatic amine problem has by no means been solved, but certain statements can be made with assurance. Veys (1972) has identified 8 main occupational groups in which an increased incidence of bladder tumours can be linked with aromatic amines, namely workers employed in the chemical and dyestuffs industries, in pigment manufacture, in rubber and cable making, in textile dyeing and printing, gas workers, laboratory workers, rodent control operatives (rat catchers), and patent fuel, tar and pitch workers.

There is irrefutable evidence that certain aromatic amines are carcinogenic, i.e. capable of producing malignant tumours. Those which are beyond doubt carcinogenic include 2-naphthylamine, benzidine, 4-aminodiphenyl, 4-nitrodiphenyl, *o*-dianisidine, *o*-tolidine and dichlorbenzidine, while there is uncertainty about the carcinogenic potential of 1-naphthylamine, auramine and magenta.

In 1974, Fox *et al.* published a survey of occupational cancer in the rubber and cable-making industries in which they analysed data for the 5-year period 1967–71. In 1976, Fox and Collier published a further study of occupational cancer in the rubber and cable-making industries in which they analysed deaths occurring in the period 1972–74. Records of nearly 41 000 men employed for at least 1 year in the rubber and cable-making industries had been under observation for 8 years. The mortality pattern for 1972–74 was compared with that for the earlier period 1967–71. Analysis indicated a significant excess of deaths from cancer of the bladder throughout the industry. Men who had not been exposed to the acknowledged bladder carcinogens were found among those with cancer of the bladder. The excess was in deaths occurring in 1973 and 1974 in the 45–64 and 65 + age groups. Excess to a statistically significant degree was observed in the foot-wear, footwear supplies except adhesives and tyre sectors of the industry.

Fox and Collier's study confirmed an earlier observation of an excess of lung cancers for certain groups of men particularly in the 55–64 year age group and those who entered the industry between 1950 and 1960. They drew the conclusion that the tyre sector of the industry has a relatively high rate of lung cancer and that further investigations are required into the continuation of bladder cancer. They drew attention to possible weaknesses in their data (they compared standardized mortality ratios for different periods but recognized that such comparisons have to be made with care). They also drew the guarded conclusion that their data related not to the magnitude of the hazard but to the existence of the hazard.

CYANOSIS

The blueness of the skin and mucous membranes, first observed among chemical workers by Rehn, is caused by a change in haemo-globin which can be induced by the aromatic amines and other chemical substances. The haem iron (spelt 'heme' in American sources) can be oxidized from the ferrous to the ferric state, the result being a disordered haemoglobin, greenish-brown or chocolate-coloured, called methaemoglobin, which cannot combine with oxygen. Thus the formation methaemoglobin represents an inter-ference with the blood–oxygen transport system, the symptoms of which, generally speaking, are related to the anoxia resulting from the impairment of oxygen transport. The victim of such an interference looks blue when 20% circulating haemoglobin has been changed to methaemoglobin, although if contact from the causal substance is removed recovery occurs rapidly. In cases where more than 20% has been changed the cyanotic colour becomes progressively deeper and the symptoms of anoxia, weakness and headache more apparent. If absorption continues, cerebral anoxia and heart failure may result. Provided that all further exposure ceases and that the methaemoglo-bin concentration has not risen to the point where the anoxia causes permanent tissue damage, recovery will be complete.

The cyanosis described by Rehn was therefore an indication of methaemoglobin caused by exposure to and absorption of quantities of aromatic amines.

In the red blood cells of people exposed to aromatic amines, small granules have been described. These are called Heinz bodies and they are believed to consist of denatured haemoglobin, possibly sulph-haemoglobin. Many substances appear to be capable of causing Heinz bodies though, unlike methaemoglobinaemia, the Heinz bodies appear not to be associated with such profound effects on the oxygen transport system. However, they are plainly an indication that substances are affecting haemoglobin. Predictably, methaemoglobin and Heinz bodies have both been explored as possible bases for the medical screening of people exposed to chemical substances.

HISTORY OF ACTION

The interplay between research, law and industrial relations has been partly documented as far as the aromatic amines are concerned, although much of the detail remains to be researched. Case's account showed that in the 1930s the recognition of bladder cancer in chemical workers in Britain led to pressure of the trade unions for the disease to be scheduled under the Workmen's Compensation Act. The chemical industries thought that this would be a premature step and that confirmation in the form of an epidemiological survey was required. Employers offered a 'gentlemen's agreement' that men with the disease would be 'not worse off' than if the disease had been scheduled. This plan was accepted by the trade unions and an epidemiological survey was proposed, but World War II led to its postponement until 1948.

In 1953, as a result of reports arising from the survey, papilloma of the bladder in chemical workers was included in the schedule of diseases under the National Insurance (Industrial Injuries) Act. The Workmen's Compensation Act had by this time been replaced, and the schedule of diseases transferred intact from the old to the new legislation and subsequently added to from time to time.

The survey (reported by Case et al., 1954) examined aniline, 1-naphthylamine, 2-naphthylamine and benzidine. The American findings—that aniline is not a carcinogen—were confirmed.

The role of benzidine and 2-naphthylamine as tumour-producing agents was clearly established, although the role of 1-naphthylamine remained, and still remains, a matter for discussion, since exposure to 1-naphthylamine appeared always to involve some exposure to 2-naphthylamine as well. The survey revealed 127 cases of death from tumours of the bladder in a male population where 4 deaths would be expected on epidemiological grounds. In addition a further 135 cases of bladder tumours were reported.

The risk gradient (see Appendix 18.1) was shown to extend from 10% to 100%, the dramatic 100% incidence occurring among workers who had been involved in 2-naphthylamine distillation. Even more remarkable was the discovery that victims were drawn from a variety of positions, and even included managing directors. Previous observations about the tumours' latent period were confirmed, examples cited in the survey ranging from 5 to 50 years, with a modal value of 18 years (see Appendix 18.1).

Following the survey, action included, as already mentioned, the scheduling of tumours of the bladder under the National Insurance legislation. As early as 1949, ICI had ceased to manufacture 2-naphthylamine, because the company recognized the serious risk to employees, and in 1952 its manufacture in Britain was totally discontinued. Five years later a voluntary Code of Practice recommending control and medical procedures was introduced by the industry (Scott, 1962). In the mid-1960s the manufacture of 1-naphthylamine and benzidine was discontinued, although both substances continued to be imported for some time afterwards.

Another result of the survey was of a quite unexpected kind. Case required a control group for his study and, because of its well-run cancer registry, he chose Birmingham, identifying rubber workers as a suitable control group. To his surprise, Case discovered an undue incidence of bladder tumours among the rubber workers, and further inquiry soon incriminated Nonox S, a substance which contained 1-naphthylamine and 2-naphthylamine mixed together. Nonox S was used in the rubber industry as an antioxidant, its function being to prevent the deterioration of rubber in tyres and other products. The principal rubber manufacturer in Birmingham at this time was Dunlop, and their supplier of Nonox S was ICI.

According to Case (1966) the Rubber Manufacturing Employers' Association decided against a survey along the lines of that done in the chemical industry, apparently for fear of unduly alarming the employees (the evidence for this comes from a statement in Parliament by R. J. Gunter on 15 February, 1965—cited by Case, 1966). Growing concern about the problem led the Ministry of Labour, in 1964, to publish draft regulations aimed at prohibiting certain substances and controlling others; regulations finally came into force in 1967.

THE INFLUENCE OF COMPENSATION

At this point it is necessary to draw attention to the 14-year interval between the scheduling of the disease, for compensation purposes, under the National Insurance (Industrial Injuries) Scheme and the introduction of regulations aimed at control of the disease. Compensation plainly took preference over prevention, at least as far as legal requirements were concerned.

The Robens Committee (1972) commented unfavourably on this general tendency (paragraph 424): 'Prevention and compensation are two sides of one problem. Most people would say that prevention should be the first consideration. In practice the situation in this country is such that whenever there is a degree of conflict between the development of better accident prevention provisions on the one hand, and compensation principles and procedures on the other, it is the latter which appears to dictate what can be done.'

In the case of bladder tumours caused by aromatic amines, compensation certainly seems to have been the dominating issue.

In 1965 a coroner's inquest showed bladder tumour disease in a cable worker. It then became apparent that the cable industry—which for years had used rubber to sheathe cables—constituted a further source of dangerous exposure to 2-naphthylamine and to other aromatic amines used in connection with rubber sheathing. In 1965 the mounting concern led the Department of Employment and Productivity to launch a campaign intended to trace and screen men thought to have been at risk in the rubber and cable industries. Case regarded this step as no more than a conciliatory gesture. The full account of what this campaign achieved has yet to be reported.

In 1971 two employees of Dunlop Ltd, Thomas Cassidy and Christopher Wright, sued ICI and Dunlop in consolidated actions (1972 13 KIR 255), which succeeded mostly against ICI. The Court of Appeal in 1972 upheld the original decision, in which O'Connor J. held that:

(1) Alpha-naphthylamine* and beta-naphthylamine were known to be carcinogens by 1935.

(2) By 1942 ICI knew that the free naphthylamine content of the dried Nonox S was 1.8% and that it presented a hazard to ICI workers.

(3) ICI knew that beta-naphthylamine was part of the impurity.

(4) ICI knew how Nonox S was used in the rubber industry.

(5) ICI knew that there was no safe level of exposure to alpha-naphthylamine and beta-naphthylamine.

---

*O'Connor J. used the older terminology for 1- and 2-napthylamine.

(6) Dunlop had enhanced the risk to the plaintiffs by failing to institute screening of their urine.

(7) There was a duty for the manufacturer, the discharge of which '. . . requires that he should inform himself of the circumstances in which the product will be used, the quantities in which it will be used, and the possibilities of exposure to which men may be subjected. If in the state of knowledge at the time he cannot say positively that the proportion of carcinogen present in the product will be harmless, then I hold that he cannot market it without giving adequate warning of the presence of the carcinogenic materials.'

In his judgment O'Connor J. noted that, by 1970, 450 cases of bladder cancer had been revealed in the rubber industry.

STATUTORY CONTROL

At the time of writing two statutory instruments are current, the Carcinogenic Substances Regulations, 1967 (SI 1967, no. 879) and the Carcinogenic Substances (Prohibition of Importation) Order 1967 (SI 1967, no. 1675). However, as these regulations apply only to premises subject to the Factories Act 1961, it is clear that they will have to be revised in order to match the scope of the Health and Safety at Work Act 1974. This is important, because some of the occupations mentioned by Veys are outside the jurisdiction of the 1961 Act but, of course, within the 1974 Act's scope.

The Carcinogenic Substances Regulations prohibit (subject to a power to grant exemptions) 2-naphthylamine, benzidine, 4-aminodiphenyl, 4-nitrodiphenyl and their salts. Substances which contain any of these compounds, unless the impurity is less than 1%, are also prohibited.

In addition to *prohibited substances*, the regulations also recognize *controlled substances*, namely 1-naphthylamine, *o*-tolidine, dianisidine, dichlorbenzidine and their salts, auramine and magenta. The precautions to be taken with the controlled substances are specified in the regulations.

The Carcinogenic Substances (Prohibition of Importation) Order 1967, prohibits importation into the UK of 2-naphthylamine, benzidine, 4-aminodiphenyl, 4-nitrodiphenyl and their salts, and any substance or article containing these compounds (except below the specified concentration). Exemptions can be granted for research and the material can be brought into a dock for the sole purposes of re-export.

Medical screening, including exfoliative cytology (see Chapter 16) is a legal requirement in connection with the regulations.

The problem of occupational cancer of the renal tract has yet to be eliminated entirely from the rubber industry. Several important lessons are to be learned from the history of this disease. In due course, no doubt, the disease will be eliminated but the question arises whether the problem could be repeated with some other compound or class of compounds, and whether carcinogenicity can be predicted, without waiting for work people to develop disease.

APPENDIX 18.1

Epidemiological data relating to occupational cancer of the renal tract presented in graphical form. The data illustrate the following points:
  (1) latent period (Fig. 18.A1),
  (2) risk gradient of certain occupations (Fig. 18.A2), and
  (3) age and sex distribution of cancer of the bladder (Figs. 18.A3–18.A5).

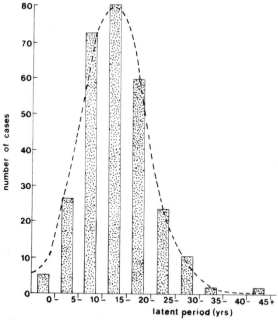

FIG. 18.A1. The latent period of tumours of the bladder in workers employed in the dyestuffs industry. The data relate to 281 cases and show the distribution of latent periods. The dotted curve is a normal distribution for a mean of 17.8 with a standard deviation of 7.18 years. Redrawn, with acknowledgements, from Case *et al.* (1954).

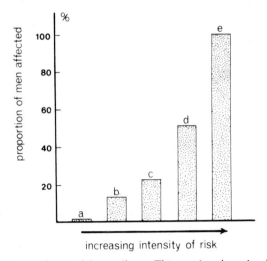

FIG. 18.A2. Occupational risk gradient. Figure showing the incidence of bladder tumours in men engaged in certain occupations, compared with the general male population. (a) general male population; (b) benzidine manufacture; (c) mixed exposure to aromatic amines; (d) exposure to 2-naphthylamine, and (e) distillers of 2-naphthylamine. Redrawn, with acknowledgements, from Case (1966).

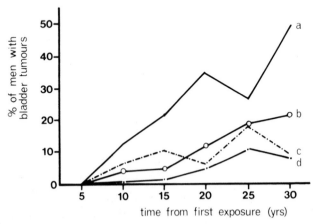

FIG. 18.A3. Incidence of disease rising with time, in men employed on different processes. (a) men exposed to 2-naphthylamine; (b) mixed exposure; (c) benzidine; and (d) 1-naphthylamine. Redrawn, with acknowledgements, from Case (1966).

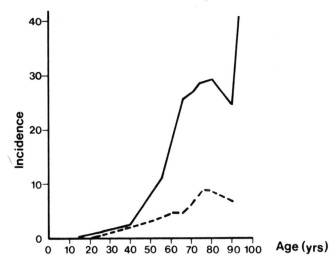

FIG. 18.A4. Annual incidence rate per hundred thousand population from papilloma of the bladder plotted by age. Male data are indicated by the solid curve, female data by the dashed curve. The data relate to the Birmingham area in about 1961. The very large sex difference is likely to be a reflection of the occupational link. Redrawn, with acknowledgements, from Waterhouse, J. A. H. (1974), *Cancer Handbook of Epidemiology and Prognosis*, Edinburgh: Churchill Livingstone.

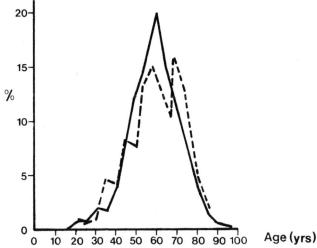

FIG. 18.A5. Distribution of bladder papilloma by sex and age. Male data shown by the solid curve, female data by the dashed curve. Data have the same origin as those shown in Fig. 18.A4. The difference in age distribution is not marked but the generally lower age distribution for the men is, presumably, a reflection of the occupational link. Redrawn, with acknowledgements, from Waterhouse (1974).

REFERENCES

BONSER, G. M. (1947). Experimental cancer of the bladder, *British Medical Bulletin*, **4**, 379–381.

BONSER, G. M., CLAYTON, D. B. and JULL, J. W. (1951). Experimental enquiry into cause of industrial bladder cancer, *Lancet*, **2**, 286. (Cited by Scott, 1962.)

BRIDGE, J. C. (1927). Report of the Senior Medical Inspector for the year 1926, in *Annual Report of the Chief Inspector of Factories and Workshops for the year 1926*, London: HMSO.

CARCINOGENIC SUBSTANCES REGULATIONS, 1967. S.I. 1967: 879 amended by S.I. 1973: 36.

CASE, R. A. M., HOSKER, M. E., MCDONALD, D. B. and PEARSON, J. T. (1954). Tumours of the urinary bladder in workmen engaged in the manufacture and use of certain dyestuff intermediates in the chemical industry, *British Journal of Industrial Medicine*, **11**, 75–104.

CASE, R. A. M. (1966). Hunterian Lecture. Tumours of the urinary tract as an occupational disease in several industries, *Annals of the Royal College of Surgeons*, **39**, 213–35.

Cassidy v. *Dunlop Rubber Co. Ltd, Wright* v. *Dunlop Rubber Co. Ltd, Cassidy* v. *Imperial Chemical Industries Ltd, Wright* v. *Imperial Chemical Industries Ltd.* II KIR 311 and XIII KIR 215.

CLAYSON, D. B. (1950). Some differences between the metabolism of β,2-naphthylamine and 2-acetamidonaphthylamine in the dog and cat, *Biochemical Journal*, **47**, 46. (Cited by Scott, 1962.)

FOX, A. J. and COLLIER, P. F. (1976). A survey of occupational cancer in the rubber and cable-making industries: Analysis of deaths occurring in 1972–74, *British Journal of Industrial Medicine*, **33**, 249–64.

FOX, A. J., LINDARS, D. C. and OWEN, R. (1974). A survey of occupational cancer in the rubber and cable-making industries: Result of five-year analysis, 1967–71, *British Journal of Industrial Medicine*, **31**, 40–51.

GEHRMANN, G. H., FOULGER, J. H. and FLEMING, A. J. (1949). Occupational carcinoma of the bladder, *Proceedings of the IXth International Congress of Industrial Medicine*, London and Bristol: Wright, p. 472. (Cited by Scott, 1962.)

GRANDHOMME (1896). *Die Fabriken des AG Farbwerke Vormals Meister Lucius and Brunig in Sanit. u Soziale Beziehung.* (Quoted in International Labour Office Report, 1921). (Cited by Scott, 1962.)

HUEPER, W. C., WILEY, F. H. and WOLFE, H. D. (1938). Experimental production of bladder tumours in dogs by administration of β,2-naphthylamine, *Journal of Industrial Hygiene*, **20**, 46. (Cited by Scott, 1962.)

INTERNATIONAL LABOUR OFFICE (1921). Cancer of the bladder among workers in aniline factories, Geneva. (Cited by Scott, 1962.)

LEGGE, T. (1913). Report of the Senior Medical Inspector for the year 1912, in *Annual Report of the Chief Inspector of Factories and Workshops for the Year 1912*, London: HMSO.

LEGGE, T. (1914). Report of the Senior Medical Inspector for the year 1913, in *Annual Report of the Chief Inspector of Factories and Workshops for the Year 1913*, London: HMSO.

LEICHTENSTERN, O. (1898). Harnblasentzündungen und Geschwulste bei Arbeitern in Farbfabriken, *Dtsch. Med. Wschr.* **24**, 709. (Cited by Scott, 1962.)

LUEUNBERGER, S. C. (1912). Die Unter dem Einfluss der Synthetischen Farbenindustrie beobachtete Geschwülstentwicklung, *Beitr. Klin. Chir.* **89**, 208. (Cited by Scott, 1962.)

MINISTRY OF LABOUR (1964). Draft Statutory Instrument, The Carcinogenic Substances (Prohibition of Importation) Order, and S.I. 1967: 1675.

NATIONAL INSURANCE (Industrial Injuries) (Prescribed Diseases) REGULATIONS, 1953, Amendment No. 2, S.I.

PARKE, D. V. (1968). *The Biochemistry of Foreign Compounds*, Vol. 5, Oxford: Pergamon.

PERKIN, W. H. (1856). British Patent No. 1984. (Cited by Scott, 1972.)

REHN, L. (1895). Blasengeschwulste bei Fuchsinarbeitern, *Arch. Klin. Chir.*, **50**, 588. (Cited by Scott, 1962.)

REHN, L. (1904). Weitere Erfahrungen über Blasengeschwulste bei Farbarbeitern, *Verh. Dtsch. Ges. Chir.*, **33**, 231. (Cited by Scott, 1962.)

REPORT BY THE COMMITTEE ON SAFETY AND HEALTH AT WORK (1972). (Chairman Lord Robens) London: HMSO.

SCOTT, T. S. (1962). *Carcinogenic and Chronic Toxic Hazards of Aromatic Amines*, Amsterdam: Elsevier.

VEYS, C. A. (1972). Aromatic amines: the present status of the problem, *Annals of Occupational Hygiene*, **15**, 11–15.

WIGNALL, T. H. (1929). Incidence of disease of the bladder in workers in certain chemicals, *British Medical Journal*, **ii**, 258–259.

Chapter 19

Metals in the Disservice of Man

OBJECTIVES OF THE CHAPTER

This chapter provides:
  (1) a framework for knowledge about the harmful effects and certain associated modes of action of metals,
  (2) an exemplification of the point that specific substances are pathogenic at relatively high levels of intake but essential micronutrients at 'dietary' levels,
  (3) emphasis of the importance of inhalation as a mode of input for the metals,
  (4) emphasis of the point that identification of the target organ of a particular substance can be a useful first step in identifying its modes of action and effects, and
  (5) brief summaries of the principal effects associated with 27 metals and substances resembling metals.

INTRODUCTION

This chapter examines certain metals and other substances which, though not metallic elements, possess certain properties which resemble those of metals. Metals such as sodium and potassium are not included because they are so closely connected with metabolism that their inclusion in a chapter concerned with the pathogenicity of substances would be misleading. The substances mentioned here are: aluminium, antimony, arsenic, barium, beryllium, cadmium, chromium, cobalt, copper, gold, iron, lead, lithium, magnesium, manganese, mercury, molybdenum, nickel, phosphorus, platinum, selenium, silver, thallium, tin, tungsten, vanadium, and zinc. Compounds of some of these substances are included in the discussion.

Luckey and Venugopal (1977) pointed out that the study of metals harmful to humans has in the past been compartmentalized into: (1) metals essential for mammals, (2) toxic metals, (3) radioactive metals, and (4) stimulatory compounds and non-reactive or non-toxic compounds. But now they see these as merging into a single topic for study—the interplay between mammals and metals. Ideas are changing rapidly; metals hitherto regarded as toxic are now being reconsidered as possible essential metals—for example selenium, nickel, vanadium, tin, and even arsenic. Evidence of pathogenicity is building up against certain essential nutrients—such as cobalt, chromium, iron, and zinc.

'Toxic metals' is less and less an appropriate concept. All metals (and, indeed, all chemical substances) are pathogenic in excess, whilst some are undoubtedly essential for life and their absence or deficiency is a cause of disease. Hence it is possible for a metal to render both biological service and disservice, according to the quantity present in the target organ(s).

As an aside, it is interesting to note that Luckey and Venugopal observed from their review of metal pathogenicity, generally, that many non-essential metals were capable, in small quantities, of causing stimulation of biological processes, an effect they referred to as 'hormology'. Such effects represent yet another complication in the setting up of dose–effect and dose–response relations. (It seems possible that many pathogenic substances possess hormological capabilities.)

NUTRIENTS AND NON-NUTRIENTS

Davies (1972) has reviewed the clinical significance of the 'essential' biological elements. He has adopted strict criteria for 'essential', and considered that six elements were essential in trace amounts: chromium, cobalt, copper, manganese, molybdenum, and zinc. The six can be referred to as micronutrients. Of the six, chromium, cobalt, copper, manganese and zinc are definitely the cause of harmful effects from industrial exposure (see later). Molybdenum is not associated with industrial disease although excess of this element causes disease in animals.

Biologically active groups including, *inter alia*, sulphydryl (—SH) are recognized as a basis for enzyme action. For example, acetyl coenzyme A is involved in acetylation, a reaction which is important in detoxication as well as energy cycles (Yudkin and Offord, 1973). Sulphydryl groups readily bind a number of metals. Thus detoxication or energy metabolism (or both) could be interfered with at the cellular level by the 'poisoning' of the sulphydryl group by an alien metal. In this way the detoxication of substances quite unrelated to the alien metal can be disrupted.

Other metals participate in metabolism in various important ways, although they do not satisfy Davies's (1972) strict definition of essential micronutrients. These other metals can be labelled nutrients and they may be pathogenic by deficiency or by excess. Examples are iron and selenium. A further group of metals appears on present knowledge, to have no nutrient role and hence they can be labelled non-nutrients, although, as previously stated, this knowledge is under review. This group comprises arsenic, cadmium, lead, mercury, nickel and platinum.

Davies underlined the point that most of the stable elements are found in minute quantities in the human body. The quantity actually found depends on numerous factors including the concentrations in food and the external milieu generally. For many of the stable biologically inactive elements there are no homeostatic processes and consequently accumulation occurs unless bio-dumping is adequate. To an extent, therefore, people's elemental make-up reflects the external milieu they inhabit.

The six micronutrients cannot be labelled 'toxic metals' because deficiency of them causes disease. The point was made in Chapter 8 that the concepts of 'toxicity' and 'pathogenicity' need careful separation. The pathogenicity of the essential micronutrients is clearly established but they cannot be labelled toxic substances.

Metallic micronutrients are necessary for the functioning of numerous enzymes (such enzymes are termed metallo-enzymes but not all enzymes belong to this family). In some instances the metal is closely incorporated in the enzyme protein molecule. In other instances the binding of the metal to the protein is not firm. 'Coenzyme' is used to describe *any* small molecule loosely associated with enzymes but necessary for their normal functioning. Metallic micronutrients form an important category of coenzymes.

The knowledge that micronutrients are closely associated with enzyme function provides an important set of clues to the pathogenicity of metals and similar substances. Enzyme function can be disrupted if the micronutrient is replaced by another metal; cadmium probably competes with zinc in this way. Enzyme production is subject to regulatory processes at the ribosomes and metals may interfere with the regulatory processes. An example is the inhibition by inorganic lead of a number of important enzymes associated with energy production in mitochondria.

A number of the non-nutrients occur widely, lead, for example, being ubiquitous and its presence in the human body inevitable. Indeed, there appears to be a 'normal' range for concentrations of lead in the blood, even though lead appears to have no role in life processes. The average dietary intake of lead by persons in the UK is about 200–250 μg per day, an amount which does not appear to have increased over the past 25 years (Department of the Environment, 1974).

The bodily intake of lead has in recent years excited much controversy, and has led to reports such as that just mentioned. There can be little doubt that the case against lead has been overstated by some commentators. Others, however, point out that the ubiquity of lead is not a justification for contaminating man's external milieu, at work or otherwise. Its very ubiquity requires that man's normal lead burden is not increased. Much concern centres on the apparently narrow margin between the 'normal' concentration of lead in the blood (10–30 μg/100 ml) and the level above which further intake is regarded as excessive (80 μg/100 ml). For a detailed discussion on this issue see the monograph by Waldron and Stöfen (1974).

Concentration in the blood is not necessarily a satisfactory indicator for lead (or, for that matter, for other substances). Arguments based purely on blood concentrations need cautious appraisal. Blood concentrations, in general, may be misleading because the substance is stored in one of the body compartments and the quantity stored in such a compartment will be related to concentration gradients. However, mobilization of substances may take place against gradients, perhaps in an extent sufficient to cause the critical concentration for target organs to be exceeded.

The blood concentration of lead is kept within 'normal limits' by the buffering effect of the lead bone store. The result is the accumulation of lead in bone. Sudden mobilization of the store can result in high concentrations of lead being passed into the circulating blood and then distributed to the target organs, perhaps years after the *cessation* of exposure to lead. An increase in body temperature (caused by a pyrogen, for example, in non-occupational disease) may be sufficient to mobilize lead from the bone store.

Similar reasoning can be applied to any substances which are transported in blood and which are also accumulated in bodily compartments. In Chapter 12 the point is made that the target organ—the site where the damage is effected—may not necessarily be that in which greatest accumulation occurs. Indeed, storage may be seen as a defence process, although one with limitations.

MODES OF INPUT

In Chapter 2, modes of input at various organs were discussed. As discussed in Chapter 12, in many instances the input organ appears in no way to have been harmed by the harmful substance. Harm is effected in the target organ reached by transport of the substance. This only occurs after the substance has gained access to the bodily interior. Other substances cause harm to the input organ on contact, even though the principal target organ may be remote from the input organ. There are many other substances which attack the input organ exclusively. In these cases the input organ is the target organ.

The substances listed in this chapter offer numerous examples of this diversity. Certain of them, including antimony and cadmium (see below), attack the gut as a target organ. Lead and compounds of lead are inhaled and pass into the circulation without apparently causing harm to lung tissues. Other substances, such as cadmium, also attack the respiratory pathways, although they enter the body by this route. The respiratory pathways appear to be the sole target organ for substances such as vanadium, that is for substances where inhalation is the mode of input.

Inhalation dominates occupational exposure to metallic substances. Unless steps are taken to exercise control, processing of many of these, for example by burning, heating and machining, gives rise to substantial concentrations of aerosols in the occupational external milieu. Dusts and fumes are the most common forms of these aerosols. However, the substances may be present as liquid droplet aerosols or as vapours. In this context the ability of mercury to give off vapour at room temperature and pressure makes it a pathogenic substance. Such pathogenicity takes place in any circumstance where quite small quantities of metallic mercury are allowed to contaminate ill-ventilated spaces.

Aerosols reaching the alveoli cross the alveolar membrane by one or more of the transport systems described in Chapter 3. Lead has already been mentioned as a substance which passes the alveolar membrane without causing harm to the membrane; another substance with this capability is mercury vapour.

The clearance mechanisms from the conducting pathways deal with those aerosols which do not reach the alveoli. This mechanism has been discussed in Chapter 3, where it was seen to bring particles up the nasopharynx and thence to the gut. A proportion of aerosols in the gut is absorbed, and the remainder is excreted in the faeces. Of that absorbed, some may be excreted lower down the gut.

Aerosols reaching the gut by way of the respiratory clearance mechanisms add themselves to substances ingested in the diet. Ingestion normally dominates the input of the nutrients and non-nutrients discussed in this chapter. In general, only a portion of the ingested substances actually crosses the gut wall, the remainder being excreted in the faeces. This is seen in the study by Tati, Katgri and Kawai (Nordberg 1976) of the average daily intake of cadmium in Japan. The average daily intake was found to be 60–80 μg, whereas the amount of faecal cadmium found in the same random sample ranged from 41.1 to 79.4 μg/day.

Absorption through the skin can be an important mode of input for some of these substances. The organometallic compounds and metal salts of fatty acids are readily absorbed through the skin. Implantation of small quantities of beryllium may result in chronic inflammation and ulceration. Several of the metals penetrate the surface of the skin in their attack on it as a target organ. The attack can take the form of ulceration as with chromium, a fact evidenced by the risk to chrome platers of punched-out ulcers of the hands and perforations of the nasal septum. A number of metals, such as nickel and platinum, set up altered sensitivity, which may manifest itself in dermatitis.

TARGET ORGANS, MODES OF ACTION AND EFFECTS

Identification of the target organ or organs is an important step in understanding the modes of action and effects of nutrients and non-nutrients. Thereafter the effects on the target organ should be considered. As mentioned in Chapter 12, effects are defined as biological changes, in order to distinguish them from biologically non-participating phenomena such as accumulation.*

In Chapters 12 to 16 certain general points were made about effects from harmful inputs. It was observed that there are certain clearly defined groups of effects which are non-specific. The same effect is observed as a result of inputs widely differing in character, whereas no one effect is uniquely linked to any one input. We saw that four principal groups of non-specific effects accounted for a substantial proportion of occupational disease. These four groups are: (a) inflammation, degeneration, cell death and necrosis; (b) altered sensitivity; (c) disorder of repair; and (d) disorder of growth.

It will be recalled from Chapter 12 that the pattern of functional or morphological dysfunction can be reasonably well predicted if the target organ and the effect caused in that organ by any substance are known. On this basis, Table 19.1 provides a rough-and-ready basis for the prediction of harmful effects. It must be emphasized, however, that the table is by no means exhaustive since, in particular, it lacks information about the compounds of many of the substances listed. Generalizations may be made about those compounds not listed:

(1) Though solubility is an important determinant of biological

---

*It must be acknowledged, however, that accumulation can sometimes be biologically participating. Lead participates biologically when it is being stored in bone. Nevertheless, the distinction between effects and non-participatory phenomena is useful.

activity, since, in general, the more soluble compounds of a substance are more biologically active, this is by no means always true.

(2) Organic compounds may be more active than inorganic compounds, although the most biologically active forms are the metallic carbonyls and metallic hydrides.

Table 19.1 details substances listed in this chapter, their principal target organs and the effects on those target organs, identified under the four groups. A further column identifies accumulation and certain specific effects.

The table has been compiled from information from the following sources: Browning (1969), Waldron and Stöfen (1974), Beliles (1975), Hunter (1975) and Nordberg (1976). Included in Table 19.1 are substances which appear, from *occupational* exposure, to produce an effect in one or other of the target organs; excluded are those effects attributed to medicinal or other non-occupational input; also omitted are observations made as a result of animal experiments. The table relates only to those substances mentioned in the text, together with one or two of their compounds, its purpose being to provide a framework for summarizing the effects of those substances listed. The table allows certain general statements to be made and it can provide an outline for conceptual purposes.

Table 19.2 provides some indication of the likely effects of all the substances on target organs, although it cannot be relied upon for more than this. The various target organs, together with the principal disease patterns associated with the effects, are summarized in the table. From Tables 19.1 and 19.2 certain general conclusions can be drawn:

(1) The listed substances (principally metals) are strongly associated with irritation leading to inflammation in the skin and respiratory tract.

(2) Certain of the substances cause irritation leading to inflammation or degeneration in other target organs, notably the brain, gut and kidney.

(3) Even though the pattern of disease varies somewhat from substance to substance there are many important features in common.

Altered sensitivity in skin (dermatitis) results from continued association with several of the metals, whereas metal fume fevers, associated with zinc, copper and magnesium, have been classified as altered sensitivity of blood, plasma and blood-forming organs. As this classification is intended to show, in all likelihood the metal fume fevers are an immunopathological process, presumably involving antibody formation. Disorders of repair include the collagenous pneumoconioses, which are examined in Chapter 15, and the fibrous tissue formation observed in kidney with cadmium and in liver with manganese. The disorders of growth outlined in the tables include neoplasms of the skin and respiratory tract. They also include the effect of certain metals on the growth of red blood cells in the blood-forming organs.

SUBSTANCES

In this section each substance will be identified and the effects attributed to it highlighted; this should furnish the reader with a useful supplement to the summary given in Tables 19.1 and 19.2. Sources cited previously on p. 280 have provided the material from which the accompanying synopses were compiled.

Aluminium

The manufacture of aluminium powder in Germany during World War II was associated, according to Hunter (1975), with the development of collagenous pneumoconiosis. Prior to the war aluminium had been regarded as belonging to the non-fibrogenic dusts.

Shaver's disease (Shaver and Redell, 1947, cited by Hunter) is a collagenous pneumoconiosis associated with the manufacture of corundum (Al_2O_3) from bauxite.

Antimony

According to Beliles (1975), antimony miners have developed pneumoconiosis, but McCallum (1967, cited by Beliles) reported no fibrosis. (As pointed out in Chapter 15 'pneumoconiosis' is a term with variable meaning.)

Arsenic

The inorganic compounds of arsenic act as local irritants to the mucous membranes and skin, to the latter of which they are also carcinogenic. The organic arsenicals, a group of substances which were used in chemical warfare, have a profoundly irritant and vesicant effect on the skin, in that they cause fluid-filled vesicles to develop. Arsine (AsH_3) acts as a powerful haemolytic agent.

Barium

Barium sulphate and barium carbonate accumulate in the lung, causing non-collagenous pneumoconiosis.

Beryllium

Beryllium attacks the skin as a target organ, causing dermatitis, acute ulceration and chronic inflammatory change. Inhalation of beryllium

TABLE 19.1

SUMMARY OF METALS, TARGET ORGANS, AND EFFECTS

Target organs	Harmful inflammation	Altered sensitivity	Disorder of repair	Disorder of growth	Accumulation or other effect
Skin	Arsenic, organic arsenicals, beryllium, chromium (6+), lithium hydride, mercury, organic tin, zinc	Chromium (6+), cobalt, gold, nickel platinum		Arsenic	Accumulation of lead in gums accumulation of silver
Respiratory tract	Antimony, arsenic, arsine, beryllium, cadmium, chromium (6+), lithium hydride, manganese, magnesium, mercury, nickel carbonyl, platinum, vanadium, zinc chloride	Cobalt	Aluminium, beryllium, cadmium, cobalt tungsten	Arsenic, nickel chromates	Non-collagenous pneumoconiosis from iron oxide, tin oxide, barium sulphate, barium carbonate
Blood, plasma, blood-forming organs		Copper, magnesium, zinc, iron and other metal fume fevers		Arsenic, cobalt, lead, selenium	Arsine haemolysis and cobalt polycythaemia
Kidney	Beryllium, cadmium, lead, mercury		Cadmium		Accumulation of iron and silver
Liver	Beryllium		Manganese	Beryllium	
Brain	Arsenic, cadmium oxide, mercury, nickel carbonyl, organic lead, organic mercury, organic tin				Lead encephalopathy, manganese encephalopathy, hydrogen selenide
Bone	Phosphorus			Cadmium	Accumulation of lead
Gut	Aluminium, antimony, arsenic, cadmium				Lead colic and constipation
Special senses	Mercury—mercurialentis				Thallium: loss of vision

TABLE 19.2

SUMMARY OF DISEASE PROCESSES IN TARGET ORGANS (to be read in conjunction with Table 19.1)

Target organs	Harmful inflammation	Altered sensitivity	Disorder of repair	Disorder of growth
Skin	Irritant dermatitis, ulceration	Allergic dermatitis (eczema)		Neoplasia of epidermis, dermis, and skin appendage
Respiratory tract	Rhinitis, sinusitis, naso-pharyngitis, laryngitis, tracheo-bronchitis, alveolitis, pneumonitis and pneumonia	Bronchial asthma	Collagenous pneumoconiosis	Neoplasms
Blood, plasma, blood-forming organs	Leucocytosis (the leuco-cytosis may simply be a response to inflammation in other target organs besides blood)	Metal fume fever (this may or may not be an immune response)		Disorders of haemoglobin production, or red cell formation
Kidney	Glomerulo-nephritis and nephrosis		Fibrosis	
Liver			Fibrosis (cirrhosis)	
Brain	Encephalopathy (inclusion of encephalo-pathy under inflammation, etc. is questionable)			
Bone	Necrosis and chronic inflammation			Skeletal disorders (for example, cadmium)
Gut	Mucosal-irritation	Structural changes and functional changes		

causes inflammatory changes throughout the respiratory pathway, including pneumonia. Beryllium is also associated with collagenous pneumoconiosis and can cause chronic inflammatory changes in the kidney and liver.

Cadmium

Cadmium is highly irritant to the respiratory tract and causes loss of the sense of smell (anosmia). The irritant effect of cadmium on the lung is severe, the result being marked emphysema with minimal fibrosis. Cadmium damages kidney, resulting in the impairment of kidney filtration and the finding of protein and sugar in the urine. Cadmium also causes disorders of bone metabolism, resulting in skeletal deformities. The effects of cadmium are highlighted in a report which appeared in the *Birmingham Post* on 28 January, 1977, headlined 'Tragedy of man who shrank'.

A factory worker was said by his family to have shrunk seven inches in years of suffering. At the inquest into his death it was learned that during his 36 years as a battery plate maker, he was exposed to cadmium which left him with 'brittle bones' and a 'kidney complaint'. His bones were said to be so fragile that one fractured when he climbed out of bed. According to evidence given at the inquest he 'needed new hip joints', and had lost his sense of smell.

Tests on his kidneys, liver and bones after his death showed a 'gross excess' of cadmium. The medical adviser to the company concerned told the coroner that when he first examined the man in 1956 medical knowledge had not linked a kidney complaint with cadmium, and that it was now the company's policy to redeploy any worker showing signs of a kidney complaint to a job where there was no possibility of contact with cadmium.

Chromium

The hexavalent form of chromium is particularly active biologically and, as previously indicated, causes dermatitis, ulcers on the hands and forearms and perforation of the nasal septum. It can also result in irritation and inflammatory change of the upper respiratory tract. In this instance, dermatitis is considered to be an altered sensitivity. Hexavalent chromium is also suspected as a cause of lung cancer from occupational exposure (Levy, 1977; personal communication).

Cobalt

According to Payne (1977) cobalt bears major responsibility for the development of 'hard metal' disease, the acute form of which results in an asthma-like altered sensitivity of the upper respiratory tract. The chronic form of 'hard metal' disease results in a collagenous pneumoconiosis, which develops 20 or more years after initial exposure and usually without acute symptoms. Workers in toolrooms are particularly at risk, since the tungsten carbide dust created when tools are ground and sharpened also contains cobalt.

Copper
Apart from causing metal fume fever, copper is not associated with occupational disease.

Gold
Dermatitis resulting from altered sensitivity has been observed in a worker electroplating gold with a gold salt solution as the electrolyte.

Iron
Long-term exposure to iron results in the accumulation of iron in the interstitial spaces of the lungs. It causes non-collagenous pneumoconiosis which is reversible once exposure ceases.

Lead
The target organs for lead are the central nervous system (brain), the gut, the blood and blood-forming organs, the kidney, the gums, the skin and the extensor muscles of wrist or foot. Changes in the nervous system include lead encephalopathy, although this is now rare. The florid manifestations include attacks of coma, convulsions and severe mental disorder. Encephalopathy has been observed in children who have absorbed excess quantities of lead, for example, from lead-containing paints. The gut effects of lead include colic, loss of appetite, nausea, vomiting and constipation. Lead accumulation is also associated with loss of weight, which may be the result of gut effects or a direct effect of the lead. Lead colic is one of the commoner symptoms of chronic lead poisoning, and may be followed, if the disease is allowed to progress, by the weakness of the extensor muscles, a weakness manifested as wrist or foot drop.

Haemoglobin and red cell production are also affected by lead, the result being anaemia and changes in the microscopic appearance of circulating red cells. Lead impairs the numerous enzymes involved in haemoglobin synthesis, causing the anaemia of lead poisoning. Renal changes resultant from lead intake have been a matter of dispute, although there have been suggestions that lead workers suffer from raised blood pressure, a condition possibly related to renal changes.

As previously discussed, lead accumulates in bone and also as a blue line in gum margin.

Lithium
Lithium hydride is intensely corrosive to the skin and respiratory tract.

Magnesium
Magnesium, particularly in the oxide form, causes metal fume fever.

Manganese
Manganese is highly irritant both to lung and gut. More serious, however, is the specific effect on the central nervous system, since manganese is known to affect certain parts of the brain. The resultant effects resemble Parkinson's disease, and the condition generally is marked by psychiatric disorder involving irritability. External manifestations include difficulty in walking, speaking and movement, all of which arise from specific damage to the central nervous system.

Mercury
Mercury vapour is inhaled, but mercury may also be absorbed through the skin. It accumulates in the brain and, at high concentrations, is associated with severe haemorrhage of the brain and of other organs. Long-term exposure to mercury vapour results in a characteristic pattern of central nervous system degeneration manifest in increased irritability, in aggressiveness and in a marked tremor which interferes with writing. Before medical knowledge was more advanced these symptoms often led to mistaken diagnosis. The phrase 'mad as a hatter' for instance is supposed to derive from the unusual behaviour often associated with those in the felt hat trade. Such peculiarities are now generally accepted as having been caused by the effects on the brain of long-term exposure to mercury, which was used in the treatment of felt.

Exposure to mercury vapour has also been observed to cause a brown discoloration in the lens of the eye, a condition known as mercurialentis.

Apart from the brain, the kidneys are other principal target organs for mercury. People exposed to mercury at work are known to excrete protein in the urine, a condition indicative of kidney damage.

The chemical form of the mercury has an important influence on the pattern of disease it induced. The alkyl mercury compounds, of which methyl mercury is the most dangerous, behave differently from other organic mercury compounds. Methyl mercury is readily absorbed through lungs, gut and skin. Like other alkyl mercury compounds it can be transformed from mercury by bacteria in the external milieu.

Transformation appears to have been involved in the notorious

Minamata disease. From 1953 people living near Minamata Bay, an area of sea off the south-west coast of Kyushu, one of the southern Japanese islands, fell prey to a mysterious neurological disease, a disease ultimately traced to factory effluent discharged into the bay. This factory made vinyl chloride, for which mercuric chloride was used as a catalyst, the spent mercuric chloride being carried out into the bay. It seems likely that micro-organisms in the mud of the bay there turned it from inorganic into organic mercury including, possibly, methyl mercury. (For a discussion of Minamata disease see Tucker, 1972.)

For compounds of mercury apart from the elemental form and alkyl mercury, the target organ is normally the kidney. Workpeople exposed to mercury—especially to mercury vapour—excrete mercury in urine. However, levels in urine are considered a poor guide to individual body burden. On the other hand, grouped values for mercury in urine may provide a useful indicator of group exposure.

Molybdenum
No occupation-linked disease appears to have been reported.

Nickel
Nickel carbonyl is extremely dangerous, as instanced by Hunter (1975) in his description of those cases of poisoning from nickel carbonyl which occurred in Britain in the early part of this century. It seems that nickel carbonyl divides into nickel and carbon monoxide, and that the nickel is deposited in a fine layer on the respiratory units, resulting in intense irritation, bleeding and, finally, pulmonary oedema. The carbon monoxide may also contribute to the harmful effect, but the nickel is regarded as the primary danger.

Prolonged exposure to nickel and to nickel carbonyl is reported to have produced cancer in the lungs and nose of exposed workers (Sunderman, 1970). Altered sensitivity to nickel is common, the result being a dermatitis known as nickel itch, whereas altered sensitivity in the tract, producing an asthma-like condition, has also been reported.

Phosphorus
Phosphorus is the primary element in the metabolism of life. It does, however, present dangers in its elemental form, yellow phosphorus, and as phosphine gas and organophosphorus insecticides. There exists a whole family of these biologically highly active substances, some of which are well known. They include parthion, malathion and triorthocresyl phosphate.

Elemental phosphorus causes chronic inflammation of the lower jaw, a particularly unpleasant condition known as phossy jaw. This disease, classically associated with match dipping, was described by Hunter as 'the greatest tragedy in the whole story of occupational disease'. During the last century many people were affected by phossy jaw, and it was finally eliminated from match manufacture only by the prohibition of the use of white phosphorus. Not only was an occupational disease in a particular group of workers thereby prevented, but the search for an alternative led to the discovery of harmless, red phosphorus. This was used to produce the safety match, the desirable feature of which (i.e. that it could only be ignited when struck against a chemically appropriate surface) made the discovery all the more beneficial. It is perhaps the best historical example of effectiveness in substitution.

Phosphine (PH_3) is evolved in the accidental wetting of zinc phosphide. In this respect it resembles arsine. It is extremely dangerous and can cause fatal disturbances of the central nervous system and gut. It apparently differs from arsine in not causing haemolysis.

Platinum
Platinum is associated with altered sensitivity affecting skin and, occasionally, the respiratory tract.

Selenium
The metal selenides have effects on the central nervous system. For instance, animals affected by excess selenium suffer from a disease known as 'blind staggers'.

Silver
Excessive exposure to silver causes accumulation in the tissues. The resulting discoloration is called argyria, a condition in which the skin goes black and sometimes acquires a metallic lustre.

Thallium
According to Browning (1969) there is a risk in the manufacture of fused halides of thallium in connection with special glass manufacture: excessive exposure results in loss of vision.

Tin
Inhalation of metallic tin is associated with a non-collagenous pneumoconiosis. Organic tin compounds are very biologically active, some, such as triethyl tin, producing severe effects on the central nervous system.

Tungsten
Tungsten is a constituent of hard metal. According to Payne (1977)

cobalt is the major factor incriminated in 'hard metal' disease, though tungsten has also been associated with the disease.

Vanadium

Vanadium has an irritant effect on the respiratory tract. Exposure to vanadium pentoxide (from, for example, flue-cleaning of oil-fired burners using certain types of oil) causes a green discoloration of the tongue.

Zinc

Zinc is an important biological metal. Exposure to zinc fumes causes zinc fume fever, which follows the inhalation of freshly formed fumes of zinc oxide and of other forms of zinc. Metal fume fever may be experienced on the first contact with zinc fumes but the fever can also develop after a period of absence from exposure, for example after a holiday.

REFERENCES

BELILES, R. P. (1975). *Metals in Toxicology*, L. J. Casarett and J. Doull, eds, New York: Macmillan.

BROWNING, E. (1969). *Toxicity of Industrial Metals*, 2nd Edition, London: Butterworth.

DAVIES, I. J. T. (1972). *The Clinical Significance of the Essential Biological Metals*, London: Heinemann.

DEPARTMENT OF THE ENVIRONMENT (1974). *Lead in the Environment and its Significance to Man*, London: HMSO.

HUNTER, D. (1975). *The Diseases of Occupation*, London: English Universities Press.

LUCKEY, T. D. and VENUGOPAL, B. (1977). *Metal Toxicity in Mammals*, Volume 1. New York: Plenum Press.

NORDBERG, G. F. (Ed.) (1976). *Effects and Dose-Response Relationships of Toxic Metals*, Amsterdam: Elsevier.

PAYNE, L. R. (1977). The hazards of cobalt, *Journal of the Society of Occupational Medicine*, **27**, 20–5.

SUNDERMAN, F. W. (1970). Nickel poisoning, in *Laboratory Diagnosis of Diseases caused by Toxic Agents*, F. W. Sunderman and F. W. Sunderman eds, St Louis: Warren Green.

TUCKER, A. (1972). *The Toxic Metals*, London: Ballantine.

WALDRON, H. A. and STÖFEN, D. (1974). *Sub-Clinical Lead Poisoning*, London: Academic Press.

YUDKIN, M. and OFFORD, R. (1973). *Comprehensible Biochemistry*, London: Longman.

For Further Reading:

LUCKEY, T. D., VENUGOPAL, B. and HUTCHESON, D. (1975). *Heavy Metal Toxicity, Safety and Hormology*, New York: Academic Press.

Chapter 20

Asbestos

OBJECTIVES OF THE CHAPTER

This chapter provides:
  (1) an outline of asbestos, its modes of action and its effects,
  (2) an outline of the history of action on asbestos, and
  (3) comment on the British Occupational Hygiene Society's Hygiene Standard for Chrysotile Asbestos Dust and its relation with the Asbestos Regulations 1969.

INTRODUCTION

Asbestos is a term used to describe several mineral, fibrous silicates. In Britain, the Asbestos Regulations 1969 (Statutory Instrument, 1969) recognize four types of asbestos: crocidolite, amosite, chrysotile and fibrous anthophyllite. The fibrous nature of asbestos and its relative chemical inertness make the class of substance useful for a wide range of purposes, including those of thermal insulation and 'fire-proofing', and as building material.

The first cases of disease from asbestos were observed in Britain in 1900, and were reported by Murray in 1907. Since then evidence of the capacity of these inert, inoffensive-looking substances to cause serious disease has been accumulating.

The four types of asbestos differ in their chemical composition but their capability for causing disease, a capability which differs according to type, is now viewed more as a matter of fibre morphology than of chemistry.

Of the four types, three belong to the straight-fibre family of asbestos substances, only chrysotile possessing fibres which are curly. All forms of asbestos separate into fibres of a wide range of dimensions. All four types possess a proportion of fibres small enough to be respirable, and it is these fibres which are the prime concern of this chapter. According to Harries (1973) asbestos corns (warts) are attributable to the penetration of the skin by asbestos fibres or to the implantation of asbestos fibres by sewing needles which carry minute fibres.

290

AIRBORNE PARTICLES OF ASBESTOS

Timbrell (1973), in his explanations of asbestos-induced disease, emphasized the importance of fibre dimension. It is dimension which determines the aerodynamic behaviour of fibres. For example, Timbrell showed that an asbestos fibre of 3 μm diameter has the same falling speed as a particle, of unit density, of 10 μm equivalent diameter (in other words, the EUDS is 10 μm). It will be recalled from Chapter 2 that most particles of EUDS 10 μm and above are entrapped by respiratory filtration, so, on this evidence, greatest suspicion must fall on fibres with diameters of 3 μm and below.

Timbrell also concluded that fibres longer than 200 μm were likely to be intercepted in the upper reaches of the conducting airways. These findings were in many ways corroborated by those of Muir in 1976. In a review of health hazards from thermal insulation products Muir concluded that there was a risk of mesothelioma from the inhalation of any fibrous substance with a diameter of less than 0.5 μm and a length in excess of 10 μm. According to existing knowledge, therefore, it seems that fibres with diameters of 3 μm and less are associated with asbestos-induced disease, and that a particular category of asbestos disease is associated with fibres of a diameter less than 0.5 μm. Because of their small aerodynamic diameter, fibres less than 10 μm in length may remain airborne within the lung, or the body defence mechanisms may be able to cope with them. Fibres longer than 10 μm but shorter than 200 μm appear to present the risk of asbestos disease.

It should be pointed out that the above-cited figures on fibre dimensions of importance in respiratory disease do not entirely correspond with those presently advocated by the Health and Safety Executive in Guidance Note EH/10, 1976 for the purposes of interpreting the 1969 Asbestos Regulations.

The Guidance Note defines fibres as follows (page 2): 'Fibres means particles of length greater than 5 micrometers, a diameter of less than 3 micrometers and having a length to breadth ratio of at least 3 to 1, observed by transmitted light under phase contrast conditions at a magnification of approximately 500×.'

The whole question of asbestos standards is at present under review and the definition of fibre could well be revised in order to take account of current knowledge.

ASBESTOS-INDUCED DISEASES

(1) Asbestosis

This disease belongs to the group of diseases of the chest known as the pneumoconioses, and to the sub-group characterized by excessive collagen formation (see Chapter 15). Fibrosis arises from the excess collagen formation present diffusely throughout the lungs, and extends throughout the lung tissues. It involves the pleural lining, which at post-mortem has a shaggy appearance (see Fig. 20.1).

The fibrosis starts in the interstitial parts of the lung which are reached by the fibres of asbestos either by direct penetration of the alveolar membrane or by phagocytic migration. Once into the interstitial sites, by whatever route, the fibres are the subject of attempts at phagocytosis by macrophages.

A laboratory experiment by Beck *et al.* (1972) showed that in cell culture (i.e. experiments carried on in the test-tube rather than in a living animal) particles of glass and chrysotile asbestos small enough to be phagocytosed were not toxic to macrophages, unlike crystalline silica in the form of quartz which was highly toxic. Fibres whose

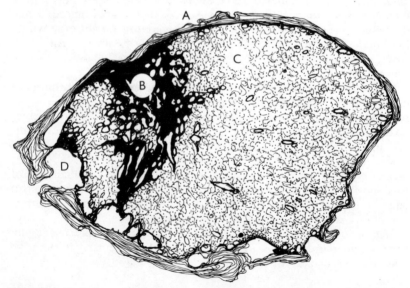

FIG. 20.1. Drawing of section of lung with asbestosis. (A) Pleural lining thickened; (B) area of fibrosis (this is exaggerated for clarity); (C) lung tissue showing evidence of emphysema; and (D) large space created by shrinkage of fibrous tissue.

length greatly exceeded the girth of the macrophages were also phagocytosed, but the prolonged 'spearing' of the macrophages resulted in an increase in membrane permeability, which allowed leakage from the macrophage body.

These findings provide the basis for a feasible explanation of asbestosis. It will be remembered from Chapter 15 that death or disruption of macrophages appears to initiate the excess production of collagen. In the case of quartz, the macrophages are poisoned, but in the case of *long* asbestos or other fibres the macrophages may simply leak to death. Either way, there is macrophage death followed by fibrosis. But how the macrophage death actually leads to the fibrosis is still a matter of conjecture.

Alveolar macrophages are probably 15–20 μm in diameter. It is reasonable to assume that the apparent harmlessness of inert fibres of less than 10 μm in length is due to their being small enough for macrophages to cope with. Fibres much in excess of 10 μm cannot possibly be wholly enclosed by macrophages; indeed the study by Beck and his colleagues shows photographs of macrophages apparently 'impaled' on long fibres.

It should be emphasized that all four types of asbestos are known to have caused asbestosis. The attention given to crocidolite in relation to mesothelioma (infra) overshadows the fact that it is also known to be a cause of asbestosis (Sluis-Cremer and du Toit, 1973).

(2) Carcinoma of the Lung

Carcinoma of the lung was first reported by Lynch and Smith in the USA in 1935. Several subsequent studies have suggested links between asbestos exposure and more than one type of malignant tumour of the lung. The cancer appears to arise in the vicinity of fibrosis (Jacob and Anspach, 1965). Parkes's account (1973) suggests that lung cancer does not occur in the absence of asbestosis, although certain American studies, for example that of Selikoff *et al.* (1964) suggest that lung cancer associated with asbestos exposure *can* occur independently of asbestosis. Strong evidence exists that cigarette smoking and asbestos exposure interact multiplicatively. Selikoff *et al.* (1968) calculated that the risk of contraction of lung cancer by asbestos workers who smoked cigarettes was 90 times greater than that of non-smokers who had never worked with asbestos. This link between severe asbestos exposure and cigarette smoking was also observed by Berry *et al.* in 1972.

According to Hill *et al.* (1966), rigorous control of the use of mainly chrysotile asbestos in a textile factory reduced the mortality rate from bronchial carcinoma to the national average. It is likely that some of the increase in lung cancer is accounted for by the decrease in mortality from tuberculosis, which was associated with asbestosis (Smither, 1966). The decline in mortality rate from tuberculosis (for asbestos workers as well as for the population as a whole) is attributable to the improvement in social conditions as well as to antituberculous therapy.

(3) Mesothelioma of Pleura and Peritoneum

Mesothelioma (sometimes called malignant mesothelioma) of the pleura and peritoneum is uncommon. Wyers first noted an association between mesothelioma and asbestosis in 1946. Fourteen years later, in a study of inhabitants of the north-west of Cape Province in South Africa, Wagner *et al.* reported mesothelioma among those with a history of industrial or residential contact with crocidolite. Present evidence particularly implicates crocidolite, but there is evidence that chrysotile and amosite have also caused mesothelioma in humans (MacDonald and MacDonald, 1975).

Mesothelioma is highly malignant and invariably fatal, death occurring within a year or 18 months of first diagnosis. The disease is also characterized by a latent period of 20–40 years. Particularly clear evidence of this long latent period is to be found in a study by Jones (1974). His subjects of study were workers engaged, for a 4-year period during World War II, in the assembly of gas masks. This work involved the use of crocidolite, and, though exposure ceased in 1945, cases of mesothelioma did not begin to appear until 1964. Cases continued to come to light in the early 1970s, clear evidence of a latent period of 20–30 years.

(4) Pleural Plaques

Jones (1974) has also summarized knowledge relating to pleural plaque. This is a yellow-white, patchy thickening of the pleura and is associated with asbestos exposure. Pleural plaques, which are present on the pleura of the chest wall, range in size from specks to wide areas up to 10 mm thick. Pleural plaques contain calcification which results from degeneration of fibrous tissue. They are plainly visible on X-ray photographs. They are not associated with adhesions between adjacent surfaces of the pleura. Though no dose–response relation has been established, it appears that once a minimal initial exposure has been exceeded prevalence depends on age. Pleural plaques are not harmful *per se*, but they do indicate asbestos exposure. Their presence on the *chest wall* indicates that, after penetrating the membrane of the respiratory unit, asbestos fibres cross the pleural space.

(5) Clubbing of the Fingers

A number of diseases of the chest are associated with deformity of finger ends, a deformity which causes the finger ends to appear rounded, rather like Indian clubs. This 'clubbing' varies in degree, but is associated with asbestosis, its presence being clearly indicative of asbestos disease.

(6) Asbestos Bodies

Asbestos bodies are found on microscopic examination of lung, following exposure to asbestos. Parkes (1973) described them as structures sometimes as much as 80 μm in length, golden yellow or brown in colour. They consist of asbestos fibres wholly or partly covered with an iron-containing protein. The significance of asbestos bodies—and indeed whether they contain asbestos at all—is in dispute. However, one body of opinion regards asbestos bodies as indicators of asbestos exposure and not necessarily of asbestos disease.

It is tempting to hypothesize that the iron-protein covering represents an inactivation and the harmless imprisoning of asbestos fibres in the lung interstitial tissue. Moreover, there is evidence, from animal experiments, that the formation of asbestos bodies is associated with phagocytosis by alveolar macrophages and that the iron-protein may derive from the phagosomes of the macrophages (see, for example, Suzuki and Churg, 1969).

(7) Other Cancers

American studies, such as those of Enterline (1965), Mancuso (1965) and Selikoff et al. (1970), have shown increased incidence of cancers of the gut in asbestos workers. However, according to a report published in 1967 by the then Ministry of Labour, there then existed no unexpected incidence of gut cancer in Britain. Suggestions have been made that cancer of the ovary and of the blood-forming organs may be associated with asbestos exposure, although, as Parkes made clear, this evidence is weak.

HISTORY OF ACTION ON ASBESTOS

Attention was first drawn to asbestosis by Cooke, in 1924. However, according to Hunter (1975) it was the case of Sieler, in 1928, which triggered the inquiry of Merewether and Price and their report in

1930. This report was followed, in 1931, by the Asbestos Industry Regulations, which were replaced, in 1969, by the Asbestos Regulations.

As is implied by their title, the Asbestos Industry Regulations were decidedly weak, in that they were limited to the asbestos industry. Asbestos appears in many forms in many industries, and consequently the scope of the regulations did not correspond to the scope of the problem. Even so, in the enforcement of the regulations (in regard to Acre Mill) the Factory Inspectorate '... did to some degree fall short of the standard of performance of the functions ... that could properly have been expected of them'. (Parliamentary Commissioner, 1975).

The Parliamentary Commissioner (the Ombudsman), in the report quoted above, was commenting on the case of a Mr Buick. It was Mr Buick's contention that he had contracted asbestosis as a result of his employment at Acre Mill, Hebden Bridge, Yorkshire, a mill operated (from 1939 until its close-down in 1970) by the Cape Asbestos Company (now Cape Industries Ltd). The mill came within the jurisdiction of the Asbestos Industry Regulations and, latterly, the Asbestos Regulations. Mr Buick maintained that his contracting of asbestosis was partly due to maladministration by the Factory Inspectorate, in that it had not adequately carried out its responsibilities in respect of Acre Mill. The Commissioner drew attention to two other persons said to be suffering from asbestosis, but their cases he could not consider, because they had never worked in the factory, and because the Factory Inspectorate's jurisdiction did not, at that time, extend beyond the factory premises. This aside in the Commissioner's Report is important, in that it suggests the possible occurrence of 'neighbourhood' cases of asbestosis as well as mesothelioma.

The Parliamentary Commissioner's Report contains a valuable summary of the history of asbestosis and its control. He emphasized the point (made by Merewether and Price) that full development of asbestosis could be prevented if the concentration of asbestos fibres at a place of work was maintained below that found in the spinning processes of a factory processing raw materials (because the incidence of disease was observed by Merewether and Price to be low in such processes).

The Asbestos Industry Regulations required the use of exhaust ventilation systems 'to ensure that no asbestos dust was allowed to escape into the air of any room in which people worked'. Where dust was unavoidable, breathing apparatus and special clothing had also to be provided. The Commissioner's report reveals that the then Chief Inspector of Factories had informed the Home Secretary that it was 'desirable to emphasize that this limit is clearly provisional and is subject to alteration in the light of further medical experience'.

No progress had been made towards a safe limit during the 30 years following the report of Merewether and Price. In 1960, however, the American Conference of Governmental Industrial Hygienists had set a Standard for asbestos. It nevertheless emerged (according to the Commissioner) that in 1964, in the USA, evidence had been provided of an abnormally high incidence of lung cancer among asbestosis sufferers. Instances of the development of mesothelioma following only short periods of exposure to certain types of asbestos had also been provided.

As we have seen, however, the first report of a link between asbestos and carcinoma of the lung had been made by Lynch and Smith, in the USA, as early as 1935. A near-contemporary report in Britain, by Merewether, had suggested that 17.7% of a male population with asbestosis had died of lung malignancy.

Only in the mid-1960s did the Ministry of Labour begin consultations designed ultimately to revise the 1931 Asbestos Regulations. In 1965 a special medical advisory panel was instituted under the chairmanship of the Senior Medical Inspector. This panel concluded that the American Standard was unsatisfactory, that crocidolite was more dangerous than other kinds of asbestos and that fibre-counting was the preferred method of assessing airborne asbestos concentration. Their major conclusion was that no 'safe' limit could be established for asbestos. In 1968 the British Occupational Hygiene Society published its Hygiene Standard for chrysotile asbestos (see below).

The Commissioner's report continued with a detailed investigation of the Factory Inspectorate's approach to the problem at Acre Mill. It is clear from the Commissioner's account that the Factory Inspectorate did not show the concern that would be evident today. Attitudes towards asbestos have changed radically since the BOHS published its Standard in 1968 in that the previous complacency has been replaced by anxiety. The Inspectorate's activities have to be judged against the background of this profound shift in opinions.

As has already been indicated, the Commissioner did criticize the Factory Inspectorate's attitude towards their responsibilities in respect of Acre Mill. He found that there was little direct contact between inspectors and the general workforce at the mill. It appears that this contact was not the custom and practice at Acre Mill, since neither Inspectorate nor workforce ever consulted. One can only speculate whether consultation would have reduced the incidence of asbestosis. However, section 28(8) of the Health and Safety at Work Act 1974, requires an inspector to keep employees, or their representatives, adequately informed about matters affecting their health and safety. In this respect, therefore, history is unlikely to be repeated.

The Parliamentary Commissioner found that for many years both employers and the Factory Inspectorate generally were unaware of the full extent of the dangers from asbestos dust. 'Not until the early 1960s had the development of scientific knowledge shown that the dangers were greater than had been apparent at the time of the Merewether report in 1930.' One might argue that Doll's evidence in 1955 provided sufficient corroboration of the findings of Lynch and Smith 30 years before, and that a vigorous campaign should have been mounted earlier. That it was not mounted is evidence, I believe, of an attitude of complacency towards asbestos in medical and official circles.

Relating to Acre Mill and the Inspectorate, the earliest record to be discovered by the Commissioner was dated 19 May, 1949. The then visiting inspector recorded that arrangements for handling the crude asbestos were 'criticized strongly to the management' and that some of the methods being used to eliminate dust were 'unsatisfactory'. A number of visits followed, and individual inspectors were certainly concerned at the apparent risk. However, it was thought by the Commissioner that the Inspectorate did not subject the management to as much pressure as it might. It appears that the Inspectorate failed to make full use of available statistics, a failure which might have had a bearing on their attitude to prosecution. It is suggested that the Inspectorate's attitude might have been that advice, help and persuasion were just as effective as prosecution. According to the Commissioner, 'the Department now acknowledges, with hindsight, that in general they did not take a sufficiently firm line in regard to asbestos dust prior to the introduction of 1969 Asbestos Regulations. Prosecution for breaches of the 1931 Regulations during the 40-odd years they were in force were extremely rare (by contrast, I understand some 51 firms have been prosecuted for some 120 separate contraventions) since the 1969 Regulations came into force.'

British Occupational Hygiene Society's Hygiene Standard for Chrysotile Asbestos Dust

The British Occupational Hygiene Society produced the Hygiene Standard for chrysotile asbestos dust in 1968. The Standard was based on the proposition that the risk of '... the earliest demonstrable effects on the lung due to asbestos' would be less than 1% for an accumulated exposure of 100 fibre years per ml. This corresponds, for example, to a concentration of 2 fibres per cm^3 for 50 years or 4 fibres per ml for 25 years, and so on. A confidence limit of 90% was attached to the estimate. The data were derived from one factory processing chrysotile asbestos, where fibres were defined as particles longer than 5 μm and a length to breadth ratio in excess of 3 to 1.

The Hygiene Standard noted the significant risk of lung cancer associated with asbestosis, and the risk of mesothelioma of pleura and the peritoneum particularly associated with inhalation of crocidolite dust. It specifically stated that there was no known quantitative relation between asbestos and cancer risk, and that it was not possible at that time to specify a safe limit for asbestos in regard to the cancer risk or, by implication, the mesothelioma risk.

Asbestos Regulations 1969

Section 2(3) defines asbestos dust as 'dust consisting of or containing asbestos *to such an extent as is liable to cause danger to the health of employed persons.*' Guidance Note EH/10 (HSE, 1976), as has been seen, gives guidance on how Factory Inspectors should interpret this phrase until a revision is forthcoming, because the whole matter is at present under review.

For chrysotile, amosite and fibrous anthophyllite the regulations are not be be enforced if the concentration of asbestos dust over any 10-min sampling period is less than 2 fibres per ml. Where the concentration is above 2 fibres per ml but less than 12 fibres per ml, over the 10-min sample, a further period of sampling is carried out over a period of 4 h. The sampler should draw air at a rate between 200 and 500 ml/min. If the 4-h sample exceeds 2 fibres per ml the inspector requires the standard of control to be improved. The extent to which this is required depends upon the amount by which the average concentration, over the period of exposure, exceeds the 2 fibres per ml standard. If the average concentration of asbestos over any 10-min period exceeds 12 fibres per ml, inspectors, according to Guidance Note EH/10, should normally seek to confirm the measurement and then take action to enforce the requirements, in particular those relating to exhaust ventilation.

The regulations apply in full whenever workpeople are engaged in processes involving crocidolite. An approved form of respirator is required unless the concentration in the worker's breathing zone is maintained below 0.2 fibres per ml, when measured as the average concentration over a 10-min sampling period.

It is explicitly stated that the Guidance Note standards are related to the 100 fibre years per ml stipulated in the BOHS Standard. There is a further resemblance, in that in the Guidance Note fibres are defined as particles of length between 5 μm and 100 μm, having a length to breadth ratio of at least 3 to 1. This corresponds to the definition of fibre used in the BOHS Standard. The BOHS Standard recommends sampling over a full working shift (with certain exceptions). The Guidance Note, on the other hand, stipulates 10 min as the minimum continuous sampling period. This was plainly designed with an eye to the practicalities of an inspector carrying out measurements in a factory, but it does, in fact, represent a weakness in the standard of control. A great deal depends on the manner in which inspectors approach the problem of measurement. The 10-min sampling period has to be chosen with care, in order that the inspector might decide whether the Asbestos Regulations apply or do not apply to a particular premises. According to the Guidance Note 'ideally the sample should be taken during a period of maximum dust emission so as to determine peak exposure level', but in practice periods of maximum emission may be difficult to detect without extensive sampling—and certainly longer than 10 min.

The Asbestos Regulations and the British Occupational Hygiene Society's Standard are all the subject of continuing discussion and review. It will be interesting to see whether the numerical value assigned to the Standard is tightened up, and whether a different strategy to sampling is adopted.

REFERENCES

ASBESTOS REGULATIONS (1969). S.I. 1969, No. 690.

BECK, E. G., HOLT, P. F. and MANOJLOVIC, N. (1972). Comparison of effects on macrophage cultures of glass fibre, glass powder and chrysotile asbestos, *British Journal of Industrial Medicine*, **29**, 280–6.

BERRY, G., NEWHOUSE, M. L. and TUROK, M. (1972). Combined effects of asbestos exposure and smoking on mortality from lung cancer in factory workers, *Lancet*, **2**, 476–9.

BRITISH OCCUPATIONAL HYGIENE SOCIETY (COMMITTEE ON HYGIENE STANDARDS OF THE BRITISH OCCUPATIONAL HYGIENE SOCIETY) (1968). Hygiene standards for chrysotile asbestos dust, *Annals of Occupational Hygiene*, **11**, 47–69.

COOKE, W. E. (1924). Fibrosis of the lungs due to the inhalation of asbestos dust, *British Medical Journal*, **2**, 147.

DOLL, R. (1955). Mortality from lung cancer in asbestos workers, *British Journal of Industrial Medicine*, **12**, 81–86.

ENTERLINE, P. E. (1965). Mortality among asbestos products workers in the United States, *Annals of the New York Academy of Sciences*, **132**, 156–65.

HARRIES, P. G. (1973). Clinical signs, in *Biological Effects of Asbestos*, W. Davis, ed, Lyons: International Agency for Research on Cancer.

HILL, I. D., DOLL, R. and KNOX, J. F. (1966). Cohort analysis of changes in incidence of bronchial carcinoma in a textile asbestos factory, *Annals of the New York Academy of Sciences*, **132**, 526–35.

HUNTER, D. (1975). *The Diseases of Occupations*, London: English Universities Press.

JACOB, G. and ANSPACH, M. (1965). Pulmonary neoplasia among Dresden asbestos workers, *Annals of the New York Academy of Science*, **132**, 536–48.

JONES, J. S. (1974). Pathological and environmental aspects of asbestos-associated diseases, *Medicine, Science and the Law*, 14, (3), 152–8.

LYNCH, K. M. and SMITH, W. A. (1935). Pulmonary Asbestos III. Carcinoma of lung in asbesto-silicosis, *American Journal of Cancer*, **24**, 56–64.

MACDONALD, A. and MACDONALD, C. (1975). Epidemiology of mesothelioma from estimates of incidence, Paper presented at the *XVIIIth International Congress on Occupational Health*, Brighton.

MANCUSO, T. F. (1965). Discussion of asbestos and neoplasia epidemiology, *Annals of The New York Academy of Science*, **132**, 589–602.

MEREWETHER, E. R. A. and PRICE, C. W. (1930). Report on effects of asbestos dust on the lungs and dust suppression in the asbestos industry, London: HMSO.

MINISTRY OF LABOUR (1967). Problems arising from the use of asbestos, H.M. Factory Inspectorate, London: HMSO.

MUIR, D. C. F. (1972). *Clinical Aspects of Inhaled Particles*, London: Heinemann.

MURRY, M. (1907). Departmental Committee on Compensation for Industrial Diseases, Minutes of Evidence. Cmnd. 3496, London: HMSO, pp. 127–8.

PARKES, W. R. (1973). Asbestos-related disorders, *British Journal of Diseases of the Chest*, **67**, 261–300.

PARLIAMENTARY COMMISSIONER (1975). Report Supplement (Department of Employment) Case No. C253/V, Danger to Health from Asbestos, London: HMSO.

SELIKOFF, I. J., CHURG, J. and HAMMOND, E. C. (1964). Asbestos exposure and neoplasia, *Journal of The American Medical Association*, **188**, 22–6.

SELIKOFF, I. J., CHURG, J. and HAMMOND, E. C. (1968). Asbestos exposure, smoking and neoplasia, *Journal of The American Medical Association*, **204**, 106–12.

SELIKOFF, I. J., HAMMOND, E. C. and CHURG, J. (1970). Mortality experiences of asbestos insulation workers, in *Pneumoconiosis, Proceedings of the International Conference*, Johannesburg 1969, H. A. Shapiro, ed.

SIELER, H. E. (1928). A case of pneumoconiosis, result of the inhalation of asbestos dust, *British Medical Journal*, 2, 982.

SLUIS-CREMER, G. K. and DU TOIT, R. S. J. (1973). Amosite and crocidolite mining and milling as causes of asbestosis, in *Biological Effects of Asbestos*, W. Davis, ed., Lyons: International Agency for Research on Cancer.

SMITHER, W. J. (1966). Secular changes in asbestosis in an asbestos factory, *Annals of the New York Academy of Sciences*, **132**, 166–83.

TIMBRELL, V. (1973). Physical factors as aetiological mechanisms, in *Proceedings of the Conference on Biological Effects of Asbestos*, Lyons, 2–5 October 1972.

SUZUKI, Y. and CHURG, J. (1969). Structure and development of the asbestos body, *American Journal of Pathology*, **55**, 79–107.

WAGNER, J. C., SLEGGS, C. A. and MARCHAND, P. (1960). Diffuse pleural mesothelioma and asbestos exposure in the North Western Cape Province, *British Journal of Industrial Medicine*, **17**, 260–71.

WYERS, H. (1946). Thesis presented to the University of Glasgow for the degree of Doctor of Medicine.

Further Reading
BRADFIELD, R. E. N. (1977). *Asbestos, Review of Uses, Health, Effects, Measurement and Control*, Epsom: Atkins Research and Development.

HEALTH AND SAFETY EXECUTIVE (1977). *Selected Written Evidence Submitted to the Advisory Committee on Asbestos*, 1976–77, London: HMSO.

ZIELHUIS, R. L. (1977). Ed, *Public Health Risks of Exposure to Asbestos*, Commission of the European Communities. Oxford: Pergamon Press.

Chapter 21

Case Study: Respiratory Disease in the Coal Industry

OBJECTIVES OF THE CHAPTER

This chapter:
(1) highlights the interplay of political and scientific factors in relation to respiratory disease in coal miners and standards for respirable dust control in mines,
(2) outlines the natural history of coal-workers' pneumoconiosis, distinguishing between simple coal-workers' pneumoconiosis and progressive massive fibrosis,
(3) discusses briefly views on the role of quartz in the aetiology of coal-workers' pneumoconiosis, and
(4) mentions other theories of causation,
(5) notes forms of respiratory disease other than pneumoconiosis experienced by coal workers,
(6) discusses the social consequences of coal-workers' pneumoconiosis, highlighting questions of compensation,
(7) outlines the action taken on pneumoconiosis after 1946 in the British coal industry,
(8) outlines the controversy which surrounded the report by the Industrial Injuries Advisory Council on pneumoconiosis and byssinosis published in 1973,
(9) identifies the importance attached to compensation, and comments on the dust levels in coal mines in relation to the standards laid down in the Coal Mines (Respirable Dust) Regulations,
(10) summarizes the main provisions of the Coal Mines (Respirable Dust) Regulations, and
(11) comments on the rationale of the Regulations.

POLITICAL AND SCIENTIFIC FACTORS

The prominent part played by coal in Britain's economic development is acknowledged and documented in almost every relevant history textbook (see, for example, Briggs, 1959; Wood, 1960). The term 'King Coal' is not just a pun on a nursery rhyme. However, the industry's influence on British history has not been confined merely to the economic and industrial domain in that (as will be mentioned in Chapter 23) some of the strategies most central to occupational health and safety have been evolved as a result of the struggle against the dangers of coal mining. Disasters in coal mines continue to be of profound emotional significance nationally, as well as for the industry.* In the years before 1946 the frequently bitter relations between coal owners and coal workers influenced the political and social life of the nation, while more recently at least one Government has paid the price for its alienation of the miners. No industry has produced so many political leaders, and it is not surprising that health and safety should have figured so largely in political debate.

Of all the famous Labour Party miner-politicians, Aneurin Bevan (1897–1960) personifies the inseparability of health and politics. Son of a miner, one of thirteen children, Bevan was born at Tredegar in South Wales. He began work in the local mines at 13, and before the age of twenty became Chairman of a Miners' Lodge of more than 4000 members. He developed nystagmus (a rhythmical side-to-side motion of the eyes attributed to work in poor illumination underground). The condition was responsible for his rejection on medical grounds for military service during World War I. In connection with the nystagmus he received a lump sum compensation of £60 (Foot, 1962).

In 1926 his influence among the trade unionists of South Wales was such that he became leader of the Welsh miners during the General Strike of that year. He was elected to Parliament in 1929, as Member for Ebbw Vale, becoming a member of the Labour Party two years later. Although frequently in opposition to Churchill's policies, he held a number of important posts in the wartime coalition government.

The General Election of 1945 returned a Labour Government and Bevan was appointed Minister of Health; his prime responsibility was the introduction of the National Health Service, long an important element of the Labour Party's social policy manifesto. Contemporary with the introduction of the NHS came a change in attitudes towards compensation for injury or disease occasioned by work. This change was given concrete form by the replacement of the Workmen's Compensation Acts by the National Insurance (Industrial Injuries) Act of 1946. Dangers to the health and safety of coal miners had been a politically important force in the reform of the compensation system.

---

*A readable account of coal mine disasters from 1700 onwards is to be found in Duckham, H. and Duckham, B. (1973). *Great Pit Disasters*, Newton Abbot: David and Charles.

The Workmen's Compensation Act had, in its time, attempted to introduce a measure of social justice within the framework of 19th century ideas. Although it was not entirely unsuccessful, it had certainly produced a number of undesirable side-effects, not least for coal workers (see below). One of the most significant was the ill-feeling engendered between the coal-mining community and the 'compensation doctors' (medical referees) whom that community saw as agents of the coal owners attempting to minimize the legitimate claims of miners injured or diseased by their work.

The 'compensation doctors' are now no more than subjects for historical study, though they remain as a lingering influence on attitudes. The National Coal Board's medical service has worked hard and successfully to achieve the confidence of the coal miners. However, respiratory disease in coal workers—'the dust'—remains a political and an emotional question, as well as one of human safety. This needs to be borne in mind in any discussion of the nature, extent and social consequences of respiratory disease in coal workers.

COAL-WORKERS' PNEUMOCONIOSIS

As has been so often remarked, terminological difficulties exist in the definition of any disease or condition. The term 'coal-workers' pneumoconiosis' has gained favour because it is descriptive, it is consistent with the ILO categorization (described in Chapter 15) and it implies almost no theories about causation. Terms such as 'black lung' are deplored by Parkes (1974) as being uninformative, while 'anthraco-silicosis' implies theories of causation which, by and large, are now discredited.

The National Coal Board's medical service (McLintock, 1976) recognizes two principal terms: 'simple (coal-workers') pneumoconiosis' and 'progressive massive fibrosis' (synonymous with complicated pneumoconiosis). 'Simple coal-workers' pneumoconiosis' corresponds to the ILO's 'non-collagenous pneumoconiosis' (described in Chapter 15), while 'progressive massive fibrosis' involves a collagenous reaction resultant upon an altered tissue response to the coal dust (Fig. 21.1).

The extent to which simple coal-workers' pneumoconiosis is associated with respiratory disability is a matter of debate (see below). There is, however, general agreement that progressive massive fibrosis has more severe effects than does simple pneumoconiosis, and that it is attended by a greater tendency to life shortening and respiratory disablement. Rheumatoid pneumoconiosis ('Caplan's Syndrome') describes another disabling condition in which rheumatoid disease, with or without arthritis, is observed. In rheumatoid

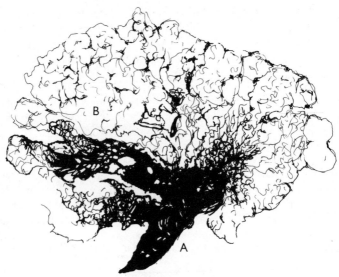

FIG. 21.1. Drawing of lung section showing progressive massive fibrosis. (A) Area of fibrosis, blackened by incorporation of coal dust in collagenous fibrous tissue; and (B) severe disruption of alveolar architecture, with advanced emphysema.

pneumoconiosis the lungs have large nodules of necrotic tissue and collagen. This disease, first described by Caplan in 1953, occurs in 2–6% of men with whatever category of pneumoconiosis (Lindars and Davies, 1967). For further reading see the description of rheumatoid pneumoconiosis by Parkes (1974).

Simple coal-workers' pneumoconiosis and progressive massive fibrosis both result from exposure to coal dust. The 'simple' condition is non-collagenous, whereas progressive massive fibrosis (PMF) is indisputably collagenous. As will be recalled from Chapter 15, the ILO classification holds that coal dust is non-fibrogenic in character and that PMF is an altered reaction of lung tissue to non-fibrogenic dust. In this view PMF results from a factor, or constellation of factors, in the host as well as from the quantity of coal dust present in the respired air, or residing in the lung.

According to Parkes (1974) four principal theories are identifiable in connection with PMF and are as follows.

(1) The Action of Silica in the Form of Quartz

From what is known about the effect of silica in provoking a disorder of repair (see Chapter 15), suspicion that silica is the agent responsible for PMF is natural. It is tempting to assume that, except when silica is present, non-fibrogenic coal dust results in simple non-collagenous pneumoconiosis, and that where silica is present the collagenous response of PMF is provoked. Unfortunately, epidemiological findings (McLintock et al., 1971) are not consistent with a quartz hypothesis because there was no evidence that exposure to respirable quartz was high in men who developed PMF. The work of Nagelschmidt et al. (1963) showed that the amount of quartz in the areas of lung affected by PMF was 'insignificantly higher' than that in the rest of the lung, and insignificantly higher than in lungs affected by simple coal-workers' pneumoconiosis. Parkes cited the observations of Policard et al., in 1967, who found that coal dust from 13 French collieries contained substances which were capable of *inhibiting* the effects of quartz on rats' lungs. These observations led them to virtually exonerate quartz from playing any part in coal-workers' pneumoconiosis.

In 1977, Walton et al. reported a study of the attack rate (defined below) of coal-workers' pneumoconiosis among 3154 face workers at 20 collieries. They observed that no generalization was possible from their data about the effects of quartz. Mass concentration of respirable dust (discussed in Chapter 3) unadjusted for composition remains, in their view, the most suitable index of risk (of coal-workers' pneumoconiosis) for use in British coal mines when the quartz content does not exceed 7.5%.

(2) Lung Burden of Coal Dust

According to this theory, there is a threshold for the accumulation of coal dust in lung beyond which the collagenous fibrogenic reaction is initiated. The work of McLintock et al. (1971) showed that the attack rate (defined as the number of cases of disease diagnosed expressed as a rate per thousand men exposed over a specified period) of PMF increased with the degree of simple pneumoconiosis, and that the rate was associated with progression of simple pneumoconiosis. Age appeared to influence the attack rate, though in a way not fully identifiable. Geographical variations were also observed in the attack rate of PMF, but these were thought largely to reflect differences in the prevalence of simple pneumoconiosis, even though other factors were unquestionably involved.

(3) Tuberculosis

The relation between coal-workers' pneumoconiosis and tuberculosis has yet to be fully evaluated. Tuberculosis, as a 'social' disease, has declined markedly in importance, owing to improvements in social conditions and in therapy. Although James, in 1954, reported tuberculosis in 40% of nearly 500 cases of coal-related fibrosis at postmortem examination, the evidence against tuberculosis as a significant factor in PMF is strong. In particular, Cochrane (1962), by means of a test based on the immune response to the tubercle bacillus, found no evidence of a difference between men with PMF and those with simple pneumoconiosis. Moreover, McCallum in 1961 (cited by Parkes, 1974), and Ball et al. (1969), showed that the use of antituberculus drugs did not prevent the progression of PMF.

(4) Immunological Factors

Immunological factors may well be involved (see, for example, Wagner, 1971). The evidence, however, is by no means conclusive.

The current view of the aetiology (cause) of simple coal-workers' pneumoconiosis is that it is related to the concentration of dust in the respired air (see later) while it is held that progressive massive fibrosis is also related to concentration of dust. Other factors may be involved in the aetiology of PMF but their role has yet, to be fully elucidated.

OTHER TYPES OF RESPIRATORY DISEASE ASSOCIATED WITH THE COAL INDUSTRY

Emphysema is rather like a ballooning of the lung tissue. Emphysema was defined by Reid (1967) as a condition of lung tissue characterized by an increase beyond the normal size of the respiratory units. It is not a disease per se, but rather a description of the state of lung tissue associated with several different forms of lung disease. For example, lung fibrosis, of whatever cause, by its shrinkage brings about distortion of the lung architecture, producing emphysematous change. Pulmonary function may not be affected by small degrees of emphysema, but its further progress may be marked by respiratory disability.

Emphysema, according to Parkes (1974), is a variable finding in simple coal-workers' pneumoconiosis. Its role in the production of disability is disputed, as is the very question of disability associated with coal-workers' pneumoconiosis.

Chronic bronchitis is a condition in which there is persistent and increasing bronchial secretion (from the goblet cells in the bronchial epithelium) of a degree sufficient to cause expectoration. The condition may be complicated by a narrowing of and eventual obstruction of the airways, in which case the term 'chronic obstructive bronchitis' is used. Cough is a further complication. As the condition advances, breathlessness increases to a disabling degree. The downward progression often follows a series of steps, each marked by a bout of acute upper respiratory illness, which is frequently initiated by a common cold. Smoking, especially cigarette smoking, atmospheric pollution, and respiratory irritants are known factors in the causation of chronic bronchitis. However, the extent to which work in the coal industry is a cause of chronic bronchitis remains the subject of debate. The study by Rogan *et al.* (1973) indicated that once early bronchitic symptoms are present, the disease may progress independently of factors initiating the disease process. This particular observation is of interest in the light of the discussion which now follows, concerned with the question of disability from respiratory disease among coal miners.

THE SOCIAL CONSEQUENCES OF COAL-WORKERS' PNEUMOCONIOSIS

The Industrial Injuries Advisory Council's Report, 'Pneumoconiosis and Byssinosis' (published, in 1973, by the Department of Health and Social Security), dealt with two principal questions: the definition of pneumoconiosis, and the account which should be taken of respiratory conditions accompanying any pneumoconiosis.

On the matter of the definition the Report recommended the following:

> [pneumoconiosis is] permanent alteration of lung structure due to the inhalation of mineral dusts and the tissue reaction of the lung to its presence but does not include bronchitis and emphysema.

This definition is narrower than that adopted in 1971 by the International Labour Office (see Chapter 15); under the ILO definition bronchitis and emphysema would not necessarily be excluded. For reasons which will be mentioned later the Industrial Injuries Advisory Council attached great importance to the exclusion of these two diseases from the compensation scheme for pneumoconiosis and byssinosis. On this issue (and certain others) there were four members of the council who supported a 'Note of Dissent' which was included with the report and which expressed views against the exclusion of the two diseases.

On the matter of disability, the Report recommended no change in the existing scheme, namely that where bronchitis and emphysema co-exist with pneumoconiosis and, overall, the disability is severe, the effects of the bronchitis and emphysema should be treated as though they are effects of the pneumoconiosis. This implied, of course, that the less severe conditions, and those without evidence of pneumoconiosis, would continue to be excluded.

The report prompted vigorous argument in medical and non-medical circles, principally about the implications of the decisions in connection with coal-workers' pneumoconiosis (even though the Report had been concerned with pneumoconiosis generally and not only that of coal workers).

In any discussion of these implications two powerful influences are involved, whether implicitly or explicitly:

(a) the political implications of any decision affecting the coal industry, and

(b) questions of compensation.

Analysis of the political importance of coal-workers' pneumoconiosis is not the writer's present concern; that it is important, however, is unquestionable, if only because of the powerful links between the coal industry and the nation's political life.

QUESTIONS OF COMPENSATION

The role of compensation in influencing the medical debate is pervasive, though perhaps not as immediately apparent as the political influence. Hugh-Jones and Fletcher, in 1951, published a Medical Research Council Memorandum on the social consequences of pneumoconiosis among coal miners in South Wales. This restrained commentary demonstrated with great clarity the results of legislative attempts to achieve a measure of social justice through compensation. Such attempts created a social problem which the authors called 'an example of the unfortunate results that can follow the most well-intended legislation'.

Prior to 1928, as Hugh-Jones and Fletcher showed, partly by economic circumstances and partly by ignorance of the nature of their condition, coal workers suffering from pneumoconiosis were forced to remain at work until they were severely affected. Workmen's compensation legislation had been revised in 1925, and three years later a scheme was instituted whereby silicosis was recognized as a compensable disease. Under this scheme, miners who had become totally disabled and who were known to have worked in silica rock could apply for compensation to the Silicosis Medical Board. So stringent were the requirements that not one application for compensation was made until 1931. In that year possible compensation also became available to the partially disabled, and thereafter (until July 1948, when the National Insurance—Industrial Injuries—Act of 1946 came into force) the scheme for coal workers underwent numerous revisions each aimed at reducing the stringency of the requisite criteria. Between January 1931 and July 1948, as a result of this reduced stringency, in South Wales alone 19 000 men were certified under the Workmen's Compensation Act (Hugh-Jones and Fletcher, 1951). All were suspended from further work in the mining industry.

The reason for these suspensions was humane enough, in that it was hoped that the suspended men would be protected from further physical deterioration. However, the social consequences of this misdirected humanity were serious, since the certified men were forced to compete for alternative employment with the normal (i.e. non-certified) labour force. For this competition they were manifestly ill-equipped—by their disability (because physical exertion left many of them abnormally breathless), by their age (many were between 40 and 60) and by geographical accident, in that the valleys of South Wales in which coal is mined are difficult of access and remote from centres of alternative employment.

Prior to 1939 few of the certified men found work, though during World War II, when labour was short, perhaps half were in employment. Those who remained unemployed were invariably the older and more severely disabled. In the latter stages of the war, and in the years immediately afterwards, certification was extended to the less severely disabled. No suitable work was available for the many men newly certified, so that by 1951 (Hugh-Jones and Fletcher, 1951) 5000 sufferers from pneumoconiosis were unemployed through suspension from work in the coal mines of South Wales. Of these probably 75% were capable of medium or light industrial employment and there is evidence that they could have given satisfactory service in jobs which were not too strenuous. In terms of their sick rate and absenteeism the men's record was no worse than that of comparable groups of non-certified men. This picture should, however, be qualified by the following statement (Hugh-Jones and Fletcher, 1951), a statement of importance in the light of subsequent debate '... since their disability tends to increase in the course of time, some men will gradually become incapable of work which they could formerly manage'.

Most men suffered considerable loss of income as a result of certification, while the psychological effects of certification were often equally harmful. In addition, coal-workers' pneumoconiosis represented a serious loss of skilled manpower to the mining industry and imposed a considerable financial burden in terms of compensation, not just in South Wales, but nationally.

Hugh-Jones and Fletcher noted that certification often caused considerable mental depression. A contemporary Social Survey found that many men were reluctant to leave the mines, that they resented their changed life-style and that they feared loss of income and the resultant insecurity. The National Joint Pneumoconiosis Committee's working party of 1948 reportedly received bitter complaints from certified men to the effect that the regulations 'took their jobs away from them'.

Many workers found to have pneumoconiosis were seriously frightened by the diagnosis. Often they had seen friends or relatives die of the disease, so that diagnosis of pneumoconiosis naturally carried the implication of fatality. Hugh-Jones and Fletcher thought that these anxieties worsened the actual disability which the men experienced.

As part of its social programme, the Labour Government of 1945–51 took the coal industry into public ownership. The newly-nationalized industry thus inherited not only dusty coal mines but also the after effects of a well-intentioned though socially-crippling compensation scheme. The Coal Nationalization Act of 1946 (effective from 1 January 1947) placed on the National Coal Board a general responsibility for the health and welfare of its employees, even though no specific recommendation was made about how this responsibility should be discharged (McLintock, 1977).* Though the dust problem had long been recognized (dust control was practised in pits before nationalization), the main impetus for that dust control was the recognition of the part played by coal dust in the propagation of coal dust explosions.

In 1959, by means of mobile X-ray units, the periodic X-ray scheme for miners was introduced. The results of this scheme, together with results from the National Coal Board's pneumoconiosis field research which linked X-ray survey data with dust measurements made in the pits (Jacobsen et al., 1971), provided the authorities with perhaps the most comprehensive data ever available for the purposes of standard setting. How these data were in fact formalized into regulations is dealt with below.

In 1936, as a result of the increasing number of cases of coal-workers' pneumoconiosis reported, the Medical Research Council had undertaken an investigation of the disease. Because it featured so prominently in South Wales this area was chosen as the centre for research. Reports dealing with chronic pulmonary disease in South Wales coal miners (Reports 243 and 244) appeared in 1942 and 1943, the latter year being that in which coal-workers' pneumoconiosis became compensable. The Medical Research Council formally established a pneumoconiosis research unit in South Wales in 1945,

---

*The Coal Mines (Respirable Dust) Regulations 1975 specifically require 'chest radiographs' as part of the duty of medical supervision imposed on mine owners; see also Appendix 21.1.

appointing Dr C. M. Fletcher its first Director. From then on pneumoconiosis research burgeoned. The result has been the creation and development of several highly specialized teams investigating respiratory disease in the coal mining industry. Thus was provided the final ingredient for the controversy surrounding the Industrial Injuries Advisory Council's Report, the powerful political pressures, the social and economic pressures, and, finally, a multiplicity of research teams.

CATEGORIES OF PNEUMOCONIOSIS

In order to follow the debate, it is necessary to outline the categorization of coal-workers' pneumoconiosis. To speak of 'categories', is however, somewhat misleading, since clear differentiation is detectable only on X-ray films. This X-ray categorization has become the predominant factor against which all others are tested.

Several systems of classification have been used, and the description which follows is a simplification. 'Simple coal-workers' pneumoconiosis' presents a picture in which the collections of coal dust cast shadows on the X-ray film. These shadows are variously referred to as 'nodules', or 'shadows'. The simple condition is divided into three categories (1, 2 and 3) according to the dimensions and populations of the nodules. For some purposes the three categories are then subdivided according to nodule size, these sub-categories being: p (pinhead), m (micronodular), and n (nodular), in increasing order of size. Progressive massive fibrosis presents shadows which are very much larger than those of simple coal-workers' pneumoconiosis. The PMF shadows are divided into three categories (A, B and C), A being the smallest and C the largest.

Even without any knowledge of the differentiation between simple coal-workers' pneumoconiosis and PMF, is obvious that this X-ray categorization gives rise to questions of an epidemiological nature; for example, is the natural history of the disease to be represented by the progression from 1/2/3 to 1/2/3 + A/B/C? If not, and if the sequence 1/2/3 represents a different disease from A/B/C, does cross-over occur from one disease to another? Does, for instance, 2 become B or C, or does A become 3? Jacobsen (1975) has discussed aspects of this question, and his study should be consulted for further discussion of epidemiological questions.

Further sets of questions can readily be generated by the addition of further variables such as expectation of life, dust exposure and age. Other questions arise from the search for relations between X-ray categories and emphysema, or chronic bronchitis, or impairment of respiratory function (as measured by the battery of tests now available for exploring respiratory physiology) or, the most important but most elusive feature of all, disablement.

Cochrane (1973) reported a 20-year follow-up study of part of South Wales, the male population of which had been first examined in 1950–51. Cochrane found that survival rates for miners and ex-miners appeared to be independent of the X-ray category of pneumoconiosis, except for category B and/or C where the survival rates were much reduced. The survey was so effective that all but 10 of the original sample of 6062 were traced. From this data Cochrane felt he could 'quite reasonably put forward the suggestion' that, other than in categories B and C, no loss of expectation of life (in 20 years) is associated with coal-workers' pneumoconiosis. He also cited a previous study carried out with his colleagues on the basis of which it was suggested that appreciable disability is associated only with categories B and C.

The Industrial Injuries Advisory Council Report (1973) concluded that it is possible to have simple coal-workers' pneumoconiosis with neither measurable disablement nor appreciable effect on life expectancy, thereby effectively accepting Cochrane's findings.

In 1974, Davies published a paper which disputed a number of the Industrial Injuries Advisory Council's conclusions. In particular he took exception to their conclusions about disability and life shortening, pointing out the major difficulties in assessing disability as well as the problem of taxonomy. Davies's thesis was that at least some of the researchers had used ventilatory (lung function) tests in such a way as to distort the true picture. His criticisms were (a) that the results of ventilatory tests, if normal, would be interpreted as proving disability to be absent, and (b) that abnormal findings would be attributed to a condition other than simple pneumoconiosis (because of the preconception that simple pneumoconiosis is not disabling).

In a letter published in the *British Medical Journal* in September 1974, B. Curry, of the Department of Economics, University College, Cardiff, criticized Cochrane's findings on the basis that survival rates were distorted as a result of the small amount of data available for the older groups, and that certain groups showed a lower survival rate in categories A and 3. He also drew attention to the possible distorting effect caused by the reduction in manpower in the coal industry since 1951.

In August 1974, W. K. C. Morgan, writing to the *British Medical Journal* from West Virginia, took issue with Davies. While conceding that British coal miners do have a decreased life expectancy, Morgan suggested it was not attributable to occupation, *per se*. His proposition was that decreased life expectancy could also be attributed to the fact that, historically, coal miners and their families had been poorly housed, had lived in polluted areas, and had been inadequately provided with medical and social facilities. These factors, in Morgan's view, also accounted for the high incidence of bronchitis contracted by British miners and, more significantly, by their wives.

Davies's article commented on the work of the Pneumoconiosis Medical Panels. These comprise medical experts responsible for diagnosing and grading pneumoconiosis for the purposes of the National Insurance (Industrial Injuries) Scheme. In August 1974, McGowan et al. (from the Department of Health and Social Security) expressed the view that cigarette smoking rather than coal dust was the major factor in the production of any chronic obstructive bronchitis in coal miners. They pointed out that no correlation between chronic bronchitis and the category of simple coal-workers' pneumoconiosis had yet been demonstrated.

In that same month (August, 1974) McLintock et al., on behalf of the National Coal Board and the daughter organization, the Institute of Occupational Medicine, expressed their support for some, if not all, of Davies's contentions. They disagreed with Cochrane's view that simple pneumoconiosis is not associated with disability, and suggested that further investigation was required in order to explore the disability question. They also emphasized the fact that an association had been established between 'X-ray pneumoconiosis' and respiratory symptoms.

In October 1974 Muir reviewed the IIAC Report for the *British Journal of Industrial Medicine.* He commented on the failure of the report to define precisely what constituted an assessable disablement. It will be recalled that the IIAC Report took the view that, in effect, the mere diagnosis of simple coal-workers' pneumoconiosis carried no implication of disability.

Chronologically, the next important aspect of the debate came in November 1974 with Davies's reply to the various comments of McGowan, McLintock and others. In particular, Davies pointed out the need to avoid use of the word 'bronchitis' to describe symptoms of cough and sputum. 'Bronchitis' in Davies's view, constitutes a specific disease, whereas cough and sputum are but symptoms common to many respiratory diseases. This, in fact, is a crucial point; the Industrial Injuries Advisory Council in its inquiry and Report having specifically *excluded* bronchitis (and also emphysema) in the absence of pneumoconiosis.

The reasons for excluding bronchitis are plain enough, in that the disease is still widespread in Britain and, because of the multi-causal aetiology, it is difficult to attribute in individual cases to single factors—whether it be cigarette-smoking, general atmospheric pollution, or exposure to irritants or other harmful inputs at work. Understandably, the IIAC has always avoided accepting chronic bronchitis as an occupational disease, no doubt largely from fear of precipitating a rush of additional disablement claims, related to non-occupational conditions. Disablement benefit, in the IIAC's view, is obtainable only by persons suffering from occupational disease and not from disease caused by the general, non-occupational milieux. Given this viewpoint, it is easy to see the logic of Davies's arguments that, in effect, chronic bronchitis, disablement and simple pneumoconiosis were being artificially separated at the stage of diagnosis. On the other hand, Cochrane's report (see above) is also very convincing, notwithstanding Curry's criticism about the small number of data for the older age groups.

AGREEMENT ON COMPENSATION

In September 1974 an agreement was drawn up between the National Coal Board and the mining unions. This provided benefit (financial compensation), as an alternative to common law damages, for men certified to be disabled by pneumoconiosis or for the dependants of men who have died from the disease. Although the agreement involved the National Coal Board and the mining unions only, the Government contributed £100 M to help meet the cost of benefit payable to men certified to be suffering from the disease prior to 1 October 1974. McLintock (1977) noted that in 1973–75 the total of certifications by the pneumoconiosis panel had increased, from 2 per 1000 in 1973, to 2.2 per 1000 in 1974 and to 2.8 per 1000 in 1975. As McLintock commented: 'there can be little doubt that this represents a continuing interest in pneumoconiosis by miners and more particularly ex-miners resulting from the agreement between the Government, Board and Unions.'

DUST STANDARDS FOR BRITISH COAL MINES

The Regulations paraphrased in Appendix 21.1 meet the requirements (the 'three methods of approach') outlined by Walton (1969), namely
  (1) to ensure that the dust levels nowhere exceed a specified maximum,
  (2) to control employment between dusty and less dusty places of work so that, for each man, the overall average exposure does not exceed the 'safe' limit, and/or
  (3) to subject men to regular medical examination (biological

monitoring) so that susceptible individuals showing early symptoms of disease can be detected and warned of the risk. They (the Regulations) also marked the culmination of nearly two decades of increasing concern, debate and collection of data (i.e. since the conception in 1951 of the comprehensive surveillance scheme and the recommendation of the (British) Medical Research Council, in 1959, that the most suitable method of measuring dust concentrations was by aerodynamic selection, as in the lungs, of the respirable mass of dust).

Efforts to reduce airborne dust had been made long before 1959, though, because of the long latent period of dust associated diseases, and because much of the dust control was directed at explosion risk the substantial progress made in dust control was not at that time truly reflected in the official statistics. That substantial progress had in fact been made was never in dispute. However, there is an interesting comparison to be made, for example, with South Africa, where legislation aimed at prevention of silicosis was first enacted in 1911 and where, in the gold mines, an estimated six-fold reduction in the levels of airborne dust was recorded between the years 1917 and 1937. The reductions recorded in British mines since 1945 have been less dramatic though by no means unsatisfactory, the number of newly-recorded cases of pneumoconiosis, for example, having fallen from an average of 6.05 per thousand men employed in 1954–58 to 1.86 in 1967 (Walton, 1969). These data should be compared with the trend in certifications commented upon by McLintock (1977), mentioned previously.

The mining of any brittle material inevitably results in production of airborne dust, which, in coal mining, constitutes perhaps only 1% of the dust total. *Removal* of airborne dust from a working is achieved by exhaust ventilation, or (less efficiently) *suppressed* by 'damping down' with a water spray. Far more important, however, is dust *prevention*, (likely to be achieved by strict observance of the Regulations) by improved (i.e. less dusty) operating methods and machinery. The coal industry, however, remains torn between the sometimes conflicting considerations of safety and productivity, in that increased output and increased dust concentration are both related to increased mechanization.

The need 'to measure the prevailing dust concentrations and to specify "standards" for the maximum possible levels' (outlined by Walton, 1969) has been largely met by the 1975 Regulations, but the multiplicity of standards still used internationally does hinder co-operative research, in that it makes comparability of results difficult. Different methods of measurement inevitably produce different assessments, though assessment is in itself complicated by the fact that, even within the same working, the dust concentration levels are subject to considerable daily, even hourly, variation. For many years research progress was hindered by the difficulty of isolating for weighing particles of respirable dust less than 5 μm in diameter. This difficulty has to some extent been overcome by improvement in sampling methods, such as those by which particles are measured according to their 'falling speeds'. Even this method is not entirely satisfactory, however, in that equal 'fall speed' does not necessarily mean equal particle size (see Chapter 3).*

Constant improvement of sampling and measuring apparatus is essential to the success of efforts to prevent, remove, or suppress dust. Over the 25 years various instruments have been produced, both in prototype and commercially, in order to meet the dust sampling requirements of the (British) Medical Research Council. During the early days of the pneumoconiosis field research, standard environmental sampling was based upon the standard thermal precipitator (STP). The hexhlet, the first horizontal elutriator introduced around 1954, proved generally unsuitable for use in British coal mines, so other instruments were developed. By the mid-1960s two portable dust samplers were generally available—the NCB Mining Research Establishment's Gravimetric Dust Sampler type 113A and the Safety in Mines Research Establishment's 'Simgard'. With the former some 4000 shifts were sampled in the period 1965–71. In the late 1960s experiments began with a personal dust sampler mounted on the miner's cap lamp, while by 1970, with the approval of the National Joint Pneumoconiosis Committee, efficient gravimetric samplers (like type 113A) became standard, and routine microscopical (particle) counting was discontinued. About this time the Coulter Counter also began to be generally used as a means of size characterization of the gravimetric respirable dust samples.

Not surprisingly, coal material is the major constituent of airborne dust in coal mines, though the nature of the dust varies greatly according to the nature of the parent rock and the nature of the coal mined. Both concentration and composition of dust may be factors which interact.

The mineral composition of the samples measured by the 113A sampler showed that dust of a type normally found during face work contains quite sizable amounts of kaolin (11% average), mica (8%) and quartz (4–5%) (traceable to the non-coal strata), as well as ash and lesser quantities of calcium, magnesium and other carbonates (Dodgson *et al.*, 1971). Figures of course vary greatly, often between different faces in a single seam, according to the ash content of the

---

*Detailed experiments into particle characteristics and methods of measuring total dust concentration are described in Dodgson *et al.* (1971), p. 774 *et seq.*

samples, the size distribution of the particles and the carbon content of the coal. The single most important factor in producing such variation is mining method. Power loading, for example, is much dustier than hand-getting.

Another reason for statistical variation in dust assessment is the fact that the relation between pneumoconiosis (and other 'dust diseases') and the milieu varies according to locality and circumstances. This relation was examined in some detail by the Conference on Technical Measures of Dust Prevention and Suppression in Mines, held in Luxemburg in October 1972. For the purposes of his 'General Report' on the subject, Walton took 'cumulative dust exposure' as the primary environmental factor. Other factors, each subject to great variation, include time or concentration of exposure to dust, and individual susceptibility to disease. This latter factor is of especial importance, in that until exhaustive research has been undertaken it is impossible to disprove the seemingly unlikely suggestion that 'radiological pneumoconiosis' especially PMF (see p. 111) strikes randomly. Far more probable and scientific is the causal thesis; that 'pneumoconiosis' can be produced by certain environmental or biological causes as yet unidentified, apart from causes already identified. Since 'individual susceptibility' is really synomymous with 'individual response' it is essential, if all the causes of pneumoconiosis are to be known, that researchers should identify completely the factors (= inputs) to which individuals respond. Since these factors are so complicated and multifarious (for example, whether the individual 'nose breathes' or 'mouth breathes') it is an almost impossible task. Obviously, it is highly probable that susceptibilities differ according to variations in tissue response and to the varying ability of bodies to clear dust from the lungs. As Walton pointed out, 'the pre-employment detection of individuals who are abnormally susceptible to dust ... would greatly help to bring pneumoconiosis under control'. Such control would, however, have features of a 'safe-person' strategy, as discussed in Chapter 24.

It might equally be argued that further attention to factors of superficially peripheral importance to mining could help control pneumoconiosis. Among factors other than dust which might have a bearing on the incidence and severity of the disease are such inputs as climatic conditions and cigarette-smoking, both of which are of course recognized as contributory causes of chronic bronchitis.

NUMERICAL STANDARDS

The following paragraphs provide a brief outline of a complex subject. In September 1975 the Coal Mines (Respirable Dust) Regulations were introduced under the Health and Safety at Work Act 1975. These regulations specified 'permitted' dust standards in quantitative terms (see below). Prior to these Regulations there had been a requirement under the Mines and Quarries Act (S74) that 'dust of such character and in such quantity as to be likely to be injurious to the persons employed' should be minimized. This requirement had largely been met by the adoption, in about 1970, of voluntary standards based on gravimetric methods, standards agreed upon by the National Coal Board, the mining trade unions, and the Mines and Quarries Inspectorate (Jacobsen, 1977). Mention of 'dust standards' in the coal-mining industry prior to 1975 is to these *agreed standards*, whereas post-1975 references presumably are to the statutory standards, even though this is not always explicitly stated.

Chamberlain *et al.* (1971) gave the data relating to the agreed standards; these are presented in Table 21.1 and should be compared

TABLE 21.1

STATUTORY AND AGREED STANDARDS FOR DUST IN COAL MINES. COMPARISON OF STANDARDS LAID DOWN IN THE COAL MINES (RESPIRABLE DUST) REGULATIONS 1975 WITH THE AGREED STANDARDS DESCRIBED BY CHAMBERLAIN *et al.* (1971); DATA IN MG/M³.

| Work place | Statutorily permitted amounts by 1975 Regulations | | Agreed standards |
	Regulation* 9(1)	Regulation* 9(2)	
Longwall face (LWF)	8		8
†LRDI > 12		12	
LRDI < 12		8	
Drivages (D)			3
Quartz > 0.45	3		
LRDI > 9		9	
LRDI < 9		6	
Other operations (0)	6		
LRDI > 9		9	
LRDI < 9		6	

*See Appendix 21.1 for an explanation of these Regulations.
†LRDI means last *respirable dust index*; *respirable dust* means dust which may be retained in the respiratory system, *respirable dust index* means (*inter alia*) the average respirable dust content of samples; and *respirable dust content* means the weight in milligrammes present in each cubic metre of air where the sampling device retains all dust with aerodynamic diameter not exceeding 1 μm (EUDS), half of the dust of aerodynamic diameter 5 μm and none in excess of 7 μm.

TABLE 21.2
WORKING PLACES (COAL MINES) MEETING DUST STANDARDS (TAKEN FROM McLINTOCK, 1977)

Working places	March 1973		March 1975*		March 1976	
	Number	Percentage meeting standard	Number	Percentage meeting standard	Number	Percentage meeting standard
Longwall coal faces						
Production shift	904	92.3	787	93.2	775	98.7
Preparation shift	172	97.1	98	100.0	86	100.0
Cutting shift	70	95.7	53	94.4	29	100.0
Board and pillar operations	—	—	—	—	63	100.0
Drivages						
In coal	499	93.4	444	96.2	819	95.97†
In stone	296	62.5	392	70.7		
Intake roadways to coal faces	—	—	—	—	775	100.0
Transfer points						
In return airways	138	98.6	142	100.0	183	100.0
Loading points						
In return airways	16	100.0	15	100.0	12	100.0

*Because of industrial action, figures for March 1974 are not comparable and were omitted by McLintock.
†As from September 1975 the arbitrary distinction which had been made between stone and coal drivages was dropped.

with the standards required under the 1975 Regulations. An outline explanation of the requirements of the 1975 Regulations is given in Appendix 21.1, but for the purposes of discussion of Table 21.1 it should be pointed out that Regulation 9(2) is crucial in its requirement

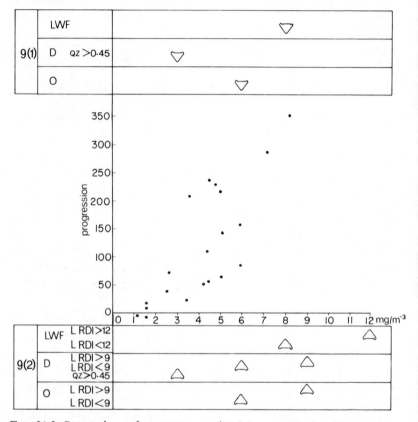

FIG. 21.2 Comparison of statutory permitted dust standards with data reported by Jacobsen *et al.* (1971) showing the relation between a certain index of progression of coal-workers' pneumoconiosis and the mean dust concentrations in 20 collieries. The index of progression was obtained from calculations based on X-ray findings of coal miners' chests. Firstly, the X-ray classification scheme was elaborated to give a 12-point scale. Serial X-rays were then placed on the scale by a team of X-ray film readers. From these data steps of progression were obtained, these steps then being expressed as a rate per million shifts worked at the coal face. The mean dust concentrations correlated well with coal-face exposure (r = 0.94). For an explanation of the data in the left-hand panels see Table 21.1.

that work must cease where the stipulated levels are exceeded. It can be seen from the table that as far as the dustiest longwall faces are concerned the Regulations represent a relaxation of the agreed standard (from 8 to 12 mg/m^3). Relaxation has also been applied to operations in drivages, except where the quartz content exceeds 0.45 of the total dust. (Doubtless, the statutory permitted amounts will be lowered in due course.) McLintock (1977) reported the proportion of working places meeting what are termed the current dust standards (Table 21.2), a proportion universally high. At longwall faces there was a greater improvement in meeting the standard between March 1975 and March 1976 than in the period March 1973 to March 1975 (averages 3.7% and 0.83% respectively), an improvement presumably attributable, at least in part, to relaxation of the standards for the dustiest longwall faces.

Figure 21.2 presents a comparison between data on progression of coal-workers' pneumoconiosis reported by Jacobsen *et al.* (1971) and the statutory permitted dust standards. The data of Jacobson *et al.* relate to 4122 men and come from the comprehensive surveillance scheme begun, as mentioned previously, in 1959. Progression was judged by side-by-side comparison/contrast between the X-ray films of 4122 men, the films being repeated after an interval of up to 10 years (see also the legend to Fig. 21.1). From other data given by Jacobsen *et al.* it can be shown that the probability is about 0.4 of an individual progressing from the 0 to the 2 category of pneumoconiosis with 35 years of exposure to a mean dust concentration of 12 mg/m^3.

CONCLUSION

An overall conclusion which can be drawn from this account of respiratory disease in the coal mining industry is that in occupational health and safety there is a highly complex interplay between science, medicine, and social issues.

APPENDIX 21.1: A PARAPHRASE OF EXTRACTS FROM THE COAL MINES (RESPIRABLE DUST) REGULATIONS 1975 (SI 1975 No 14331)

These controls were enacted in order to check the amounts of respirable dust present in coal mines. They apply to all mines at which more than 30 coal-face workers are employed.

The Regulations direct that samples of the air below ground should be taken and examined by competent persons and that the results should be tabulated in such a way as to constitute an easily consulted record. In addition, the Regulations require, unless respirators are worn, that work should cease when levels of respirable dust rise above levels considered safe. Further, it is required that respirators be within reach, that competent medical supervision be provided, and that every effort be made to prevent and control dust.

Comprehensive definitions are included in the Regulations; for example, 'respirable dust content' is defined with scientific precision (Regulations, *Part I*, 'Interpretation').

Each sample, according to the general provisions outlined in *Part II* of the Regulations, must be taken under normal working conditions during a normal working shift below ground. This is of obvious importance if samples are to be taken in conditions which approximate to the norm. Samples must be representative of the respirable air at the place of work.

In relation to the taking and examining of samples the Regulations are explicit: competence is required of all persons supervising or actually taking samples. Sampling equipment must be of an approved type, while the laboratories, in which the samples are examined and respirable dust content and quartz content are recorded, must similarly be approved by the Health and Safety Executive (HSE). The manner in which these records are kept and indexed is also open to scrutiny by the HSE. It is a further condition of the Regulations that the mine owner should notify the mine manager of the results of tests within four working days of the sample being taken. The manager is required to display these results for a period of thirty days after notification.

The Regulations leave to the mine owner the choice of which scheme shall be adopted for the taking of samples, though in relation to the chosen scheme he must specify what equipment is to be used, and by what means it is to be used. In addition, lest equipment be improperly maintained, details must be provided of how samples are tested.

If the dust exceeds the permitted level, it is the manager's duty, within three working days, to inform the HSE of what measures he has adopted, or intends to adopt, to reduce the level. In the meantime all work in the 'unsafe' area must cease. It remains the manager's responsibility, throughout that and the succeeding month, to ensure that work is discontinued immediately in areas where the level is above that permitted. Only after the level has been successfully reduced, and after the district inspector has been informed of this, can work recommence—usually seven working days after notice has been given. Work may recommence earlier, but only with the consent of the HSE (Mines and Quarries Inspector).

The manager must ensure that the level of respirable dust in the affected working(s) does not exceed that permitted throughout the three months following the giving of notice that levels are 'safe'. Each mine manager must also ensure the provision, and supervise the maintenance, of a sufficient quantity of approved and efficient dust respirators.

Part III of the Regulations, entitled 'Medical Supervision', refers to the mine owner's responsibility for providing adequate medical supervision. This supervision, which applies to all employees affected or liable to be affected by the inhalation of dust, is to be provided without charge and is to include the taking of chest X-rays.

It is the duty of each mine manager, according to the requirements of *Part IV* of the Regulations, that apparatus to reduce and suppress dust be always provided and specified, and that it should be systematically tested and properly maintained. The scheme for prevention of dust must be comprehensive and be supervised by a person of competence appointed by the mine manager, on whom falls the added responsibility of keeping an accurate record of results.

The final part of the Regulations, *Part V*, 'Miscellaneous Provisions', ties up any possible loose ends. Other than in an emergency, (it reads) mine equipment can only be touched by persons who are expressly permitted to do so by the Regulations or who have obtained prior written instruction from the manager. Control over personnel using mine machinery is entirely the manager's responsibility, as is the provision of dust prevention apparatus. It is further provided that the approved dust prevention scheme, or a copy of the scheme, should be kept at the mine office or other approved place, and that a copy should be dispatched to the district inspector. The duties of the mine owner and manager in relation to enforcement of the scheme, appointment of competent supervisors, and recording of results of tests are again and further outlined.

Schedule 1 (Regulation 8), 'Sampling: further provisions', and Schedule 2 (Regulation 9) 'Permitted amounts of respirable dust index' are, as their titles suggest, but further extensions of the requirements laid down previously (see Table 21.1 and Fig. 21.2). These sections, comprising definitions of comparability of operations and regulations concerning frequency of sampling and permitted levels of respirable dust, are too complex and of too statistical a nature to permit a reasonable paraphrase. The reader, should he/she wish or need to know all the details contained therein, is advised to examine the Schedules (which are appended to the Regulations) at first hand. Moreover, there is the prospect that numerical values set out in the Regulations will be revised periodically.

REFERENCES

BALL, J. D., BERRY, G., CLARKE, W. G., GILSON, J. C. and THOMAS, J. (1969). A controlled trial of anti-tuberculosis chemotherapy in early complicated pneumoconiosis of coal workers, *Thorax*, **24**, 399–406.

BRIGGS, A. (1959). *The Age of Improvement* 1783–1867, London: Longman.

CAPLAN, A. (1953). Certain unusual radiological appearances in the chest of coal miners suffering rheumatoid arthritis, *Thorax*, **8**, 29–37.

CHAMBERLAIN, E. A. C., MAKOWER, A. D. and WALTON, W. H. (1971). New gravimetric dust standards and sampling procedures for British coal mines, *Inhaled Particles III*, W. H. Walton, ed., Old Woking: Unwin.

COCHRANE, A. L. (1962). The attack rate of progressive massive fibrosis, *British Journal of Industrial Medicine*, **19**, 52–64.

COCHRANE, A. L. (1973). Relation between radiographic categories of coal workers' pneumoconiosis and expectation of life, *British Medical Journal*, **2**, 532–4.

CURRY, B. (1974). Survival in coal workers' pneumoconiosis, *British Medical Journal*, **3**, correspondence, 633–4.

DAVIES, D. (1974). Disability and coal workers' pneumoconiosis, *British Medical Journal*, **2**, 652–5.

DAVIES, D. (1974). Disability and coal workers' pneumoconiosis, *British Medical Journal*, **4**, correspondence, 289–90.

DEPARTMENT OF HEALTH AND SOCIAL SECURITY (1973). *Pneumoconiosis and Byssinosis*, Report by the Industrial Injuries Advisory Council in accordance with Section 62 of the National Insurance (Industrial Injuries) Act, 1965 on Pneumoconiosis and Byssinosis, Cmnd. 5443, London: HMSO.

DODGSON, J., HADDEN, G. G., JONES, C. O. and WALTON, W. H. (1971). Characteristics of the airborne dust in British coal mines, in *Inhaled Particles III*, W. H. Walton, ed., Old Woking: Unwin.

FOOT, M. (1962). *Aneurin Bevan: A Biography*, Vol. 1, London: Macgibbon and Kee.

HUGH-JONES, P. and FLETCHER, C. M. (1951). The social consequences of pneumoconiosis among coalminers in South Wales, *Medical Research Council Memorandum No. 25*, London: HMSO.

JACOBSEN, M., RAE, S., WALTON, W. H. and ROGAN, J. M. (1971). The relation between pneumoconiosis and dust-exposure in British coal mines, in *Inhaled Particles III*, W. H. Walton, ed., Old Woking: Unwin.

JACOBSEN, M. (1975). Effects of some approximations in analyses of radiological response to coal mine dust exposure, in *Recent Advances in the Assessment of the Health Effective Environmental Pollution*, Luxemburg: Commission of the European Communities.

JACOBSEN, M. (1977). 'Scientists and Safety', Letter to the Editor, *New Scientist*, 7 July.

JAMES, W. R. L. (1954). The relationship of tuberculosis to the development of massive pneumoconiosis in wool workers, *British Journal of Tuberculosis*, **48**, 89–101.

LINDARS, D. C. and DAVIES, D. (1967). Rheumatoid pneumoconiosis. A study in colliery populations in the East Midlands coalfield, *Thorax*, **22**, 525–32.

McGOWAN, R. M., PARKES, W. R., PHILLIPS, T. J. G. and WILLIAMSON, R. G. B. (1974). Disability and coal workers' pneumoconiosis, *British Medical Journal*, **3**, correspondence, 521–2.

McLINTOCK, J. S. (1976). *Annual Report 1974–75 National Coal Board Medical Service*, London: National Coal Board.

McLINTOCK, J. S. (1977). *Annual Report 1975–76 National Coal Board Medical Service*, London: National Coal Board.

McLINTOCK, J. S., MUIR, D. C. F., JACOBSEN, M. and WALTON, W. H. (1974). Disability and coal workers' pneumoconiosis, *British Medical Journal*, **3**, correspondence, 343–4.

McLINTOCK, J. S., RAE, S. and JACOBSEN, M. (1971). The attack rate of progressive massive fibrosis in British coalminers, in *Inhaled Particles III*, W. H. Walton, ed., Old Woking: Unwin.

MORGAN, W. K. C. (1974). Disability and coal workers' pneumoconiosis, *British Medical Journal*, **3**, correspondence, 343.

MUIR, D. C. F. (1974). Pneumoconiosis and byssinosis, *British Journal of Industrial Medicine*, **31**, 322–5.

NAGELSCHMIDT, G., RIVERS, D., KING, E. J. and TREVALLA, W. (1963). Dust and collagen content of lungs of coal workers with progressive massive fibrosis, *British Journal of Industrial Medicine*, **20**, 181–91.

PARKES, W. R. (1974). *Occupational Lung Disorders*, London: Butterworth.

REID, L. (1967). *The Pathology of Emphysema*, London: Lloyd-Luke.

ROGAN, J. M., ATTFIELD, M. D., JACOBSEN, M., RAE, S., WALKER, D. D. and WALTON, W. H. (1973). Role of dust in the working environment in development of chronic bronchitis in British coalminers, *British Journal of Industrial Medicine*, **30**, 217–26.

WAGNER, J. C. (1971). Immunological factors in coal workers' pneumoconiosis, in *Inhaled Particles III*, W. H. Walton, ed., Old Woking: Unwin.

WALTON, W. H. (1969). Surveillance of airborne dust concentrations, in *Proceedings of 9th Commonwealth Mining and Metallurgical Congress*, London 1969: Offices of the Congress under the Institute of Mining and Metallurgy.

WALTON, W. H. (1973). Pneumoconiosis and environmental factors, in *Proceedings of the Conference on Technical Measures and Dust Prevention and Suppressions in Mines*, Luxemburg: Commission of the European Communities.

WALTON, W. H., DODGSON, J., HADDEN, G. G. and JACOBSEN, M. (1977). The effect of quartz and other non-coal dusts in coal workers' pneumonoconiosis, in *Inhaled Particles IV*, W. H. Walton, ed., Oxford: Pergamon.

WOOD, A. (1960). *Nineteenth-Century Britain*, London: Longman.

Chapter 22

Gassing Accidents

OBJECTIVES OF THE CHAPTER

This chapter provides:
  (1) complementary systems of classification for gassing accidents linked to the mode of action of certain gases, and circumstances of gassing accidents,
  (2) an outline explanation of certain biological processes underlying the mode of action involved in gassing accidents,
  (3) a discussion on the nature of 'nitrous fumes' and a discussion of the question whether these are associated with chronic as well as acute respiratory disease,
  (4) an explanation in biological terms of the first-aid treatment, and
  (5) 12 outline case histories of gassing accidents illustrative of biological processes and circumstances under which gassing accidents take place.

INTRODUCTION

Under this general heading are included accidents which result in exposure to gases, vapours, or fumes. Whether transitory or prolonged, such exposure can be physically harmful and may cause acute or chronic ill-health, or even death.

For the purposes of this study 'anoxia', or oxygen deficiency, is a more apposite description of the effects of gassing than is 'asphyxia', a term commonly associated with suicide or murder by strangulation. However, by no means all gassings are caused by deficiency of oxygen. The physiological results of exposure vary, the properties of the substance to which the victim is exposed and the mode of action being two determining factors, and the circumstances of the gassing being another. This complexity of cause and effect is best illustrated by the classification of gassing accidents according to (a) their natural history and/or (b) their physical nature.

Although this chapter is primarily concerned with the classification outlined above, instances of gassing accidents examined by HM Factory Inspectorate are appended. It is hoped that the reader will be prompted to categorize these examples according to the suggested classification and consider how the knowledge gleaned from the findings could be applied to the prevention of similar recurrences. A broad framework for such prevention will be considered in a later chapter so that the examples presented are not only instances of various types of gassing accident but also background to the evolution of safety strategies. Hitherto, the subject matter of this study has been concentrated on diseases caused by comparatively lengthy exposure. Gassing accidents represent diseases in which acute and chronic effects, as well as death, result from brief exposure; these also have to be provided for in strategies for prevention.

CLASSIFICATION ACCORDING TO THE NATURE
OF THE ACCIDENT

Several sub-categories are immediately identifiable:

(a) *The unexpected* (HM Factory Inspectorate's 'surprise gassing'), where the interaction of two individually harmless substances results in the emission of a noxious gas, for example acids and hypochlorites produce chlorine; nitric acid, certain organic materials and copper produce oxides of nitrogen (discussed below); acids and sulphides produce hydrogen sulphide.

(b) *The confined area*; a sub-category into which a lamentably large number of gassing accidents fall. As the following account illustrates, all too often such accidents result in multiple deaths, in that co-workers succumb while attempting to rescue colleagues. Under the general heading 'Died when they didn't smell danger', the journal *Industrial Safety* (1974, January p. 4) reported the inquest into the deaths of Frederick Charles Whiting (43) and Basil Grace (32), workers at the Exchange Brewery in Bridge Street, Sheffield. The jury, which returned verdicts of accidental death, heard how the two men's over-reliance on their 'experience' and sense of smell precipitated the tragedy. While clearing out a giant vat both had been overcome by carbon dioxide fumes, their 'experience' having failed to warn them that the fumes had reached a dangerous level. A third co-worker, himself overcome and taken to hospital, told the court how the younger man (Mr Grace) had gone to the assistance of his colleague and thereby sacrificed his own life. In fact both men had needlessly lost their lives since the necessary safety equipment (testing and breathing apparatus, and safety harness) was on hand but was 'rarely used', such was the men's faith in their own abilities of fume detection. The coroner, Dr Herbert Pilling, remarked that the case was an obvious example of familiarity having bred contempt.

However accurate a press report, for our purposes it is hardly an authoritative primary source. The reported 'findings' therefore require amplification and further interpretation. The following list should provide this, as well as informing the reader of other factors contributory to gassing accidents and the circumstances under which these occur.

(i) The slow build-up (insidiousness) of odourless gases, like carbon dioxide or carbon monoxide, is often deceptive and causes the unsuspecting to underestimate the dangers.

(ii) If the constituents of the contaminant can be ascertained, 'atmospheric' testing should present no complications. Where the contaminating gas is unidentified or where the possibility of oxygen deficiency has not been considered, testing and evaluation of results, especially by non-specialist personnel, can be difficult and sometimes misleading.

(iii) An important factor is that work people may be unaware of the nature of potential damage from gassing.

(iv) Respiratory protection may be imperfect through defect or be unsuitable for the purposes to which it is being put. For example, canister respirators rapidly become saturated when called upon to cope with large quantities of noxious gas.

(v) Gassings are often caused by process failures, or by a variety of accidental bursts, leaks or spillages.

(vi) Certain substances, notably plastics, undergo chemical breakdown under conditions of extreme heat. Breakdown is accompanied by the giving off of gases such as hydrogen chloride and hydrogen cyanide, or, where fire is present, carbon monoxide (Bowes, 1974). Death by gassing is thus a frequent adjunct to fire.

CLASSIFICATION ACCORDING TO THE PHYSICAL EFFECT(S) OF THE GASSING

Under this heading are included the following sub-categories of gassing: (i) those resulting in anoxia, (ii) those connected with irritation of the respiratory tract, (iii) those which promote arousal of the central nervous system (CNS) or, more often, a general reduction in CNS activity, (iv) those which assume the distinctive features of the fume fevers, and (v) those miscellaneous conditions which fall into none of these sub-categories, or those which are the consequence of disparate or combined physical effects and which cannot be simply classified.

(i) Gassings Resulting in Anoxia

This sub-category can be further sub-divided into (a) simple anoxia, and (b) toxic anoxia.

(a) *Simple Anoxia*
This is a term descriptive of oxygen deficiency in inhaled air, whatever the cause. Simple anoxia gassings frequently follow the same pattern, namely displacement of oxygen by gases like methane, helium, carbon dioxide, nitrogen, and hydrogen, all of which are comparatively inert.

Simple anoxia gassings are not uncommon, a particularly severe example being that in Abilene, Kansas, reported by the *Guardian* of December 1974 under the heading 'Killer Leak'. The report carried details of the deaths of 6 workmen engaged in the repair of a pipeline leak. All had been rendered unconscious by methane gas and had drowned in crude oil which had escaped from the pipe. These bare facts suggest that methane given off from the oil, though lighter than air, had been of sufficient quantity to displace the oxygen and, despite its biological inactivity, had 'poisoned' the men.

(b) *Toxic Anoxia*
This is a term used to describe the condition in which parts of the body's oxygen transport or utilization systems are harmed by the adverse action of biologically-active substances. The impairment of oxygen transport by carbon monoxide poisoning, which causes stable carboxyhaemoglobin to preferentially form instead of oxyhaemoglobin in the erythrocytes (red blood corpuscles), is an important example of toxic anoxia. So too is the harm occassioned by arsine poisoning, which similarly impairs oxygen transport through haemolytic destruction of the erythrocytes. It should be noted that the haemolytic destruction causes the kidneys to be overburdened with part-destroyed haemoglobin, often with fatal consequences. Hydrogen cyanide gas impairs oxygen utilization in tissue. Impairment of oxygen utilization in the central nervous system eventually leading to respiratory paralysis results from inhalation of hydrogen sulphide.

(ii) Gassings Connected with Irritation of the Respiratory Tract

As has been emphasized hitherto, the organ (or organs) most vulnerable to a particular input is known as the target organ. In the case of gassings, the site of action in the target organ is partially decided by both the aggressiveness and solubility of the gas(es).

'Deep' gases—a term which can be used to describe gases such as ozone, phosgene, and oxides of nitrogen (see below)—penetrate the conducting airways and sometimes give rise to delayed pulmonary oedema. The membranes of the respiratory units, expecially alveoli, suffer increase in permeability owing to the effects of the irritant gas, causing fluid to accumulate in the respiratory units. That the oedema is delayed is a typical feature of inhalation of 'deep' gases and one which makes it all the more dangerous. The delay reflects the relatively slow processes of permeability increase and fluid accumulation.

Included among these 'deep' gases are those which can be 'aggressive' such as chlorine and ammonia, but under the general heading 'respiratory irritants' should also be included the acidic mists of, for example, hydrochloric acid.

(iii) Gassings which Upset the Equilibrium of the Central Nervous System

Numerous volatile fat-soluble hydrocarbons, especially the halogenated hydrocarbons, have a narcotic effect on the nervous system. The fat-solubility of the hydrocarbon compound(s) is an important feature of such gassings, since nervous tissue possesses a high proportion of fat. The narcotic effect varies not only according to the quantity of substance inhaled but also according to its concentration within the cells of the target organ (Gerarde, 1960).

At this point it is appropriate to refer to Chapter 11 in which bio-dumping of substances is discussed. Recovery from narcosis involves the transportation and excretion via the respiratory system of volatile substances, accompanied by detoxication (see Chapter 8). Where there is repeated exposure or, possibly, short-term excessive exposure, chronic lesions may be induced. In the liver, for example, cell degeneration may be followed ultimately by fibrosis.

Chapter 19 makes reference to the dangers of phosphine (phosphorus trihydride) with its capacity for causing convulsions, a gas which should be included under this heading.

(iv) Gassings which Assume the Distinctive Features of Fume Fevers

Reference is made in Chapter 19 to the metal fume fevers, to which should be added the closely related 'polymer fume fevers' (see appended case study ix), the characteristics of which were first set out by Harris in 1951 in reference to fluorine-comprising polymers. The study reported by Malten and Zielhuis in 1964 helped further identify the dangers inherent in polymer fume fever, fumes being given off at 400–450°C, at which temperature polymer products begin to disintegrate.

Fume fevers resulting from exposure to fumes are not always classified as gassing accidents because the part played by the fumes may not be recognized either by the victim or by the doctor who gives treatment.

(v) Gassings which are the Consequence of Combined Toxic Effects

Under this heading are included gassing accidents the result of which are manifold. Gassing from nickel carbonyl (described by Hunter, 1975) is an appropriate example, in that the consequences are two-fold, namely (i) extreme irritation of the respiratory system caused by the deposit of particles of metallic nickel, and (ii) formation of carboxyhaemoglobin from the separated carbon monoxide. Benzene (C_6H_6) (synonym benzol(e), not to be confused with benzine, which is a hydrocarbon *mixture*) is an aromatic liquid hydrocarbon of high-lipid solubility, a characteristic which facilitates narcosis. The narcotic effects of benzene, however, pale beside its effect on the blood-forming system: severe bone-marrow aplasia (discussed by Browning, 1965). The pathogenic effects of vinyl chloride inhalation are now considered manifold (see Chapter 23), though until recently narcosis was considered its only consequence. As has been mentioned, carbon monoxide poisoning is now thought to have chronic as well as acute ill-effects.

'NITROUS FUMES'—THEIR NATURE AND EFFECT

For the purposes of the National Insurance (Industrial Injuries) Act, 1965 'poisoning by nitrous fumes' is Prescribed Disease 17 which is recognized for any occupation involving the use or handling of nitric acid or exposure to nitrous fumes. The nature of the disease recognized for the purposes of prescription is described in *Notes on the Diagnosis of Occupational Diseases* (Department of Health and Social Security, 1972). In the Notes it is pointed out that 'nitrous fumes' (their nature is discussed below) are *the most insidious of all irritant gases*. (This is why they have been picked out for discussion here.) Upper respiratory tract irritation tends to be slight, although exposure may be accompanied by a cough. In many cases exposure is followed by a latent period from 2 to 20 h during which the individual may continue with his work and return home 'as if nothing were amiss'. After the latent period there may be sudden onset of tightness in the chest, breathlessness, and cough with clear, frothy phlegm which may develop blood staining. The breathlessness may become very marked and there may also be cyanosis (this describes a blue tinge of the finger nails, lips and ear lobes which is seen when the haemoglobin circulating in red cells is incompletely saturated with oxygen). Apart from respiratory symptoms there may be nausea, pain in the abdomen with vomiting, and a rise in body temperature. The acute inflammation of the smallest conducting airways and respiratory units results in pulmonary oedema, which obstructs oxygen exchange. Not surprisingly, the medical examination may lead to the diagnosis of pneumonia, unless the doctor elicits a history of exposure to 'nitrous fumes'.

Anoxia is undoubtedly the cause of death in a number of cases. In non-fatal cases slow recovery can be expected and, according to the *Notes on Diagnosis of Occupational Diseases*, there is no satisfactory evidence that exposure to concentrations of nitrous fumes insufficient to cause acute pulmonary oedema results in chronic ill-health.

Kennedy, in 1972, disputed that conclusion. From a detailed study of acute chest illnesses and emphysema among a hundred coal miners he drew the conclusion that both acute and chronic illness may result from excessive exposures to fumes from certain types of shot firing, fumes which contained, *inter alia*, 'nitrous fumes'.* According to Kennedy, short, heavy exposures to fumes result in acute chest illnesses which can be expected to recover fully. Low grade exposure to fumes, however, over a period of months or years appears, in Kennedy's view, to be associated with development of an emphysema-like condition. In his study many of the miners who developed acute pulmonary oedema or severe acute chest illnesses also had a history of prolonged exposure to fumes, and they later developed emphysema. Kennedy was of the view (1970) that the description used in the *Notes on Diagnosis of Occupational Diseases* should be amended to take account of the possibility of chronic illness from 'low grade' repeated exposure.

Apart from his own findings, Kennedy was able to adduce evidence from reports of animal experiments which showed that emphysema does develop as a result of prolonged, low grade exposure to 'nitrous fumes'. Kennedy summarized the toxicology of 'nitrous fumes':

Concentrations in excess of 100 ppm—death from anoxia associated with pulmonary oedema occurs with exposures to concentrations of 500 ppm or more, for one hour or less, and from concentrations of 100 ppm or more for a period of a few hours.

Concentrations less than 100 ppm; short exposures—exposure to

*Shot firing is disappearing from underground mining.

more than 50 ppm occasionally results in death, whereas exposure to less than 50 ppm does not result in death. However, it should be emphasized that these data relate to animal and not human exposures.

Concentrations less than 50 ppm; long exposures—animal experiments show that long exposure to concentrations in the range 10 to 50 ppm of 'nitrous fumes' results in emphysema in rabbits and rats.

Morley and Silk (1970) defined 'nitrous fumes' as any airborne oxide of nitrogen other than nitrous oxide (N_2O) or any mixture of these oxides in sufficient concentration to have an adverse biological effect. They pointed out that in practice these oxides always occur as mixtures. Moreover, the oxides react with each other and with atmospheric oxygen; the resultant mixtures have no constant composition. As Morley and Silk pointed out, variability in the effects on exposed subjects is to be expected. It is not surprising, therefore, that they could not attribute specific biological effects to the individual oxides. Because of the uncertain nature of 'nitrous fumes' it is evidently preferable to follow Morley and Silk's terminology and use 'oxides of nitrogen' in place of 'nitrous fumes'.

Morley and Silk reported a case of a man who, following exposure to oxides of nitrogen, developed pulmonary oedema which appeared to recover satisfactorily with treatment. However, he was found dead in bed 43 days later, having retired the previous night apparently in good health. At the post-mortem examination the cause of death was attributed to acute viral pneumonia.

Morley and Silk's case may represent an interaction between inputs of the model described in Chapter 2. The irritant effects of oxides of nitrogen might impair, in some unknown way, the respiratory defence processes against micropredators. It is not possible to identify the importance of irritant fumes as predisposing factors to bacterial or viral diseases of the chest, although it could very well be important.

Kennedy has summarized, in a chart, the diverse circumstances under which oxides of nitrogen have been encountered. These include: welding of many kinds (Morley and Silk pointed out that oxides of nitrogen result from welding flames even where the flame is not in contact with the metal); in the aftermath of shot firing and explosion of certain types of explosive; fire and combustion; in the production of silage; in various chemical reactions (previously described); with exhaust fumes from internal combustion engines; and from electric arc and similar processes. The list can be updated with the following space accident. During re-entry, the American astronauts of the Apollo–Soyuz mission accidentally inhaled oxides of nitrogen (Hatton *et al.* 1977). They showed signs of diffuse chemical pneumonitis, and it was thought that the considerable collagen breakdown, which apparently occurred during the re-entry, was due to the inhalation of the oxides of nitrogen.

FIRST-AID

Although the administration of first-aid is peripheral to the question of classification, it is appropriate to mention aspects of it because diagnosis and treatment are both facilitated by reference to the biological classification. A series of steps are prescribed which apply generally to all gassings in classes (i) to (v). The first and most obvious step is to convey the victim to a place free from gas. Thereafter efforts should be made to reduce the victim's oxygen demand to a level sufficiently low to allow the oxygen transport/uptake system to function normally. This system will be impaired, and the victim's condition worsened, if oxygen demand is too great. To avoid this the victim should be afforded both rest and warmth, although, where necessary and available, oxygen should of course be provided. In extreme circumstances, i.e. where the victim is quite evidently near to death, exhaled air resuscitation (popularly known as 'mouth-to-mouth' resuscitation) should be administered, an operation which is *without* danger to the first-aider so long as the victim's skin is free of chemical contamination.

Rest and warmth are appropriate for victims exposed to irritant gases because they are likely to experience difficulty in breathing and have a troublesome cough. Pulmonary oedema, as a delayed risk, needs to be considered for victims seemingly recovered from the immediate effects of a gassing.

PREVENTION

For strategies and tactics of prevention (discussed as a general topic in the next chapter) the reader is advised to consult the Factories Act, 1961, especially s.30; Statutory Instrument 1341, 1961; and the ILO Information Sheet CIS No. 6 on confined area working.

CASE STUDIES

The dozen case studies cited below have all been taken from the (British) Ministry of Labour publication *Accidents*, a publication which has been discontinued. The examples have been drawn selectively from over a number of years, the intention being that each illustrate a point (or points) made in the preceding classification. Each has been paraphrased and précied so that the salient points can be more easily highlighted.

I The taking of 'short-cuts' is all too often the cause of industrial accidents. A chargehand, in contriving to speed up the cleansing of a large dye filter pan, inadvisably took it upon himself to use sodium hypochlorite, a widely-used industrial bleach with the use of which he was acquainted. The accepted means of cleaning the pan involved washing with water and with sulphuric acid. Unfortunately for the chargehand, and for a co-worker similarly affected, interaction of sulphuric acid (a film of which remained in the pan) and sodium hypochlorite produced chlorine gas. Though the gassing in this case was not fatal, merely resulting in the loss by each man of a fortnight's work, the circumstances surrounding the accident should serve as warning against departing from standard practice, as well as an example of an unexpected or surprise gassing.

II A further example of the dangers of unauthorized improvisation and ill-advised time-saving is provided by the case of a brewery worker engaged in the cleaning of a container. Once again the victim was an experienced worker who, arguably, should have known better.

The cleaning of an 'underback' (the term used for the large copper receptacle in which the beer ferments) should normally be a relatively easy task. However, in order (as he imagined) to make the task even easier the brewery worker dispensed with the usual cleaning materials (diluted alkali and tartaric acid solvents) and used a nitric acid-based solution.

He finished his work rather earlier than usual and he went home. Later that night he had to be admitted to hospital where he was pronounced 'seriously' and eventually 'extremely' ill. A difficult diagnosis was made more difficult when a phone-call to the brewery management elicited no clues as to how the man's illness could have been caused. Only the chance mention to his wife of his use of nitric acid enabled doctors to successfully diagnose and treat his illness.

Only after the man's providential recovery were the full facts surrounding the incident known. An investigation by the brewery established that according to custom and practice dilute nitric acid was occasionally used as a substitute for tartaric acid solvent, if and when stocks of the latter were low.

The effects of nitric acid on copper involve the production of oxides of nitrogen, which the brewery worker had inhaled, and which produced delayed pulmonary oedema. That the brewery company were in apparent ignorance of the danger shows a certain inexcusable negligence.

III While dismantling a galvanized metal tank, part of a factory water heating system which needed replacing, a welder became seriously ill through fume inhalation. The welder had decided his best method of dismantling the 6 ft square by 4 ft deep tank was to step inside it and cut it into sections with oxy-acetylene equipment. Working alone through a full seven hour shift, and in the confined space of the tank within the further confinement of a boiler room proved too much for the man. Despite feeling unwell he managed to reach home, merely for his illness to become more serious. Only swift medical treatment, including the administration of oxygen, saved him from succumbing to the effects of delayed pulmonary oedema due to inhalation of nitrous fumes; symptoms of zinc fume fever were recognizable as a complication.

A delay of some hours between fume inhalation and the crisis of the disease is not unusual, a fact which makes incidents of this kind all the more dangerous. A person suspected of having breathed fumes likely to cause pulmonary oedema should therefore be hospitalized so that his condition can be observed. In the instance cited above, the danger of exposure to fumes was multiplied by the confined working space, which accelerated oxygen deficiency, and by inadequate ventilation, in that consumed oxygen was replaced only relatively slowly and in an insufficient quantity. Portable apparatus designed to ensure circulation of fresh air could and should have been utilized. The factory management were also negligent in being singularly unaware of the possibilities of exposure to fumes given off from galvanized zinc and those occasioned by use of the cutting equipment.

IV Not infrequently the repair or replacement of a chlorine cylinder is the occasion of a gassing accident. The dangers arising from such an accident can of course be minimized by the swift application of respiratory protection equipment. Where respiratory equipment is unavailable, or faulty, or improperly or belatedly used the consequences are likely to be serious.

One such accident took place in a water-processing factory during the changing of two 17 cwt chlorine drums, a manoeuvre which had been safely executed on countless previous occasions. Chlorine, an essential part of the treatment process, was seen to be seeping from a connecting pipe despite the closing of the principal safety valve. Though a slight leak was usually tolerated, such was the quantity of gas escaping in this instance that a fitter was prompted to fetch canister-style respirators. Unfortunately for the fitter, and for several of his colleagues, all of whom were badly gassed, their attempt to rectify the leak precipitated a large emission of chlorine. The outflow of gas subsequently traced to a faulty valve, was of such magnitude that the old-fashioned respirators were unable to cope. Canister respirators give minimal protection, and only for a brief period; their 'working life' is short and they are completely ineffectual unless the canisters are regularly checked and, when necessary, replaced. Complete protection from large amounts of gas is better guaranteed by the use of self-contained breathing apparatus.

V   The body of an early morning charwoman was discovered in a fume-filled kitchen by office staff arriving for work at 9 a.m. Normal emergency procedures failed to resuscitate the woman.

Death, the inquest was told, was due to inhalation of fumes from a coal-fired boiler in a kitchen at the rear of the office block. Examination of the boiler revealed it to be in perfect working order, but the duct by which fumes were expelled was found to be obstructed by portions of detached cement and by soot. This obstruction had caused a slow accumulation of noxious fumes, to which the charwoman had apparently remained oblivious and to which she had eventually succumbed—death by carbon-monoxide poisoning.

Here, once again, is a case in which the victim doubtlessly smelt noxious fumes but underestimated their danger. Perhaps she had become accustomed to the fumes or was overcome in a vain and over-conscientious effort to finish her work before the physical environment became unbearable. The lesson to be learnt from this tragedy is clear—ducts designed to disperse fumes must be regularly checked for possible obstruction, lest there be dangerous build-up of carbon monoxide due to incomplete carbon combustion.

VI   Another gassing accident involving a woman office cleaner illustrates the latent dangers of leaving work half-finished. A woman engaged in the cleaning of a urinal was interrupted mid-way through her work and, on her return, neglected to take account of the fact that she had previously administered a reduced strength bleach. Normally such an oversight would not have mattered, but on this occasion the woman's forgetfulness was brought home to her when, upon her scattering cleaning crystals into the urinal, the interaction of the two substances gave off concentrated chlorine fumes. Her unexpected gassing resulted in hospitalization and a month's absence from work.

The accident could easily have been avoided had the danger of such an admixture been recognized. Unfortunately the buying of the cleaning agent in bulk and its subsequent distribution in small amounts and in makeshift tins meant that the clear warning on the original container was not transmitted to the cleaning woman.

Similar mistakes can be prevented only if rules are strictly observed. Cleaning agents must be supplied singly lest unexpected chemical reactions occur, while care must be taken to brief cleaning personnel of the potential risks of such reactions. Had a warning been given, this accident might have been averted.

VII Well-intentioned but ill-contrived efforts to fit a condensate trap to a factory gas main precipitated a leakage of carbon monoxide. A welder to whom the work had been entrusted, though suffering from the typical impairment of his critical faculties occasioned by the leaked gas, continued to work oblivious of the danger. The inevitable explosion left him with severe facial injuries. Examination proved an inserted flange joint to have been smaller than necessary—a mistake which had allowed the carbon monoxide to leak into the factory, rather than out via the main.

VIII Three tannery workers became unconscious, one for over forty-eight hours, after collapsing beside a drum of sodium sulphide solution in which hides were being immersed to remove surface hair. Overnight build-up of fumes meant that the first worker to lift the drum lid was overcome by concentrated fumes. Such was the initial concentration of these undispersed fumes that, in attempting to rescue their colleague, two other workers were also rendered unconscious. A second 'rescue party', on the scene within minutes of the first, suffered no ill-effects, presumably because by that time the vapours had been diluted.

For some little time the combination of circumstances which had caused this near-tragedy baffled investigators. Eventually the hides of a type only newly imported were traced back to the Kenyan port of Mombasa. Further investigation showed the hides to have been wet-salted, and this provided the essential clue to the cause of the accident. Due to its acidity, the salt had interacted with the sodium sulphide solution, leading to the production of hydrogen sulphide, a gas which if inhaled in sufficiently concentrated form can be fatal.

This detailed piece of detective work provided the means of prevention of further similar accidents. Henceforth every consignment of hides was to be examined, and if acidity levels were found to be too high then the sodium sulphide solution was to be supplemented by quantities of sodium hydroxide (caustic soda) so that the solution would be neutralized.

IX Polytetrafluoroethylene (PTFE) is a multi-purpose plastic, perhaps its most common use being as the 'non-stick' protective film with which some cooking utensils are coated. As far as its use in this connection is concerned it appears not to have been associated with danger. However, when subjected to heat disintegration, PTFE does have a tendency to give off fumes which can cause polymer fume fever.

PTFE was brought to the attention of HM Factory Inspectorate when two workers contracted polymer fume fever. Both men had been engaged in the excision of a seized roller bearing from a tipper truck. Their efforts to remove the bearing, which was lubricated by a film of PTFE, had involved the use of oxy-acetylene cutting equipment. The PTFE disintegrated under the intense heat, both men being overcome by the resultant fumes. Both were hospitalized with fume fever symptoms and acute respiratory inflammation.

Here was an accident caused by lack of knowledge. Not one person in the factory apparently knew that the bearing was coated with PTFE, an ignorance in part due to the manufacturer's original warning as to the use of PTFE having been obscured by grime.

X   A common form of anoxic death, though not accidental death, is through inhalation of the carbon monoxide fumes emitted from the exhaust of a car or another internal combustion engined vehicle. Within the confined space of a car, with its engine running and with a length of rubber tubing attached to the exhaust, many people, tragically, have chosen to die. Under the heading 'accident', however, deaths from a similar cause are not unknown. Happily the accident here described was not fatal—it was not even 'reportable' in that the victims did not lose three days work—although only the providential proximity of help prevented the consequences from being much worse.

During the construction of a sewer over twenty feet below ground level a number of workers became unwell. There is no need to give the exact dimensions of the tunnel in which they were working— suffice it to say, however, that the shaft constituted the typical 'confined space'. Each morning, before further excavation could commence, the workings had to be pumped clear of water which had accumulated overnight. Usually compressed air pumps were used, but on the failure of such a machine a stationary internal combustion engined pump was substituted. For perhaps an hour this pump worked away, before four workmen climbed down to begin their day's labours. Luckily for them an observant crane driver noticed that his workmate hesitatingly began to ascend the ladder and then collapsed. A second man was quickly raised in the crane's bucket, whereupon the crane driver himself descended and brought up a third mate who had collapsed in the tunnel. A hastily organized party rescued the fourth.

Artificial respiration was administered to the two most seriously gassed, not, however, by their fellow workers, who appear not to have known the procedures, but by the police.

XI The site of this fatal accident was the basement of a cold storage plant through the brick wall of which a fitter's labourer was attempting to cut a hole. Once again one encounters the example of a confined space, the basement roof being less than five feet high, a fact which increased the danger in that the man was forced to crouch in an uncomfortable working position. Either through excusable clumsiness, or through inadvertency, he leaned against an adjacent pipe, causing a connection to come adrift and enveloping him in a cloud of ammonia gas. His cry for help was immediately answered by his workmate who, wearing a respirator, dragged the man clear. Rescue came too late, however, for the labourer subsequently died from the effects of inhalation of ammonia gas and the resultant bronchial pneumonia. The rescuer was fortunate that he did not also become a victim because the canister respirator which he used for the rescue was faulty and, moreover, did not fit properly.

XII The following case report comes, not from the journal *Accidents* but from a short report by Ross (1973) referring to three accidents involving loss of consciousness to men engaged in burning using oxy-acetylene flames in confined spaces. One of the accidents was investigated by Ross and the salient data are repeated here in order to make the point that not all gassing, despite thorough investigation, can be fully explained. Of the three men involved, one died. In all three cases the men were working with oxy-acetylene torches inside steel tanks in which the ventilation was presumably inadequate, although this could not be established beyond question in the investigations subsequent to the accidents. In Ross's case (1970) the man, after working for about 15 min, felt himself losing consciousness and just managed to put out the torch. Fortunately the foreman noticed what was happening and the man was rescued. On rescue he was found to be unconscious with his musculature in rigid spasm. The pupils of the eyes were dilated, and the pulse was irregular. Following the administration of oxygen movements returned and after about an hour the man became conscious with full movements. On regaining consciousness he complained of sickness and headache; the headache persisted after the feelings of nausea had disappeared. In the other two cases one workman was found dead and the other unconscious inside the tank; the torch was off and some acetylene gas remained in the tank.

Possibilities to be considered are anoxia, due to the using up of atmospheric oxygen by the acetylene flame, the production of carbon dioxide or carbon monoxide to excess, or, in the case involving the two men, oxygen displacement anoxia caused by the accumulation of acetylene.

REFERENCES

BOWES, P. C. (1974). Smoke and toxicity hazards of plastics in fire, *Annals of Occupational Hygiene*, **17**, 143–57.

BROWNING, E. (1965). *Toxicity and Metabolism of Industrial Solvents*, Amsterdam: Elsevier.

DEPARTMENT OF HEALTH AND SOCIAL SECURITY (1972). *Notes on the Diagnosis of Occupational Diseases*, London: HMSO.

FACTORIES ACT, 1961 9 & 10 Eliz. 2 c.34, Statutory Instrument 1345, 1961.

GERARDE, H. W. (1960). *Toxicity and Biochemistry of Aromatic Hydrocarbons*, Elsevier: Amsterdam.

HARRIS, D. K. (1951). Polymer fume fever, *Lancet*, **261ii**, 1008.

HATTON, D. V., LEACH, C. S., NICOGOSSIAN, D. I. and FERRANTE, N. (1977). Collagen breakdown and nitrogen dioxide inhalation, *Archives of Environmental Health*, **31**, 33–6.

HUNTER, D. (1975). *The Disease of Occupations*, London: English Universities Press.

INTERNATIONAL LABOUR OFFICE (1972). Entering tanks and other enclosed spaces, CIS Information Sheet No. 6.

KENNEDY, M. C. S. (1970). Emphysema in coalworkers, *British Medical Journal*, **4**, 805.

KENNEDY, M. C. S. (1972). Nitrous fumes and coal-miners with emphysema, *Annals of Occupational Hygiene*, **15**, 285–300.

MALTEN, A. E. and ZIELHUIS, R. L. (1964). *Industrial Toxicology and Dermatology in the Production and Processing of Plastics*, Amsterdam: Elsevier.

MORLEY, R. and SILK, S. J. (1970). The industrial hazard from nitrous fumes, *Annals of Occupational Hygiene*, **13**, 101–7.

ROSS, D. S. (1970). Loss of consciousness in a burner using an oxy-acetylene flame in a confined space, *Annals of Occupational Hygiene*, **13**, 159–60.

ROSS, D. S. (1973). Loss of consciousness affecting two metallizers (one fatally) in a confined space, *Annals of Occupational Hygiene*, **16**, 85.

Chapter 23

Case History of Vinyl Chloride

OBJECTIVES OF THE CHAPTER:

This chapter:
  (1) provides an outline of the events associated with the problems presented by vinyl chloride,
  (2) identifies diseases which vinyl chloride is known to cause,
  (3) provides an outline history of the development of knowledge about vinyl chloride,
  (4) outlines the actions taken against vinyl chloride danger, and
  (5) compares the strategic approaches to the problem adopted by government agencies in the UK and USA.

OUTLINE OF EVENTS

On 22 January 1974, the American company, B. F. Goodrich, informed its employees, the press, the public of Kentucky State, the Kentucky State Department of Labor and the US Department of Health, Education and Welfare's National Institute for Occupational Safety and Health that three of its workers involved in polyvinyl chloride production had died of angiosarcomas of the liver. The notification fulminated a reaction in the world's chemical industry. Within months the US hygiene limit, which had only recently been reduced from 500 to 200 ppm was further lowered to 50 ppm. Almost immediately this was again reduced to 'no detectable level' and was finally replaced by a limit of 1 ppm. The speedy response to formal recognition of the danger from vinyl chloride was remarkable, in marked contrast to the response to dangers described in other chapters.

According to Levinson (undated, c. 1976) trade union pressure was instrumental in producing the response; on the other hand, Weaver (1974) noted that all the information linking vinyl chloride to cancer had been produced by industry on its own initiative. He cited Ralph L. Harding jr, the President of the American Society of the Plastics Industry, as having stated that 'this is a unique situation. Industry financed the studies, the industry blew the whistle on itself'. According to Weaver, the problem, once identified, was seized upon by a 'loose but not unco-ordinated network of regulatory agencies, government research institutes, academic medical teams, labor unions, and other groups united by a common commitment to eradicate environmental causes of disease'. Weaver went on to complain that this 'regulatory-medical complex' was a political movement of an essentially 'anti-business' character.

In chapter 24 reference will be made to the interplay of socio-economic and strategic questions relative to occupational health and safety. Such questions have formed an integral part of discussions on the subject for at least a century and a half. It is apparent from the contrast between Levinson's and Weaver's interpretation of events, that the socio-economic and scientific aspects of occupational health and safety are still inseparable.

VINYL CHLORIDE; POLYVINYL CHLORIDE

Vinyl chloride is an alkyl halide; it contains a carbon-carbon double bond which undergoes electrophilic addition. It is manufactured by direct halogenation of hydrocarbons, such as acetylene or ethylene, or by dehydrohalogenation of ethylene dichloride. The vinyl chloride, the monomer, is polymerized between 40°C and 70°C to produce a long-molecule plastic, polyvinyl chloride, which, together with copolymers, has been increasingly exploited commercially (see Fig. 23.1).

At room temperature vinyl chloride is a sweet-smelling gas, with a boiling point at atmospheric pressure of about −14°C. In addition to its use in polymer production, vinyl chloride has been used as a propellant for 'aerosol' cans. Small quantities of the monomer are likely to be detectable in the polymer in manufactured articles, unless special processes are employed to remove the monomer.

Effects of Vinyl Chloride and Polyvinyl Chloride

The effects of vinyl chloride include: angiosarcoma sometimes called haemangiosarcoma; a collection of signs and symptoms sometimes referred to as 'sclerotic syndrome'; and a condition which Lange et al. (1974) termed 'vinyl chloride disease'. Other terminology relative to vinyl chloride disease includes Raynaud's syndrome, acro-osteolysis and narcosis.

Hemangiosarcoma
Hemangiosarcoma (variously referred to as haemangiosarcoma, or angiosarcoma), is a neoplastic disorder of growth (discussed in Chapter 16) arising in the cells lining the blood channels (sinusoids) of the liver. It should be noted that hemangiosarcomas occur in blood

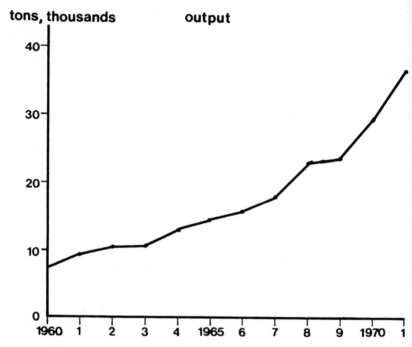

FIG. 23.1. Production of polyvinyl chloride and copolymers in Europe, USA and Japan from 1960 to 1971; average annual output in thousands of tons (data from Levinson, *op. cit.*).

vessels other than those of the liver, though such tumours can be disregarded in the context of vinyl chloride. Liver (hepatic) hemangiosarcoma is malignant, spreads invasively throughout the liver and forms secondary growths (metastasizes) widely. No report of successful treatment has been located.

Acro-osteolysis, and vinyl chloride disease
From the discussion of Ward *et al.* (1976) it appears that use of the term 'vinyl chloride disease' is preferable to that of 'acro-osteolysis', since the latter inadequately describes the full extent of the disorder suffered by vinyl chloride workers. As previously stated, vinyl chloride disease comprises:
  (a) sclerotic changes in the skin, changes marked by hardening of the skin (induration) resultant upon changes in the collagen fibres of skin. Conceptually, it can be related to disorders of repair;

(b) osteolysis, loss of calcium from the bones of fingers, hands, wrists, forearms and certain joints of the pelvis (Markowitz *et al.*, 1972);

(c) circulatory disturbances, whereby the fingers display features of Raynaud's disease, closely related in pattern but not in cause to vibration-induced white finger;

(d) thrombocytopenia, a reduction in the number of thrombocytes (also referred to as platelets), which are intimately involved in blood clotting;

(e) portal fibrosis, or fibrosis of the liver, associated anatomically with the portal vein (a vessel which carries blood, and therefore ingested nutrients and non-nutrients, from the gut to the liver);

(f) impaired liver (hepatic) function; this refers to abnormal results obtained from tests of blood and urine constituents, which reflect the performance of the liver); and

(g) impairment of lung (pulmonary) function, a change which may be associated with pulmonary fibrosis, that is a disorder of repair of lung.

Narcosis

Vinyl chloride in common with numerous other water-insoluble, fat-soluble volatile organic compounds, on inhalation, produces depressant effects on the central nervous system. The extent of these effects is closely related to the concentration of substances in the target organ, the brain.

Outline History of Knowledge about Vinyl Chloride

Narcosis was the first effect to be recognized, followed by a component of vinyl chloride disease, and finally by hemangiosarcoma.

Dublin and Vane, in about 1933 (cited by Malten and Zielhuis, 1964), appear to have been the first to report the narcotic effects of vinyl chloride. For a brief period vinyl chloride was experimented with as an anaesthetic agent for surgical purposes. However, it was found to disturb the heart rhythm, and its use was soon discontinued. In 1964, Malten and Zielhuis summarized the existing knowledge of its narcotic effects, citing Dublin and Vane's reports (1933 and 1941) of two 'slight intoxications' observed in workers exposed for about three minutes to between two and five per cent vinyl chloride. The symptoms experienced by these workers were reversible—vertigo (dizziness), light disorientation and a burning feeling in the feet. In Britain, two serious cases were reported in the *Annual Report of HM Chief Inspector of Factories for the year 1953* (Ministry of Labour, 1954). In one case a worker collapsed after a 10-min exposure near an open polymerization tank. Danzinger, in 1960, reported three cases of narcosis from vinyl chloride in Canada, of which two were fatal.

On the basis of the narcotic effects, the 'maximum allowable concentration' was set by the American Conference of Governmental Industrial Hygienists at 500 ppm, in 1962. Malten and Zielhuis thought this too high, and drew attention to the experimental data of Torkelson *et al.* (1961) which showed that repeated exposure to vinyl chloride had affected the livers and kidneys of laboratory animals. In 1963, however, the animal experiments of Lester *et al.* produced no marked change in organs, apart from a relative weight increase of liver and spleen. In the event, in 1971, the Threshold Limit Value was revised downwards by ACGIH to 200 ppm. It remained at this level until the hemangiosarcoma announcement by the Goodrich company in 1974.

From the documentation published by ACGIH (1971) it is unclear why the TLV was lowered to 200 ppm. This documentation cited the work of Kramer and Mutchler (1972) as well as that of Torkelson *et al.* (1961) and Lester *et al.* (1963). ACGIH stated that 'the multi-phasic screening tests are continuing in all exposed workers, and analysis of the findings is being made on an individual worker basis. In the meantime, a time-weighted average Threshold Limit Value of 200 ppm vinyl chloride (with a few ppm vinylidene chloride) seems appropriate to prevent adverse systemic effects from long continued daily exposure'.

Vinyl Chloride Disease

That aspect of vinyl chloride disease known as acro-osteolysis was first described in 1966 by Cordier *et al.*, in France. Further cases were reported in Great Britain in 1967 (Harris and Adams) and in that same year, in the USA, Wilson *et al.* related 37 cases in the B. F. Goodrich company. Jühe *et al.* (1972), in Germany, reported in vinyl chloride workers thrombocytopenia, liver damage, disorder of pulmonary function, and alteration in bone. Three studies sponsored by the American Manufacturing Chemists Association in 1967 were reported by Dinman *et al.* (1971), Cook *et al.* (1971) and Dodson *et al.* (1971). These three studies established the pattern of vinyl chloride disease as it was renamed, in 1974, by Lange *et al.*

Carcinogenicity with Special Reference to Hemangiosarcoma

An Italian research worker, Viola, is credited with the first report, in 1970, of carcinogenicity of vinyl chloride. He observed that rats exposed to 30 000 ppm for periods of four hours per day, five days a week, over a year, developed squamous cell carcinoma of the external auditory meatus (a common neoplastic site in rats). They also exhibited carcinoma of lung and squamous metaplasia (explained in Chapter 16), and bone changes. These findings were reported in greater detail by Viola *et al.* in 1971.

Viola's findings were doubted, and consequently Montedison and other European producers commissioned another Italian researcher, Maltoni, to undertake animal experiments to check Viola's findings. In 1973 Maltoni produced the necessary confirmation. In December 1973, Creech, B. F. Goodrich's occupational physician at their plant in Louisville, Kentucky, received notice that one of the company's workers had died of liver angiosarcoma. He recalled that two years earlier another worker had died of the same condition. In January 1974 the company therefore began an epidemiological study during the course of which Creech received a report of the death of a third worker from angiosarcoma. Then, as stated, the fateful 22 January announcement was made.

Although Viola was first to report carcinogenicity, there had been earlier supposition that the effects of vinyl chloride involved more than narcosis. For example, in the Soviet Union in 1958, Filatova *et al.* (cited by Malten and Zielhuis, 1964) had reported cases of 'chronic angioneurosis' among workers in a polyvinyl chloride factory. These workers had been exposed to between 20 and 315 ppm of vinyl chloride, and also to certain solvents such as dichloroethane and methanol. It is not exactly clear what was implied by 'chronic angioneurosis', although the condition, whatever it was, involved the liver. Other reports of a similar nature appeared soon after in various East European scientific journals, and this, with hindsight, is consistent with the findings of researchers like Torkelson (1961), published three or so years later.

By 1975, a world-wide total of 32 cases of hemangiosarcoma had been reported in polymerization workers, and a further 6 diagnosed among non-polymerization workers (Lloyd, 1975). According to Lloyd's data for the polymerization workers, the median age at diagnosis was 44, the median value for years between first exposure and diagnosis was 17 and the median years of exposure to vinyl chloride was 16. Although these data do not necessarily indicate a long latent period, it seems reasonable to assume that it is typically of the order of 17 years.

Action Against Vinyl Chloride Danger

United States of America
Levinson (*op. cit.*) has reviewed the developments in the USA. On 11 March 1974, the National Institute of Occupational Safety and Health of the US Department of Health, Education and Welfare issued a

Recommended Standard for Occupational Exposure to Vinyl Chloride, relative to the manufacture of synthetic polymer from vinyl chloride. Eleven days later the Occupational Safety and Health Administration (the section of the Department of Labor charged with administering the Occupational Safety and Health Act) issued an Emergency Standard for vinyl chloride of 50 ppm. The Emergency Standard was the subject of criticism from American trade unions, and public hearings were organized at which the case for and against a 'no-detectable level' limit was heard. Musacchio (1974) summarized the arguments put to the NIOSH. Broadly, representatives of industry testified that:

(a) there was no scientific evidence establishing the need for a 'no-detectable level' Standard,
(b) enforcement of that Standard would destroy the industry, and
(c) current control technology could not achieve the concentration levels ordered by that Standard.

The trade unions, on the other hand, pressed hard for a no-detectable level Standard, on the basis that the industry's claims of economic impairment were invalid, citing testimony that some plants had already achieved 'no-detectable' levels. Although OSHA did not accept that particular evidence it did adopt, later in 1974, a time-weighted average of 1 ppm, with a ceiling value of 5 ppm, averaged over no more than a 15-min sampling period. This Standard became legally effective on 1 January 1975 (US Federal Register 39 FR 35890; and 29 CFR part 1911; 1910.93 of part 1910 of title 29, Code of Federal Regulations).

United Kingdom

In February 1975 the Health and Safety Executive produced a Code of Practice for health precautions with vinyl chloride. This included an interim hygiene standard with a ceiling value of 50 ppm and a requirement that the average exposure over a whole working shift must not exceed 25 ppm. In October of the same year the interim standard was revised, with a ceiling value of 30 ppm and a time-weighted average of 10 ppm being set. The full background to these decisions is not accessible for public information.

Between May and December 1974 the *New Scientist* carried a number of articles by L. McGinty criticizing the 50 ppm limit. This had, in fact, been proposed by the Health and Safety Executive's predecessors, the Department of Employment. According to McGinty (1974) the ceiling of 50 ppm was worked out by the Department of Employment in conjunction with the Trades Union Congress and the Chemical Industries Association on the basis that the proposed limit was the same as that laid down in the OSHA Emergency Standard.

However, as we have seen, by 1975 the 1 ppm Standard had become law in the USA. McGinty stated that there was opposition to the British 50 ppm Standard by David Warburton of the General and Municipal Workers Union. He took the view, apparently, that the least detectable amount of exposure should be the Standard. It is possible, on this evidence, that trade union pressure led to the 30 ppm Standard ultimately adopted in the UK.

McGinty drew attention to the way in which the economic argument had been used in the USA to show that the stringent Standards were not within the economic capabilities of the industry. McGinty quoted spokesmen for the British chemical industry saying that even the 50 ppm level would be difficult to achieve and that to make all plants conform to the 50 ppm Standard would mean a great deal of modification which in any case would not be practicable for old plant.

There is a clear implication in the various articles written by McGinty that the decision makers in Britain yielded to the economic argument although, as we have seen, soon after publication the Code of Practice was tightened up. Because of the obvious importance of the economic argument it is interesting to examine the reasoning adopted by OSHA. This is to be found in the US Federal Register (1974, 39 FR 35890):

> Since there is no factual evidence that any of the VC or PVC manufacturers have already attained a 1 ppm level or in fact instituted all available engineering and work practice controls, any estimate as to the lowest feasible level attainable must necessarily involve subjective judgement. Likewise, the projections of industry, labor, and others concerning feasibility are essentially conjectural . . . .
> We agree that the PVC and VC establishments will not be able to attain a 1 ppm TWA level for all job classifications in the near future. We do believe, however, that they will, in time, be able to attain levels of 1 ppm TWA for most job classifications most of the time. It is apparent that reaching such levels may require some new technology and work practices.

Plainly, in the absence of evidence that the 1 ppm Standard was in fact achievable, there could be no question of legislation in the USA which was absolutely enforceable. Thus the Code of Federal Regulations (29 CFR part 1911; 1910.93 of part 1910 of title 29, Code of Federal Regulations) under *Methods of Compliance* requires *feasible* engineering and work practice controls to be instituted immediately; and 'a program shall be established and implemented to reduce exposures to at or below the permissible exposure limit, or to the greatest extent feasible, solely by means of engineering and work practice controls, as soon as feasible.'

In the UK in 1977 the Health and Safety Executive's Code of Practice on vinyl chloride has yet to be approved under the Health and Safety at Work Act, 1974; it thus has no legal standing. A most interesting question arises, therefore, namely how the approaches to vinyl chloride in the USA and UK compare where the one country has regulations requiring 'feasible' control and the other has an entirely voluntary code of practice; in fact, in both countries the approach to the minimization of danger from vinyl chloride has been especially thorough. The tabulation which follows (Table 23.1) compares and contrasts, in outline, the OSHA Occupational Safety and Health Standard for vinyl chloride published in October 1974, and the Health and Safety Executive's Code of Practice, published in February 1975 and amended in October of the same year. The OSHA Standard follows closely, though not exactly, the NIOSH Recommended Standard published six months earlier.

The two documents are set out side-by-side because, as previously stated, they represent a stringency of control which is unusually detailed. Other harmful factors, such as ionizing radiations, or certain aromatic amines are controlled with a similar stringency, though the detail of the requirements may not be so closely specified. The vinyl chloride requirements, therefore, provide models for control against dangerous substances generally. However, it is not suggested that all the control strategies adopted in the case of vinyl chloride need be applied to all potentially dangerous chemical substances. Indeed, criticism can be levelled at some of the strategies invoked for vinyl chloride, and the side-by-side comparison shows that neither document represents the entire gamut of health and safety strategies. (The broad question of the development of strategies and how they are chosen is dealt with in Chapter 24.)

There is relevance in Levinson's (op. cit.) point that there are shortcomings in the data linking vinyl chloride with human and animal exposure. In particular, insufficient is known about what Levinson calls the 'multiple substance' effects. He pointed out that more than vinyl chloride is involved in polymerization: use of plasticizers, solvents, phthalates and other agents and catalysts is also entailed. Residues in the reactors certainly contain vinyl chloride monomer, but they also contain a range of chemical substances which, according to one manager speaking to me during an informal discussion at a vinyl chloride plant, have so far defied complete chemical analysis. Levinson's point should be borne in mind because most reported cases of hemangiosarcoma have been contracted by persons employed to clean out the reaction vessels. While it is not suggested that vinyl chloride is guiltless, some caution is required (as Levinson has reminded us) in the acceptance of single factor causation in carcinogenesis; certain of the problems touched on in Chapter 16, such as co-carcinogenicity, necessarily cannot be excluded from the consideration of vinyl chloride.

TABLE 23.1

SIDE-BY-SIDE, SUMMARIZED COMPARISON OF THE US DEPARTMENT OF LABOR'S OCCUPATIONAL SAFETY AND HEALTH ADMINISTRATION STANDARD ON EXPOSURE TO VINYL CHLORIDE, AND THE UNITED KINGDOM'S HEALTH AND SAFETY EXECUTIVE'S CODE OF PRACTICE FOR HEALTH PRECAUTIONS ON VINYL CHLORIDE.

OSHA	HSE
STATUS OF DOCUMENTS	
Rules and Regulations printed in US Federal Register	Code of Practice produced on the assumption that it would be submitted for approval under Sections 2, 16 and 17 of the Health and Safety at Work Act, 1974. At the time of writing the Code has yet to be approved.
SCOPE AND APPLICATION	
Regulates the control of employee exposure to vinyl chloride; applies to the manufacture, reaction, packaging, repackaging, storage, handling or use of vinyl chloride but not to the handling or use of products made of polyvinyl chloride. Applies to the transportation of vinyl chloride (provides for regulation by Department of Transportation).	Limits the exposure of persons by means of the interim hygiene standard (see below), the first priority being to minimize exposure in the manufacture of vinyl chloride and its subsequent polymerization.
LIMITS	
Permissible exposure limit (1) No employee may be exposed to vinyl chloride at concentrations greater than 1 ppm averaged over any 8-h period, and (2) no employee may be exposed	*Revised interim Hygiene Standard* A ceiling value of 30 ppm and a time-weighted average of 10 ppm, allowing that wherever practicable exposure should be brought as near as possible to zero concentrations.

TABLE 23.1—*contd.*

OSHA	HSE

LIMITS—*contd.*

to vinyl chloride at concentrations greater than 5 ppm averaged over any period not exceeding 15 min.

(3) No employee may be exposed to vinyl chloride by direct contact with liquid vinyl chloride.

Action level
This is a concentration of vinyl chloride of 0.5 ppm averaged over an 8-h work day.

ENVIRONMENTAL MONITORING

(1) Vinyl chloride concentrations shall be determined by methods with specified precision; records shall be kept for at least 30 years, and shall be made available to appropriate government departments on request.

(2) Monitoring is to be performed to find out if any employee is exposed, regardless of respiratory protection.

(3) If any employees are found to be exposed, further monitoring is required:

(a) if the permissible limit is exceeded, environmental monitoring is required monthly

(b) if the action level is exceeded, environmental monitoring is required quarterly.

(4) Environmental monitoring may be discontinued when at least two consecutive measurements, with at least 5 days between them, do not exceed the action level.

(5) Environmental monitoring is required if there has been a change in the process.

(1) A monitoring plan has to be prepared according to guidelines set out in an appendix of the Code of Practice. This plan shall be available for inspection by HM Factory Inspectorate and by employees' representatives.

(2) Monitoring is required at least once per shift in all working areas where vinyl chloride is liable to be present in concentrations above the hygiene standard, and in all areas outlined in the monitoring plan.

(3) Monitoring shall be done by a person trained in environmental sampling, in analysis and in the interpretation of results.

(4) Results of the monitoring have to be recorded.

(5) Where excursions take place above the hygiene standard, investigation is required and the results must be recorded.

(6) Results of atmospheric monitoring shall be made available for inspection by personnel working in the areas where vinyl chloride is

TABLE 23.1—*contd.*

OSHA	*HSE*

ENVIRONMENTAL MONITORING—*contd.*

(6) Employees or representatives shall be afforded reasonable opportunities to observe the monitoring and measuring.

likely to be present. The results have to be summarized on a shift basis, showing the shift average level, accompanied by an analysis of excursions above the hygiene standard. The results shall normally be posted within two hours of the end of the shift and be left posted for at least two days; data processing equipment is recommended to facilitate the presentation of data.

REGULATED AREAS; IDENTIFICATION OF AREAS TO WHICH REQUIREMENTS APPLY

A regulated area shall be established where vinyl chloride or polyvinyl chloride is manufactured, reacted, repackaged, stored, handled or used and where vinyl chloride concentrations are in excess of the permissible exposure limit. Access to the regulated area shall be limited to authorized persons and a daily roster shall be made of authorized persons who enter these areas.

Access to areas where it is foreseeable that concentrations of vinyl chloride vapour could be above the hygiene standard should be restricted. Only persons whose duties require them to enter such areas will be permitted entrance.

COMPLIANCE

Exposure of employees shall be controlled to or below the permissible exposure limit by (a) engineering, (b) work practice, and (c) personal protective controls. Written plans covering (a) and (b) have to be submitted to the government authority at least every six months and feasible action is required immediately and further action is required as soon as feasible.

Not specifically mentioned, but see later.

TABLE 23.1—*contd.*

OSHA	HSE

RESPIRATORY PROTECTION

Respiratory protection is required when the exposures cannot be reduced to or below the permissible exposure limit. The employer is required to provide respirators which conform to certain, laid-down specifications. The stringency of protection is determined by the atmospheric concentration likely to be encountered in non-routine as well as routine situations. For concentrations in excess of 3600 ppm the requirement is for open-circuit, self-contained breathing apparatus, pressure-demand type with a full face piece.

There are certain restrictions:

(1) Entry into unknown concentrations or those greater than 36 000 ppm may be made only for the purposes of rescue; concentrations between 3600 and 36 000 ppm may be entered for rescue purposes and also to prevent further release of vinyl chloride.

(2) Where air-purifying respirators are used, the canisters shall be replaced on a planned basis and, where it could be reasonably expected that atmospheric concentrations of vinyl chloride might exceed the allowable concentration for the respiratory protection devices, a continuing monitoring and alarm system shall be provided.

Respiratory protective equipment is specified in all situations where concentrations of vinyl chloride vapour are known to be above the hygiene standard, or where such concentrations can reasonably be expected to be present. Respiratory protective equipment involves a full-face or half-mask continuous flow, compressed air line or self-containing breathing apparatus conforming to BS 4667: Part 3: 1974. No other respiratory protective equipment should be used without written agreement from HM Factory Inspectorate. The respiratory protective equipment requires a suitable system for cleaning, inspection and issue; it should be inspected by a competent person at least every 14 days. No person should be allowed to rely on protection by respiratory protective equipment unless he holds a respiratory protection licence renewed within the previous 12 months. This licence is only issued following a suitable course of training or a refresher course; guidance about the licence and training are set out in an appendix to the Code of Practice.

TABLE 23.1—*contd.*

OSHA	HSE

Requirements Relating Specifically to Categories of Regulated Areas/Controlled Areas

(Items summarized under this heading relate specifically to the specified category and are in addition to the general requirements applying to vinyl chloride processes overall.)

For routine operations the following are required:

OSHA	HSE
Apart from one requirement (see below) this category is not appropriate for the OSHA standard.	(1) Control of vapour shall be the fundamental objective, and all practical steps shall be taken to keep concentrations of vapour in the working atmosphere below the hygiene standard. (2) No one shall be exposed to concentrations of vapour greater than the hygiene standard. (3) Access to 'forseeable' areas shall be restricted to persons whose duties require them to enter such areas; persons required regularly to do such work must have written authorization and a record must be kept of their name(s). (4) Respiratory protective equipment must be worn at any time when the vapour concentration exceeds the hygiene standard. (5) Plant and equipment shall be designed so as to ensure maximum containment; where vents and reliefs are needed they shall be positioned so that there is no contamination of the working atmosphere or of any neighbouring locations. Legal requirements relating to atmospheric pollution for the general non-occupational milieu must be conformed to. (6) New plant and modifications to existing plant have normally to be notified to HM Inspectorate of Factories at least 6 months before work commences; plans which might affect the emission must be notified to the appropriate authority.

TABLE 23.1—*contd.*

OSHA	HSE

REQUIREMENTS RELATING SPECIFICALLY TO CATEGORIES OF
REGULATED AREAS/CONTROLLED AREAS—*contd.*

	In non-enclosed plant, or where breaches are necessary, procedures are needed to ensure that vapour is not released into the working atmosphere.
Hazardous operations:	Reactor vessel entry:
Hazardous operations include entry of vessels for cleaning. Employees engaged in hazardous operations shall be provided with and required to wear and use: (a) respiratory protection as specified in the standard and (b) protective clothing to avoid skin contacts; this clothing shall be provided clean and dry for each use.	(1) 'Confined space' is defined as any reactor, autoclave, tank, chamber, vat, pit, pipe, flue, duct, bunker or underground room in which vapour is known to have been present or is liable to be present.
	(2) Entry by personnel into confined spaces shall be kept to a minimum in terms of frequency and of time spent within the confined space.
	(3) A permit-to-work system shall be instituted for entry to any confined space, as specified under the HSE's Technical Data Note 47.
	(4) All practicable steps shall be taken to remove contamination from confined space and to test its atmosphere prior to persons entering.
	(5) Where vapour levels cannot be reduced below the hygiene standard, respiratory protection must be provided (as specified under the HSE's Technical Data Note 47).
	(6) Suitable protective clothing shall be provided and worn (specified in detail in appendix 5 of the Code of Practice).
	(7) A permit for a person to enter a confined space without respiratory protective equipment shall state positively that the person issuing the certificate is satisfied that the concentration of vapour has been reduced below the hygiene standard,

TABLE 23.1—*contd.*

OSHA	HSE

REQUIREMENTS RELATING SPECIFICALLY TO CATEGORIES OF REGULATED AREAS/CONTROLLED AREAS—*contd.*

| | and that its rising above the standard cannot be foreseen.

Maintenance and Decontamination:
(1) Proper supervision should be exercised over maintenance work, because it may involve exposure to high concentrations of vapour, and that principles of protection and control (as specified generally in the Code of Practice) should be followed in the course of maintenance work. |

PROCESSES WITHIN REGULATED AREAS/CONTROLLED AREAS

| Dealt with under other headings. | Four areas/processes are recognized: (a) the working atmosphere within the controlled area generally, (b) equipment which is not totally enclosed or where enclosure must necessarily be breached as a routine process operation, (c) entry to confined space, and (d) maintenance work. |

EMERGENCY PLAN

| A written operational plan is required for emergency situations for each vinyl chloride facility. The plan should provide for employees to be suitably equipped with respiratory protection and for the evacuation of employees not equipped. | The Code of Practice calls specifically for the formulation of an emergency contingency plan to cater for a substantial escape of vinyl chloride vapour, and for the training of personnel in dealing with such emergencies. In particular the plan should adequately provide for raising the alarm, evacuation, control of the occurrence, plant shut-down and containment, the assistance by fire and police as well as site resources. Such plans should be readily available to HM Inspector of Factories and to the Joint Consultative Committee. |

TABLE 23.1—*contd.*

OSHA	HSE

EDUCATION AND TRAINING, SIGNS AND INFORMATION

Training is required to cover:

(1) the nature of the health hazard from vinyl chloride mentioning specifically the cancer hazard,

(2) the specific nature of the operation which could result in exposure to vinyl chloride, and the necessary protective steps,

(3) the purposes, proper use, and limitations of the respiratory protective devices,

(4) the risk of fire and the acute toxicity of vinyl chloride, and the necessary protective steps,

(5) how and why atmospheric monitoring is done,

(6) how and why the medical surveillance is done,

(7) emergency procedures,

(8) how the employee might recognize a release of vinyl chloride, and

(9) review of the standard.

The material used for training should be available for inspection by the government agency.

Signs and labels:

(1) entrances to regulated areas shall display a sign:
'Cancer-suspect agent area authorized personnel only'

(2) areas containing hazardous operations or where an emergency currently exists shall display a sign stating:
'Cancer-suspect agent in this area protective equipment required authorized personnel only'

Containers of polyvinyl chloride and polyvinyl chloride resin waste shall be labelled:
'Vinyl chloride cancer-suspect agent'

(1) All employees likely to be exposed to vinyl chloride above the hygiene standard should be suitably instructed about the dangers to health and the protective measures which are necessary. In particular, the following has to be dealt with:

(a) the nature of the dangers of vinyl chloride, and the extent of the medical supervision required by the Code of Practice,

(b) the general methods outlined in the Code of Practice aimed at protecting employed persons,

(c) the role of the employee in safe working methods and personal hygiene,

(d) correct operation of the plant with which the employees are concerned,

(e) the correct use of respiratory protective equipment,

(f) the significance of the monitoring plan and the monitoring results,

(g) the works emergency procedures,

(h) works rules relating to vinyl chloride,

(i) the role of joint consultation, and

(j) procedures for reporting defects in plant and equipment and for making suggestions.

(2) A high level of awareness is to be maintained by continued instructions, supervision and training.

TABLE 23.1—*contd.*

OSHA	HSE

MEDICAL SURVEILLANCE AND MEDICALLY SUPERVISED WORKERS

Each employee exposed, regardless of the use of respirators, to vinyl chloride in excess of the action level shall be subject to medical surveillance which includes examinations and tests carried out by a licensed physician without costs to the employee.

A clinical examination shall be provided at the time of initial assignment, every six months for any employee who has been employed in vinyl chloride for ten years or more, and annually for all other employees. In addition, any employee exposed to an emergency shall be afforded medical surveillance.

The physician is required to provide a statement of an employee's suitability for work in vinyl chloride including the use of protective equipment; a copy of the statement shall be given to each employee.

An employee shall be withdrawn from possible contact with vinyl chloride if the employee's health would be materially impaired by continued exposure.

Laboratories carrying out specified analyses shall be licensed.

The requirement in the Code of Practice is to identify 'medically supervised workers', those who are occupationally exposed to vinyl chloride on a regular basis, but not 'fortuitously'. Supervised workers are required to undergo a medical examination every 12 months, this to include a full medical and occupational history, a clinical examination with particular reference to the abdomen, skin and extremities, X-rays of the hands, and such further tests as may be indicated or subsequently recommended; at the examination an official advisory leaflet is to be given to the employee. Work people who report symptoms or who are off work for more than 2 weeks should be seen by the occupational physician. If indicated, medical examinations may be undertaken at any time.

Appendix 10 of the Code of Practice sets out the difficulties and shortcomings of medical examinations, and warns against the indiscriminate carrying out of such tests, since this would impose a 'serious burden' both on the exposed person and on medical resources. Appendix 11 discusses the approach to epidemiological investigation.

REPORTS AND RECORD KEEPING

Records are required to be kept showing the name and Social Security number of each employee; the records are disclosable to a government authority. Specific data required to be kept in the records include:

(1) Registers and records required by the HSE Code of Practice include the names and addresses of all persons authorized to enter 'controlled' areas; permits to work; reports of any leakages of vinyl chloride; detailed records of examinations of

TABLE 23.1—*contd.*

OSHA	HSE

Reports and Record Keeping—*contd.*

(1) the date of monitoring, the concentrations found, and identification of the instruments and methods used,

(2) personnel rosters shall be kept as records,

(3) monitoring and rosters records shall be maintained for 30 years,

(4) if an employer ceases from business, records are to be passed to the government authority,

(5) employees or their representatives shall be provided with access to examine and copy records of monitoring and measuring; former employees shall also have this access. The medical record shall be disclosed to another physician designated by the employee upon written request by the employee,

(6) the government authority shall be notified of regulated areas, and the number of employees involved; emergencies shall be reported within 24 h, and

(7) within ten working days of any monitoring and measuring disclosing that an employee has been exposed, regardless of the use of respirators, in excess of the permissible exposure limit, the employee shall be notified in writing of the results of the measurement and the steps being taken to reduce the exposure.

ventilation equipment; results of atmospheric monitoring; details of investigations of any excursion above the hygiene standard; and of inspections of respiratory protective equipment. (All these requirements are specified in close detail in appendices to the Code of Practice).

(2) Records shall be readily available on request by HM Inspector of Factories or by the Joint Consultative Committee (required under the Code of Practice).

(3) Records of atmospheric monitoring must be retained for at least 30 years, as must records of persons who regularly work in the 'controlled areas'; the personnel records shall be capable of being linked with monitoring records.

(4) Other records shall be kept for periods of between one and three years, according to their nature; where a factory or process is closed down, records have to be deposited with the local Inspector of Factories.

Responsibilities

With a few exceptions, the numerous obligations under the standard are not specifically allocated; rather it is implicit that they fall on employers/management. The only ex-

Senior management: a member of senior management should be allocated specific responsibilities and be given appropriate support. The senior manager would be responsible

TABLE 23.1—*contd.*

OSHA	HSE

ception is a specific requirement for employees to use respiratory protection, where provided.

for ensuring that joint consultation and monitoring and other steps provided for in the Code of Practice are properly carried out.

Supervisors have the responsibility to oversee those parts of the Code of Practice which relate to their particular situation, to report difficulties, and to pass forward suggestions from people over whom they have authority.

Employees are required to comply with those parts of the Code of Practice applying to them, to directly use the facilities and services provided under the code, and to report faults to their supervisors.

GENERAL REQUIREMENTS

Dealt with previously.

(1) Certain general requirements have been detailed in previous sections. In addition, attention shall be paid to 'housekeeping', requiring the transfer of solid waste material (containing significant levels of residual vinyl chloride) in special containers located in designated areas within the works boundary. Disposal must be effected in accordance with the relevant statutory requirement. Solid and liquid waste shall be separated, and solid PVC waste be recovered as dried polymer where possible.

(2) Preventive maintenance shall be instituted so that a high standard of maintenance in the whole plant may be achieved, in order to facilitate compliance with the hygiene standard; relevant equipment shall be given a visual check by a responsible person at least once a shift, in order

TABLE 23.1—*contd.*

OSHA	HSE
	to detect leaks; checks for leaks shall be made with a leak detector, in accordance with the monitoring plans (see below); where leaks are detected immediate steps shall be taken to minimize their effects and to eliminate them; ventilation equipment shall be tested thoroughly at least once every six months, and defects corrected as soon as practicable.

JOINT CONSULTATION

OSHA	HSE
Although there are numerous requirements for employees to be kept informed, there are no specific requirements for a formal system of joint consultation.	The Code of Practice requires a formal system of joint consultation between management and employees on all aspects of the Code of Practice. The precise form of the formal system, however, is left for local decision-making. The committee in which the joint consultation takes place will include representatives of senior management, workers' safety, and representatives appointed by recognized trade unions. (The specific legal provisions relating to safety representatives come into force in October 1978.) The consultative committee is required to maintain a watching brief over the health precautions, to promote cooperation between management and employees on matters concerning vinyl chloride, and to act as a forum for the exchange of information and discussion of problems. Facilities are to be provided for the consultative committee to do its work, and the proceedings of the committee are to be published within the plant so that every employee may examine them.

TABLE 23.1—*contd.*

OSHA	HSE

EMPLOYMENT RESTRICTIONS

It is recommended that no woman who is pregnant or who expects to become pregnant should be employed directly in vinyl chloride monomer operations.

REFERENCES

AMERICAN CONFERENCE OF GOVERNMENTAL INDUSTRIAL HYGIENISTS (1971). *Documentation of the Threshold Limit Values for Substances in Work Room Air*, 3rd Edition, Cincinnati: ACGIH.

COOK, W. A., GIEVER, P. M., DINMAN, D. D. and MAGNUSON, H. J. (1971). Occupational acroosteolysis; II An industrial hygiene study, *Archives of Environmental Health*, **22**, 74–82.

CORDIER, J. M., FIEVEZ, C., LEFEVRE, M. J. and SEVRIN, A. (1966). Acroosteolysis combined with skin lesions in two workers exposed in cleaning autoclaves, *Cahiers de Medicin du Travail*, **4**, 3–13.

DANZIGER, H. (1960). Accidental poisoning by vinyl chloride—report of two cases, *Journal of the Canadian Medical Association*, **82**, 828–30.

DINMAN, D. D., COOK, W. A., WHITEHOUSE, W. M., MAGNUSON, H. G. and DITCHECK, T. (1971). Occupational acroosteolysis; I An epidemiological study, *Archives of Environmental Health*, **22**, 61–73.

DODSON, V. N., DINMAN, B. D., WHITEHOUSE, W. M., NASR, A. N. M. and MAGNUSON, H. J. (1971). Occupational acroosteolysis; III A clinical study, *Archives of Environmental Health*, **22**, 83–91.

HARRIS, D. K. and ADAMS, W. G. F. (1967). Acroosteolysis occurring in men engaged in the polymerization of vinyl chloride, *British Medical Journal*, **3**, 712–14.

HEALTH AND SAFETY EXECUTIVE (1975). *Vinyl Chloride Code of Practice for Health Precautions*, London: Health and Safety Executive.

JÜHE, S., LANGE, C. E., STEIN, G. and VELTMAN, G. (1972). Scleroderma, Raynaud's syndrome and acroosteolysis in workers in the polyvinyl chloride industry, *Deutsche Medizinische Wochschrift*, **97**, 1922–3.

KRAMER, C. G. and MUTCHLER, J. E. (1972). The correlation of clinical and environmental measurements for workers exposed to vinyl chloride, *Journal of the American Industrial Hygiene Association*, **33**, 19–30.

LANGE, C-E., JÜHE, S., STEIN, G. and VELTMAN, G. (1974). Die sogenannte Vinylchlorid-Krankheit—eine berufsbedingte Systemsklerose, *Internationales Archiv für Arbeitsmedizin*, **32**, 1–32.

LESTER, D., GREENBERG, L. A. and ADAMS, W. R. (1963). Effects of single and repeated exposure of humans and rats to vinyl chloride, *Journal of the American Industrial Hygiene Association*, **94**, 265–75.

LEVINSON, C. (undated, *c.* 1976). *Work Hazard: Vinyl Chloride*, Geneva: International Federation of Chemical and General Workers' Unions.

LLOYD, J. W. (1975). Angiosarcoma of the liver in vinyl chloride/polyvinyl chloride workers, *Journal of Occupational Medicine*, **17**, 333–4.

MALTEN, K. E. and ZIELHUIS, R. L. (1964). *Industrial Toxicology and Dermatology in the Production and Processing of Plastics*, Amsterdam: Elsevier.

MALTONI, C. (1973). Occupational Carcinogenesis, *2nd International Symposium on Cancer Detection and Prevention*, **2**, 19–26.

MARKOWITZ, S. S., McDONALD, C. J., FETHIERE, W. and KERZNER, M. S. (1972). Occupational acroosteolysis, *Archives of Dermatology*, **106**, 219–23.

McGINTY, L. (1974). *New Scientist*, 13 June, 663, 675; 11 July, 19 Dec.

MINISTRY OF LABOUR (1954). *Annual Report HM Chief Inspector of Factories for the year 1953*, London: HMSO.

MUSACCHIO, C. P. (1974). The VC Standard: the arguments are in, OSHA's decision is imminent, *Occupational Hazards*, September, 101–5.

TORKELSON, T. R., OYEN, F. and ROWE, V. K. (1961). The toxicity of vinyl chloride as determined by repeated exposure to laboratory animals, *Journal of the American Industrial Hygiene Association*, **22**, 359–61.

US DEPARTMENT OF HEALTH, EDUCATION AND WELFARE (1974). *National Institute for Occupational Safety and Health Recommended Standard for Occupational Exposure to Vinyl Chloride*, Cincinnati: National Institute for Occupational Safety and Health.

US DEPARTMENT OF LABOR, Occupational Safety and Health Administration, *Occupational Safety and Health Standards—Standard for Exposure to Vinyl Chloride*, Chapter XVII, part 1910, Federal Register, **39**, 35890–8 (see also Federal Register, **40**, 13211).

VIOLA, P. L. (1970). Cancerogenic effect of vinyl chloride (abstract), *Proceedings of the Xth International Cancer Congress*, Houston, Volume 29.

VIOLA, P. L., BIGOTTI, A. and CAPUTO, A. (1971). Oncogenic response of rat skin, lungs and bones to vinyl chloride, *Cancer Research*, **31**, 516–22.

WARD, A. M., UDNOON, S., WATKINS, J., WALKER, A. E. and DARKE, C. S. (1976). Immunological mechanisms in the pathogenesis of vinyl chloride disease, *British Medical Journal*, **1**, 936–8.

WEAVER, P. H. (1974). On the horns of the vinyl chloride dilemma, *Fortune*, October 1974.

WILSON, R. H., McCORMICK, W. E., TATUM, C. F. and CREECH, J. L. (1967). Occupational acroosteolysis, *Journal of the American Medical Association*, **201**, 577–81.

Further Reading

A comprehensive bibliography is that of HEIMANN, H., LILIS, R. and HAWKINS, D. T. (1975). A bibliography on the toxicology of vinyl chloride and polyvinyl chloride, *Annals of the New York Academy of Sciences*, **246**, 332–7.

WALKER, J. R. A. (1976). *Vinyl Chloride and Polyvinyl Chloride Health Risks*, Technical Bibliography Series No. 1, 2nd Edition, Birmingham: Central Libraries.

Section 6: STRATEGIES

This section comprises one chapter, divided into two parts. Part (A) is principally concerned with the use of hygiene limits as a strategy in occupational health and safety, while Part (B) is devoted to a history of strategies generally.

From Part (B) it will be deduced that by 1915 all the principal strategies (including that of hygiene limits), as well as all the problems had been outlined. The major strategies were thus available prior to, or soon after, the recognition of the problems discussed in Chapters 17–23. It is left to the reader to judge the effectiveness and appropriateness of strategies adopted in relation to these problems.

Chapter 24

Strategies in Occupational Health and Safety

OBJECTIVES OF THE CHAPTER

This chapter is divided into two parts: Part (A) Hygiene limits as a strategy, and Part (B) Control strategies in occupational health and safety.

The objectives of Part (A) are three-fold:

(1) to make clear that many hygiene limits are essentially compromises and that these compromises raise questions of socio–economic as well as of scientific importance;

(2) to outline the principles involved in the process of setting standards; and

(3) to highlight certain administrative and legal problems associated with hygiene limits (a) their restricted applicability outside their country of origin, (b) terminological difficulties, (c) the need to apply qualifications to hygiene limits, (d) uncertainty about the nature of the harm which limits are aimed to control, and (f) the difficulties encountered in legal enforcement of hygiene limits.

Part (B) of this chapter:

(1) provides a framework for understanding strategies in occupational health and safety,

(2) demonstrates that the principles underlying the principal strategies were evolved prior to 1915, and

(3) suggests that present-day discussion about occupational health and safety sometimes reiterates certain strategic problems identified long ago.

PART (A): HYGIENE LIMITS AS A STRATEGY

The setting and application of hygiene limits is a strategy of the utmost importance in human safety, even though no such system has yet been evolved which is entirely problem-free. In this part of the chapter, data are put forward in support of the following concepts:

(1) That many hygiene limits are compromises, and that the compromises are of a socio–economic as well as a scientific nature.

(2) That standard setting should be seen as a two-part activity, namely the elaboration of scientific data and decision making based

upon that data and upon certain, often conflicting, socio–economic considerations.

(3) That difficulties of practicability arise when systems for hygiene limits are incorporated into legal and administrative frameworks.

HYGIENE LIMITS: GENERAL COMMENTS

Hygiene limits are traceable to Germany in the 1880s. Despite this long history some people question the very concept. Levinson, for example, in reporting a conference held by the International Federation of Chemical and General Workers' Union (of which he is Secretary General) wrote (1975):

> One of the most serious problems raised by the trade unionists, industrial hygienists and medical specialists at the ICF Conference on Occupational Health was the problem of what level of exposure could be permitted in the work place for the large number of chemical substances which are toxic or otherwise hazardous to humans. There really is no answer to this question as, obviously, no exposure at all to hazardous substances is the best prevention against toxicity. This is particularly true for the substantial number of chemicals which have been shown to cause cancer in humans when workers are exposed to these substances in the work environment.

The 'no-exposure' approach is adopted in some countries in cases of certain proven carcinogens but current thinking holds that this approach is only appropriate for a limited number of substances. It is assumed that the use of all other pathogenic substances involves some degree of unavoidable contamination of the milieu (for the employed or for the population at large). That this assumption is widely accepted is instanced by reference to the paper of two Soviet scientists, Roschin and Timofeevskaya, published in 1975. They wrote:

> ...the hygienic standard is not the optimum for the environment... it would be better to do away with pollution altogether. But complete elimination of contaminants in the foreseeable future is unrealistic, since under conditions as they exist in industry we fail to have at our command an ideal technology, devoid of waste materials.

Elsewhere in the paper they expressed the opinion that:

> the establishment of the maximum permissible content of harmful substances in the environment is of immense importance,

since it forms a backbone for the environment and protection against pollution.

That industry frequently ignores some of the established standards, they suggested, in no way prejudices 'their (the standards) vast importance for the whole cause of protection of the environment against all sorts of pollution'.

THE SETTING OF LIMITS

The process of setting limits presents numerous difficulties, not least the fact that as yet there exists no internationally standardized methods. Soviet and American approaches to limits were compared by Roschin and Timofeevskaya, from which comparison it is clear, as also from the reviews by Hatch in 1972 and 1973, that limit setting in the USSR and USA relies on markedly contrasting scientific ideas.

Scientific Data Needed for Standard Setting

The prime requisite for the setting of limits should be the dose–response relation and dose–effect relation (discussed in Chapter 12). In theory, a committee charged with decision making of a socio–economic nature should be provided with numerical data showing the quantitative relation between the contaminant and its effects. As has been mentioned in previous chapters, these quantitative relations are difficult to establish—at least with any certainty. Human beings cannot be used in experiments designed to validate dose–response relations. Instead, attempts have to be made to relate the pattern of human disease to patterns of exposure to harmful inputs at work and elsewhere. In most instances the necessary data do not exist. Other possibilities include chemical analogy or data derived from animal experiments. There are difficulties with analogical data (see below) while perhaps even more formidable are the problems involved in the use of animal (or other biological) data for predicting human disease patterns. Nevertheless, as far as ethical experimentation and obser-vational difficulties permit, the relation between the effect and the harmful input has to be studied *both* in qualitative and quantitative terms.

Socio–Economic Decision Making

Decision making becomes socio–economic in nature when it involves questions of cost which are expressed in social or economic terms. Increasingly, decision making is a matter of choice of strategy.

Influencing the choice nearly always will be the question of cost of implementation; in general, committees will be looking for 'the best buy' among the possible strategies with which to deal with a particular problem. Up to a point decision making can be based on rational and scientific criteria, but overriding these will be emotional and political factors such as those identified in previous chapters.

In 1975 the US National Academy of Sciences published a report on decision making in relation to the regulation of chemicals in the environment. Within the report's conclusions are to be found clear statements of the principles upon which future decision making for codes of practice and hygiene limits should be based. As the report commented, there is no scientific formula for making decisions, since each decision always involves values about which the affected parties disagree. Hence, the values of the decision makers must play a crucial role in the outcome. No satisfactory way exists to summarize all the costs or benefits of a strategic option, and there are no terms which can be mathematically added, subtracted or compared. As the report further stated, there is no substitute for an experienced decision maker exercising good judgement.

The decision maker is encouraged by the NAS recommendations to follow four cardinal principles:

(1) The decision making process should be as open as possible to outside participation. The process of making the decision should include the groups in society who will experience the impact of the decision; the consideration given to the interests of potentially affected individuals should be proportional to the anticipated effects of the decision on those individuals.

(2) Adequate and reliable information is fundamental to sound decision making.

(3) The regulatory agency, that is the government agency charged with the responsibility, should be as explicit as possible about the factors which have entered into the making of the decision both during and after the decision making process: relevant factors should clearly be displayed for the decision maker and for the public.

(4) Existing knowledge about the principles of decision making should be utilized. Even though no mathematical technique for decision making exists, techniques developed by decision theory and benefit–cost analysis can provide the decision maker with a useful framework and language on which to formulate the decision making process.

The NAS recommendations represent a departure from established practice in most, if not all, spheres of decision making in human safety. Stokinger (1969) described the composition of the 'TLV

Committee' (TLVs are discussed later) showing that it is an expert committee. No complete picture has so far been given of the decision making procedure but it seems unlikely that the principles advocated by the NAS were utilized, in the past, in the setting of TLVs.

Neither, it seems, are the principles utilized in the setting of radiation protection standards. Pochin et al. (1976) identified two main considerations in the setting of numerical radiation limits:

(a) the type and incidence of harm caused by the radiations, and
(b) the degree of safety that should be regarded as necessary, taking into account economic and social factors.

With these considerations in mind, presumably, the International Commission Radiological Protection (ICRP), a committee of international experts, makes recommendations which, according to Pochin et al. (1976), are intended to:

conform to a level of safety likely to be generally acceptable, e.g. by limiting total occupational risks to levels at least as low as those in occupations which are widely regarded as safe.

The ICRP leave 'National' authorities to decide whether the recommendations are appropriate for their countries' circumstances. It would be at this stage, presumably, that the NAS decision making procedure (or its equivalent outside the USA) could be instituted in decision making about radiation limits.

PROBLEMS ASSOCIATED WITH HYGIENE LIMITS

As previously stated, problems associated with hygiene limits become prominent when systems of hygiene limits are incorporated into administrative or legal frameworks. Sometimes problems arise because the hygiene limit is being used for an administrative or legal purpose for which it was not intended. Other problems occur as a result of the complex nature of hygiene limits and problems also would be encountered as soon as the limit was applied to any administrative or legal framework. As stated previously, five such are identified for discussion in this study:

(a) the restricted applicability of hygiene limits outside their country of origin,
(b) difficulties of terminology,
(c) the need to apply qualifications to hygiene limits,
(d) uncertainty about the nature of the harm which hygiene limits are intended to control, and
(e) difficulties encountered in legal enforcement.

By way of example, the American Conference of Governmenta Industrial Hygienists' TLV system has been selected for discussion* because of its wide use in the UK.

(a) Restricted Applicability of Hygiene Limits Outside the Country of Origin

The ACGIH hygiene standards are used for *advisory purposes* by the Health and Safety Executive in the UK. The ACGIH system ha been adopted in its entirety—because this is a condition imposed by ACGIH. At present the position is that the Health and Safety Executive annually reprint the ACGIH hygiene standards as a Guidance Note (HSE EH15/76 was current in 1977). Whether Britain will continue indefinitely to import *en bloc* the ACGIH system is debatable in that problems arise from the detailed differences in approach to particular dangers adopted by the USA and UK. In 1977, the approaches differed for: asbestos, carbon monoxide, cotton dust, 4,4'-methylene-bis-(2-chloraniline), mica and talc, other non-siliceous mineral dusts, noise, vinyl chloride, and chromium metal and its insoluble compounds. Moreover, certain substances listed by the ACGIH (though not necessarily given a TLV) are prohibited in the UK by the Carcinogenic Substances Regulations 1967 (see Chapter 18). In some instances the UK and USA also differ in their approaches to radioactive substances. In due course it is to be expected that some of these differences may be reconciled, but noise limits, for example, are so different and the difference seems so deeply rooted that harmonization is not in prospect.

The preface to the ACGIH TLV system (1976) contains the specific statement that the system is not intended for adoption by countries whose working conditions differ from those of the USA and where substances and processes differ. Systems of hygiene limits should, in general, be seen as having restricted applicability beyond their country of origin.

(b) Terminology

It seems inevitable that there should be difficulties with terminology. The terms *Threshold Limit Value, time-weighted average*, and *Maximum Allowable Concentration* continue to cause confusion,

---

*The American Conference of Governmental Industrial Hygienists is not, as its name appears to suggest, a government organization. In fact, ACGIH resembles a learned, scientific society like its counterpart the British Occupational Hygiene Society.

even though their definitions have been established for some years.

It is important, firstly, to note that the term 'Threshold Limit Value' (TLV) is a proprietary term belonging to ACGIH. Since 1976 the ACGIH has used 'Adopted Value' to refer to the numerical values for TLVs. This is a useful step forward in avoiding confusion and *Adopted Value* should be used to refer to the numerical value published by the ACGIH under the generic term Threshold Limit Value.

The term Threshold Limit Value is widely used and is likely to remain so for many years because it has become a generic term for hygiene limits generally. In recognition of this fact it is useful to bear in mind changes in terminology of hygiene limits which have taken place over approximately the past 30 years.

The next step towards understanding a complex terminology is to note that *time-weighted average* describes a method of calculation for use in connection with certain types of limits including but not restricted to the ACGIH's Adopted Values.

The principle involved in expressing a time-weighted average is embodied in the following:

Time-weighted average =

$$\frac{\text{Average concentration during working hours} \times \text{Number of hours actually worked}}{\text{Total number of working hours}}$$

Where the extent of exposure can only be expressed in terms of a complex pattern of level and time, the calculation is extended to include each combination of level and time, as in Table 24.1.

There are certain constraints which must be observed in computing time-weighted averages. The ACGIH system invokes 'excursion

TABLE 24.1

Duration	Level	Duration × level
t_1	l_1	$t_1 l_1$
t_2	l_2	$t_2 l_2$
$\vdots$	$\vdots$	$\vdots$
t_n	l_n	$t_n l_n$
$t_1 + t_2 + \cdots + t_n$		$t_1 l_1 + t_2 l_2 + \cdots + t_n l_n$

$$\text{Time-weighted average} = \frac{t_1 l_1 + t_2 l_2 + \cdots + t_n l_n}{t_1 + t_2 + \cdots + t_n}$$

factors' which impose an upper limit on the peak value that may be used and without which the computation of time-weighted average is open to abuse; for example, a very high peak of contamination could be followed by a prolonged period of low contamination. (As was commented in Chapter 12 the time-weighted average method of calculation makes no allowance for dose-rate dependency.)

Excursion factors, described as rules of thumb, are given for TLVs not bearing a 'C' notation (see below). Excursion factors are inversely related to the magnitude of the TLV as follows:

TLV 0–1   (ppm or mg/m^3), excursion factor = 3
TLV 1–10   (ppm or mg/m^3), excursion factor = 2
TLV 10–100   (ppm or mg/m^3), excursion factor = 1.5
TLV 100–1000   (ppm or mg/m^3), excursion factor = 1.25

They are applied by multiplication; for example acetone has a TLV of 1000 ppm, the excursion factor is 1.25 and the excursion limit is therefore 1250 ppm.

The ACGIH in its first listings, published in 1946, referred to 'Maximum Allowable Concentrations' (Dinman, 1973). It soon became apparent, however, that the values published as maximum allowable concentrations were not in fact *maxima*, but were instead *average* values. Because 'Maximum Allowable Concentration' did not accurately reflect the principles involved it was replaced by the term 'Threshold Limit Value'. From time to time 'maximum permissible concentration' was used to denote not only 'true' maximum allowable concentration but also—and somewhat confusingly—average concentrations.

In 1976 the ACGIH specified three categories of Threshold Limit Values (TLVs):

(i) *Threshold Limit Value–Time-Weighted Average (TLV–TWA)*— the time-weighted average concentration for a normal 8 h working day or a working week of 40 h, to which nearly all workers may be repeatedly exposed, day after day, without adverse effect,

(ii) *Threshold Limit Value–Short-Term Exposure Limit (TLV–STEL)*—the maximum concentration to which workers may be exposed for a continuous period up to 15 min without suffering from (1) intolerable irritation, (2) chronic or irreversible tissue change, or (3) narcosis of sufficient degree to increase accident proneness, impair self-rescue, or materially reduce work efficiency. The TLV–STEL provided that no more than four excursions per day should be permitted, that there should be at least 60 min between exposure periods, and that the daily TLV–TWA should not be exceeded. TLV–STEL should thus be considered a maximum allowable concentration, or absolute ceiling, not to be exceeded at any time during the

15 min excursion period. The TWA–STEL should not be used as an engineering design criterion or considered as an emergency exposure level (EEL).

TLV–STEL are based on one or more of the following criteria:

(1) TLVs adopted by ACGIH including those with a 'C' or 'ceiling' limit,
(2) TWA–TLV Excursion Factors listed previously,
(3) Certain statutory limits enforceable in the USA.

(iii) *Threshold Limit Value–Ceiling (TLV–C)*—the concentration that should not be exceeded even instantaneously.

(c) Qualifications Needed with Hygiene Limits

The ACGIH's preface to the list of TLVs gives the general definition of TLV:

> Threshold Limit Values refer to airborne concentrations of substances and represent conditions under which it is believed that nearly all workers may be repeatedly exposed day after day without adverse effect. Because of wide variation in individual susceptibility, however, a small percentage of workers may experience discomfort from some substances at concentrations at or below the Threshold Limit, a smaller percentage may be affected more seriously by aggravation of a pre-existing condition or by development of an occupational illness.

The preface for the British readers given by the HSE in EH 15/16 contained the following caution:

> Careful attention should be given to the preface which explains the philosophy behind the list. The following points are particularly important: (1) TLVs are not sharp dividing lines between 'safe' and 'dangerous' concentrations, (2) the best working practice is to keep concentrations of all airborne contaminants as low as practicable, whether or not they are known to present a hazard, and irrespective of their TLVs, (3) the absence of a substance from the list does not indicate that it is 'safe', (4) the application of the TLV to a particular situation should be undertaken by trained occupational hygienists.

(d) Uncertainty about Harm which Limits are Intended to Control

The ACGIH list of adopted TLVs gives no detailed indication of the harm against which it is intended that people should be safeguarded. This problem was examined by Stokinger, who in 1969 discussed the ACGIH's 1968 list of 414 substances. He found that of these, 49% were based on 'systemic effects' on organs or groups of organs, 40% on irritation, 5% on narcosis (dulling of consciousness) and 2% on odour. Thus there is a range of harm to which TLVs relate—from the trivial to carcinogenesis.

The ACGIH from time to time publishes its 'Documentation of TLVs' (third printing: 1971). 'Documentation' comprises summaries of toxicological information about substances for which ACGIH have adopted TLVs. In some instances the harm is clearly identified, but there are numerous instances where the standard-setting committee has been obliged to draw conclusions in the virtual absence of any data relating to human exposure. The bald listings as set out in the ACGIH list of adopted values, for example, do not convey the uncertainty which is so obvious in the 'Documentation'.

A further source of uncertainty is revealed by Stokinger (1970): it seems that in 1968 24% of *all* TLVs published by ACGIH were based upon analogy. In this context, analogy means that the assumption has been made that chemically similar substances produce biologically similar effects. As we have seen in several previous chapters, this is by no means a universally valid assumption.

(e) Difficulties with Enforcement of Time-weighted Averages

TLVs present difficulties for enforcement authorities. Where the time-weighted average is above the TLV by a margin which exceeds both the excursion factor and any measurement error, or where the TLV is expressed as a ceiling value, the authority has the beginnings of a case. It is explicit in the concept of time-weighted average that the average must be computed for an exposure period, normally stipulated as 8 h. Thus an inspector might be required to show that the level of contaminant exceeded or was likely to exceed the TLV for the 8-h period. With measurements only slightly in excess of the TLV, the inspector may have to sample continuously for the bulk of the exposure period before it can be demonstrated to a court's satisfaction that the recommended level has, in fact, been exceeded. (This, of course, assumes that the inspector can refute a defence resting on, for example, the imprecision of his/her instruments or the uncharacteristic nature of the day on which measurements were taken.) It is going too far to suggest that TLVs are unenforceable in law but there are not many instances of prosecution succeeding against vigorous technical defence. From an enforcement authority's point of view, maximum concentrations rather than maximum average concentrations are much to be preferred. This, presumably, is why the minimum sampling time specified for the purposes of the Asbestos Regulations, 1969 (see Chapter 20), is so short.

CONCLUSIONS FROM PART (A)

Most hygiene limits are compromises. Nevertheless, hygiene limits represent an important strategy in occupational health and safety, a strategy which can be applied successfully only so long as the limitations outlined previously are recognized. The ACGIH system was selected for the purposes of the present discussion, but all the general points (and many of the specific ones as well) are relevant to other systems. Because socio–economic considerations enter into the decision making, discussions about hygiene limits have a political as well as a scientific basis.

PART (B): CONTROL STRATEGIES IN OCCUPATIONAL HEALTH AND SAFETY

Recognition of danger is an essential prerequisite for its control. The recognition marks the starting point in discussions about what action is required in order to bring the danger under control. Even the most cursory glance at the history of health and safety at work shows that there is extreme variability in the time taken for adequate control procedures to be instituted. At one extreme there are problems such as those presented by vinyl chloride which were dealt with at great speed once the extent of the danger had been fully perceived. At the other extreme are problems such as noise, over which control has yet to be fully effected even though recognition dates back many years—in the case of noise the definitive British report of occupational deafness was published in 1886 by Thomas Barr, the otologist. In the USA, a report of comparable standing was published four years earlier by Holt.

Very often, understanding of the processes of danger has not been fully achieved until long after the means for effective control have been identified, if not instituted. Consequently, failure to understand danger cannot be advanced as a common cause of tardiness in effecting control. A much more frequent cause of delay is failure to agree, on the part of decision-makers and legislators, about the strategy to be adopted.

HEALTH AND SAFETY STRATEGIES

A strategy for health and safety (called a safety strategy for brevity's sake) is a plan for projecting and directing the larger operations in a campaign against danger. Whenever a danger is recognized—publicly, at least—detailed discussions ensue in Parliament and in the press, entered into by employers' representatives and by employees' representatives, by scientists and by members of the public. In numerous instances, over almost two centuries, discussions about strategies have been protracted, often acrimonious, and—all too often—indecisive.

By identifying the origins of safety strategies, and the obstacles to them, and by analysing their strengths and weaknesses, students of health and safety may assist decision-makers in reaching conclusions and, perhaps, even hasten action. In the selective historical outline in the following paragraphs the early applications of the principal safety strategies are identified, and certain difficulties and obstacles are highlighted. As will be seen the principles and obstacles are clearly recognizable prior to 1915 so that there is no need, for present purposes, to extend the historical outline beyond that date.

HISTORICAL DEVELOPMENT OF STRATEGIES UP TO 1915

Occupational health and safety is no new study, as this outline of historical development shows. Special problems of occupational disease and safety have long been recognized, attention having been drawn to this aspect of preventive medicine by Agricola in his manual of mining technology, in which the health and safety of miners was discussed at some length (Hoover and Hoover, 1912). Hippocrates, Pliny and Paracelsius are also said to have been aware of the problems. The science of occupational health and safety cannot, however, be said to have been founded until 1700, with the publication of Ramazzini's 'De Morbis Artificium Diatriba' (Wright, 1940).

Ramazzini's study shows an extraordinary insight into the causes of occupation-linked diseases. He pointed to the inherent dangers of prolonged periods of work without rest as well as to the harm caused by 'want of ventilation' and 'unsuitable temperatures'. Ramazzini counselled against work of a dusty nature being carried on in confined spaces, and suggested that the individual should cease work immediately upon the appearance of symptoms of disease. Ramazzini was evidently a pioneer among safety strategists—but how far have his strategies withstood the test of time, and would they still be appropriate nearly three centuries later?

The problem of metal poisoning, specifically lead poisoning, interested British physicians from the earliest days of the chemical industry in the late eighteenth century, although there are instances of concern over metal poisoning in crafts in much earlier times. One of the pioneers of research into occupationally linked lead poisoning was Dr Percival of Manchester, the creator in 1796 of an unofficial Board of Health in Manchester (Hunter, 1975). In 1816 the problem of lead poisoning was conceded by Josiah Wedgwood the second, when, in evidence to a Parliamentary Select Committee on the employment of children in the manufacturing trades, he spoke of the 'unwholesome' nature of dipping, part of pottery manufacture. Pottery workers, he said, as in 'other businesses in which workmen have to do with lead . . . are, if careless in method of living, . . . very subject to disease' (Meiklejohn, 1963).

The problem was thus recognized, and in 1822 the Royal Society of Arts awarded a prize for a paper on 'glaze for vessels not prejudicial to the health of those who make them' (Meiklejohn, 1963), a glaze which was not compounded of lead. Here may well be the earliest example of a safety strategy directed towards deliberate substitution of one compound for another solely for reasons of occupational hygiene. But the commissioning of a paper, however, was as far as the matter was taken for pottery workers and, although further evidence was accumulated, concern about the effects of lead poisoning was not fully translated into control and prevention for pottery workers until legislation at the start of World War I.

M. J. Breckin (1977, personal communication) pointed out to me that the mid-nineteenth century concern about lead poisoning in the pottery industry rather overshadowed the much older problem of *potters' rot*. This is silicosis (see Chapter 15), possibly complicated in some victims by tuberculosis.

The problem was recognized by Dr Greenhow in the 1860 report but the problem continued to lack the attention given to lead until 1892 when Dr J. T. Arlidge, a physician at the North Staffordshire Infirmary, pointed out the severity of respiratory disease among Staffordshire pottery workers. In connection with potters' rot, Arlidge commented: 'in one sense, indeed, it is unfortunate that it does not for the most part awake attention by any immediate tangible consequences. Its disabling action is very slow, but it is ever progressive, and until it has already worked its . . . results . . . it is let pass as a matter of indifference—an inconvenience of the trade'.

The existence of the problem a century and a half earlier is evident in the following quotation which comes from the preamble to Patent granted by George I to Thomas Benson in November 1713. Here was an early attempt to solve the problem of respiratory disease among potters (with acknowledgement to M. J. Breckin).

Whereas, our trusty and well-beloved Thomas Benson, of Newcastle-under-Lyme in our County of Stafford, engineer, hath by his petition humbly represented to us that in Staffordshire there is a manufacture carried on of making White Pots, the chief ingredient of which is Flint Stone, and the method hitherto used in preparing whereof has been pounding or breaking it dry, and afterwards sifting it through fine launs, which has proved very destructive to mankind, insomuch that any person, ever so healthful or strong, working in that business cannot possibly survive over two years occasioned by the dust sucked into his body by the air he breathes, which, being of a ponderous nature, fixes there so closely that nothing can remove it, insomuch that it is now very difficult to find persons who will engage in the business to the great detriment and obstruction of the said trade, which would otherwise by reason of the usefulness thereof be of great benefit and advantage to our Kingdom. That the petitioner has with great pains and expence invented and brought to perfection an Engine or new method for the more expeditious working the said Flint Stone whereby all the said hazard and inconveniences attending the same will effectually be prevented; that he has, at his own charge, made several experiments of his said invention whereby he fully knows the same will answer the ends above purposed in every particular, to the manifest improvement and advantage of the said manufacture, and preserving the lives of many of our subjects employed therein, and proposes to perform the same in the manner hereinafter described—viz.: The Flint Stones are first sprinkled with water, insomuch that no dust can rise to the hurt or damage of the workmen; then crushed as small as sand by two large wheels of the bigness and shape of millstones, made of iron, to turn round upon the edges by the power of a water-wheel, and afterwards conveyed into large pans made of iron, for that purpose circular, in which there are large iron balls which, by the power of the water-wheel above named, are driven round by such a swiftness of motion that in a little time the flint stones so broken are made as fine as oil itself without the use of launs, and when so done, by turning of a cock, empties itself into casks provided for that purpose and so kept therein for the uses above-mentioned; and having humbly prayed us to grant him our Royal Letters Patent for the invention, according to the Statute in such case made and provided; We being willing to give encouragement to all arts and inventions which may be of public use and benefit, are graciously pleased to gratify him in his request.

Throughout the Victorian era, and indeed before it, concern for the amelioration of the social evils associated with the 'factory system' prompted medical men to discourse upon the ill-health of employees attributable to that system. The various committees convened to examine factory employment drew upon a great deal of medical evidence. From this evidence the link between factory employment (especially employment in the Lancashire cotton mills) and occupational ill-health became apparent. The factory system, especially that in the northern industrial areas, was indicted by some as being to blame for tuberculous disease (consumption, phthisis and scrofula) and other ailments, many of which affected the young. Although the occupational linkage with tuberculous disease was later disputed, notably by the Manchester physician Dr Daniel Noble (1843), it was generally agreed that the high incidence of ill-health among the adult working classes, especially in the new urban areas, was due to a combination of occupational and social factors.

As for strategies at the time, it is a tenable view that these were dominated by socio–economic factors to the detriment of health and, particularly, safety at work. Nevertheless, concern for conditions of employment of workers was accompanied throughout the 1830s by a concern, in some quarters at least, that greater provision should be made for safety. The early factory inspectors noticed with alarm the high incidence of accidents to the young persons whose labour they were regulating, and to other persons employed in the mills. (There was at least one common law case concerning a young girl caught by unfenced machinery.) But in the event the factory inspectors' advice did not receive legislative expression until 1844, in the Factories Act of that year. Its Section 20 required certain machinery to be securely fenced and the fencing not to be removed when the machinery was in motion.

The spirit of that section was a reflection of the factory inspectors' view that the danger lay in the machinery. This was in marked contrast to the idea, widely prevalent at the time, that misconduct, recklessness and want of caution on the part of work people were the cause of accidents (see, for example, Cooke-Taylor's account of cotton mills published in 1844).

The coal-mining industry's place in the historical development of safety is highlighted in this brief outline because the collieries were at this time perhaps the biggest category of employers of labour, health and safety in the mines has always been in the forefront of concern (see Chapter 21), and because legislation regarding coal mining is well documented and easily picked out from among the mass of contemporary parliamentary legislation. Factory legislation of the time, and the discussions of the time concerned with it, concentrated on labour regulation; health and safety were for the most part matters of secondary importance. As an industry, coal mining was also to the fore when it came to the appointment of *qualified* inspectors, appointment of which was called for by the Miners' Association of Great Britain in a petition to the House of Commons in May 1847 and acceded to by Act of Parliament in 1850 (D. B. Owen, 1977, personal communication). As a result the Home Secretary appointed four technically qualified and experienced mining engineers as inspectors, and, although they were responsible for no fewer than 2000 collieries, the Act at least provided a step in the right direction.

Chronologically, the next steps of importance to health and safety at work were made in the early 1860s. In 1861 and 1862, as a result of reports by the Medical Officer of the Privy Council (Privy Council, 1861), further examples of the link between occupation and disease were revealed. In the first of these reports (see Hunter, 1975), Dr Greenhow announced his conviction that much of the very high mortality from pulmonary disease, in different areas of England and Wales, was due to the inhalation of dust and fumes at the place of work including, *inter alia*, potteries. A Dr Knight, in 1844, had echoed numerous previous suggestions that the high death rates of grinders and cutlers in Sheffield (where the average life expectancy of a dry grinder was 29), was directly attributable to the inhalation of dust, an inevitable result of the nature of their work. The 1862 Report confirmed this speculation, and added a weight of evidence to the already bulky file on ill-health among 'dippers' in the pottery trade (Meiklejohn, 1963). What most satisfied advocates of increased health and safety provisions at work, however, was the strong language in which the conclusions of the 1861 Report was couched, namely that 'the canker of industrial disease gnaws at the very root of our industrial strength'.

Until the Act of 1864 (which resulted from a Royal Commission in 1862) factory legislation had covered only textiles and certain processes associated with textiles. The Acts of 1864 and 1867 brought earthenware works within factory legislation and required that factories be ventilated, in 'such a manner as to render harmless, so far as is practicable, any gases, dust or other impurities generated in the process of manufacture, (which were) injurious to health.' (Bridge, 1933.)

Two major strategic principles of occupational safety and hygiene were thus established:

(1) Emphasis should be put upon control of the workplace rather than individual hygiene. This is the cardinal principle of prevention—it underlies Sir Thomas Legge's axioms (see below) which have impelled practitioners since their publication, posthumously, in 1934; and it directs the strategies for occupational health and safety visualized in this book.

(2) Epidemiological research is a necessary element in the search for, and control of, occupation-linked pathogenic factors. There is, however, a third principle which belongs with these other two, even though it was enunciated at another time and in another context: that the epidemiological search cannot of itself be sufficient; knowledge is needed of *how* the pathogenicity takes place in the living body in order that its detection can be anterior rather than posterior to the events which become the epidemiologist's data.

Further Acts of Parliament in 1878 (a consolidating Act), 1883 and 1891 supplemented the somewhat limited coverage of the 1864 and 1867 Acts. The view of Fife and Machin (1976) is that the 1878 Act was the first comprehensive health and safety legislation for factories in Britain. Apart from factory legislation, however, the Coal Mines Regulation Act of 1872 was notable because it gave to the workmen the right of systematic inspection of the mines (discussed by Williams, 1960).

For this chapter, it is sufficient to note that the factory legislation extended various provisions of earlier Acts to include compulsory ventilation where dusts injurious to health were produced, and improved sanitation. Generally, these legislations furnished factory owners and managers with a more comprehensive list of dangers to which attention was to be paid.

By 1891, when powers were provided for the certification of processes dangerous or injurious to health, dramatic progress in the area of occupational health and safety had thus been made. Progress subsequently was patchy. There were notable advances, for example, in the control of white lead manufacture as well as in the formation of special rules for the manufacture of paints, lucifer matches and explosives. In the last decade of the nineteenth century particular attention was also paid to the dangers inherent in arsenic extraction, as well as to the multiple dangers likely to arise in the chemical industry (Bridge, 1933, *passim*).

Once the more obvious dangers to occupational health and safety had been tackled, attention was directed to matters of secondary importance. In the early 1890s legislation was enacted which provided for such things as increased cleanliness, by the provision of washing accommodation, the wearing of special clothing and respirators, and the provision of acidulated drinks. Such legislation, however, was largely cosmetic, in that, although the manufacturers might insist that disease was largely caused by slovenliness and disregard of personal hygiene, the occupational link was in effect being ignored.

During the decade 1890–1900 several important developments took place. Between 1893 and 1899 the Dangerous Trades Committee enquired into a vast number of industrial processes, concentrating on those involved in earthenware manufacture and those which resulted in arsenical, phosphorus and carbon disulphide poisoning, as well as on the dangers from dust and from naphtha. From 1895, medical practitioners were required to notify the Chief Inspector of Factories of any cases of arsenical and phosphorus poisoning, a statutory duty, however, of which many remained for long unaware. Notification of cases of lead poisoning also became mandatory in 1895, and thereafter for a number of years something like 400 cases were reported annually in the pottery trade alone, a figure which decreased to 4 in 1940.

The statistics undoubtedly increased public awareness of the whole problem of health and safety at work. In 1897 the passing of the Workmen's Compensation Act led to an almost immediate increase in the number of doctors in industry, although many were employed on a purely consultative basis. Inevitably, since their principal responsibility was to help employers deny liability or reduce compensation, they were generally distrusted by the workforce (see Meiklejohn, 1956 and Schilling, 1973). Such a situation provided an unhappy application of medicine to industry. Compensation's influence on health and safety at work generally is briefly referred to later in this chapter.

A further important development took place in 1898, with the appointment of Dr (later Sir Thomas) Legge as Medical Inspector of Factories. The whole approach to the problem of lead poisoning was thereby altered, blame no longer being attached to the carelessness of the worker but to the manufacturing process. Dr Legge felt obliged to draw up a leaflet, entitled 'Lead Poisoning: How Caused and How Best Prevented', pointing out the risks and best means of protection from lead poisoning. This leaflet was then distributed to workers in the relevant industries, and thereby was started a practice which was to continue down the years (Bridge, 1933).

In that same year, 1898, a Reform Committee was set up by Gertrude Tuckwell, secretary of the Women's Trade Union League, the intention of the committee being to direct especial attention to the effects of lead poisoning in women workers and the attendant high rate of still-births. More Special Rules were introduced in 1900, by which it was stipulated that all lead was to be 'fritted' (i.e. fused with siliceous matter and rendered insoluble) and that lead was to provide not more than 2% of any glaze content. It was further decreed that no woman or child should be employed in an industry where there was a risk of lead poisoning, and that monthly medical examinations of the workforce be provided. Another prerequisite, the provision of a regular supply of milk, was promulgated by the Pottery Regulations of 1913.

One example just given represents a strategy towards a vulnerable group: the total exclusion of certain people from employment. As such it is a development of Ramazzini's strategy of advising individuals to cease work immediately upon developing symptoms of disease. Whether Gertrude Tuckwell fully supported the exclusion strategy is not apparent but even if she did, events after 1913 were to demonstrate its vulnerability.

Apart from human life, World War I consumed great quantities of munitions. In 1915, David Lloyd George became Minister for Munitions and almost immediately conceded to Christabel Pankhurst, the suffragette leader, the right of women to work—which they did in their hundreds of thousands in the shell factories (Marwick, 1977). These factories handled, *inter alia*, 2,4,6-trinitrotoluene which enters the body by skin pervasion as well as inhalation and ingestion. It causes dermatitis, cyanosis, gastritis, anaemia and a toxic jaundice which has a mortality rate of 30% (Hunter, 1975).

Arguably, shell filling was just as dangerous to women as work in lead processes. Certainly, the women suffered from TNT poisoning and died also, of course, from the TNT explosions which occurred at Silvertown, East London, in 1917, and other munitions factories. The strategy of excluding the vulnerable from employment was not invoked in the instance of munitions because it would not have been in the national interest, nor would it have been politically acceptable. From this example and the other instances in this chapter can be induced an important maxim: that strategy in health and safety at work is ultimately determined by social and political pressures.

How social and political pressures arise is not within the scope of this book; however, it is evident that these are influenced, among other things, by awareness of problems and by knowledge of means of tackling problems on the part of decision-makers, as well as availability of resources. A view expressed previously in this chapter is that early British factory legislation was primarily and no doubt necessarily aimed at regulating the employment of labour and not at securing health and safety at work. The balance slowly shifted as social abuses of employment lessened and also because there were people concerned to shift the balance. Some have been mentioned already; other notable names include Leonard Horner, one of the four factory inspectors appointed under the Act of 1833. He was one of those pioneers who recognized the dangers to safety of people at work in mills. His pragmatic approach to the fencing of machinery required under the 1844 Act is documented in his reports (see for example Reports, 1846).

Another notable name is Alexander Redgrave (1818–1894), who was appointed first *Chief* Inspector of Factories under provisions first introduced in the 1878 Act. His contribution, marked by a Knight-hood, is commemorated eponymously in a standard legal text (Fife and Machin, 1976). One instance of Redgrave's far-sightedness is to be seen in his attitude to women working with lead, expressed in a report of 1882 (Hunter, 1975). He specifically advised *against* the exclusion of women on the ground that women workers would have difficulty in finding alternative employment. As has been stated, however, the legislators subsequently ignored his advice.

A historical point which has not been studied sufficiently here or elsewhere is the extent to which many of the pioneers disagreed with *their* employers—even when the employer was the Government itself. A notable example is the resignation on 29 November, 1926 of Sir Thomas Legge (1863–1932) from his post of Senior Medical Inspector of Factories because the British Government refused to ratify the Geneva White Lead Convention (aimed, *inter alia*, at prohibiting the use of white lead in paint). In an allusion to the resignation his distinguished colleague Dr S. A. Henry (1934) wrote in his posthumous introduction to Legge's book: 'The fact that he was a Civil Servant and could in no sense be responsible for any policy which a higher authority deemed advisable, did not appeal to him . . . .'

The reason for the British Government's refusal was that it wanted to assess the effectiveness of the then newly drafted Lead Paint Regulations due to be brought into force in 1927. This example highlights the main premiss of this chapter: that uncertainty over strategy is one factor, often important, which impedes progress in health and safety.

CLASSIFICATION OF STRATEGIES

A strategy in health and safety, as previously stated, is a plan in a campaign against danger. Strategies can be classified on their relevance to danger; four types of strategy are recognizable: (1) essential (2) prerequisite (3) adjunctive, and (4) collateral.

(1) Essential Strategies

Essential strategies are those which are directed at danger and bring about its elimination, reduction or restriction, or the amelioration of its immediate consequences for life and limb. Subdivisions of essential strategies can be made into (*a*) *pre-accident*, and (*b*) *post-accident* strategies; it should be noted that 'accident' is here being used figuratively to imply any event of whatever duration which causes or threatens to cause death or bodily harm of whatever nature as a result of the activities of work. Pre-accident strategies are those aimed at forestalling *eventualities*, and post-accident strategies are concerned with ensuring that *events* cause the least possible harm in terms of severity of effect or numbers affected.

One of the great nineteenth-century principles can be used to further classify essential strategies, namely that emphasis should be put on control of the work place rather than the individual. Those strategies which secure the former are dubbed *safe-place strategies* to distinguish them from *safe-person strategies* which aim at individual hygiene, used in a broad sense (for further discussion see Atherley, 1975).

Primacy of the safe-place strategies is surely self-evident yet, even in 1977, there is still debate about the relative position of particular strategies in the hierarchy. For example, Crockford, a prominent occupational hygienist, presented an irrefutable case for improvement of personal protection (Crockford, 1977) terminated by the following statement:

> ... it is apparent that in no way should personal protection be considered as a last resort. Personal protection is one aspect of environmental control and *must* (my italics) be used as part of an integrated plan for environmental control and accident prevention. The concept of the last resort has led management, unions and men to believe, erroneously, that in some way personal protection is second best.

Crockford's view received support from Walton, a distinguished research worker whose studies in connection with coal-workers' pneumoconiosis are much referred to in Chapter 21 (Crockford, 1977, discussion p. 349):

> Too often personal protection is regarded as an undesirable substitute for good control of the general environment. In fact, when other methods are nearing their limit of effectiveness due to the law of diminishing returns, personal protection can offer a substantial further advance .... A good example has been provided by a past reluctance (shared by inspectorate, management and trade unions) to actively promote the use of dust respirators in the coal mining industry (although they are made generally available to men on request).

(For further discussion on strategy towards dust in the coal industry see Chapter 21.)

Some general points need to be made about the principles which determine the relative positions of safe-place and safe-person strategies. Manifestly, a safe-person strategy protects only those towards whom the strategy is directed. The weakness of this approach is plainly to be seen in the experience of asbestosis and mesothelioma in Naval Dockyards in Britain; on this, Harries *et al.* (1972) reported:

> It is disturbing that more than half the cases of ... (asbestosis)... related to asbestos exposure were found in so-called 'neighbour-hood workers' who have not themselves worked regularly with asbestos products. The explanation lies in the widespread contamination of ships.... At Devonport *all* (my italics) of the dockyard cases of pleural mesothelioma are in neighbourhood workers; none in recognized asbestos workers.

The recognized asbestos workers were given personal protection but the other workers were not. Personal protection can at best only protect the workers who wear it and it obviously does nothing about the danger which continues to prevail generally.

Other general points about strategies are expressed succinctly in Legge's axioms (1934, p.3)

> (1) Unless and until the employer has done everything—and everything means a good deal—the workman can do next to nothing to protect himself, although he is naturally willing enough to do his share.
>
> (2) If you can bring an influence to bear external to the workman (i.e. one over which he can exercise no control) you will be successful; and if you cannot or do not, you will never be wholly successful.
>
> (3) Practically all industrial lead poisoning is due to the in-halation of dust and fumes; and if you stop their inhalation you will stop the poisoning.
>
> (4) All workmen should be told something of the danger of the material with which they come into contact and not be left to find it out for themselves—sometimes at the cost of their lives.

And further on in the same passage:

> Examples of influences—useful up to a point, but not completely effective—which are not external, but depend on the will or whim of the worker to use them, are respirators, gloves, goggles, washing conveniences, and waterproof sand-paper.

(2) Prerequisite Strategies

Strategies which are necessary as a condition for the effective intro-duction and implementation of essential strategies can be termed prerequisite strategies. There are several important examples: high level of managerial commitment; adequate provision of financial resources; information, instruction, supervision and training; and involvement of work people.* All these are explicitly required or provided for in various sections of the Health and Safety at Work Act, including strategies such as a legal requirement for a written safety policy, relating to employees' safety representatives and joint safety committees, due to come into force in the autumn of 1978.

All these strategies are manifestly prerequisite for occupational health and safety but, in the absence of essential strategies, they cannot in themselves strike at danger.

(3) Adjunctive Strategies

Adjunctive strategies are those which are added to and subordinate to essential strategies, but which though valuable are non-essential. Obviously there is room for differences in opinion about which strategies are adjunctive and which are prerequisite; indeed, there is controversy on issues such as medical care in industry, whether generally or specifically, in the form of 'biological monitoring'. Detailed discussion of the general question is not within this book's scope. However, certain points can be made on which most authorities would agree. The mere act of diagnosing a disease or measuring a biological quantity in a worker is of no benefit to the worker unless action follows which alters his circumstances for the better in some way. Various actions are feasible—including the Ramazzini strategy of removal of people from their employment. But this is an unwelcome sequel to medical intervention if it results in loss of livelihood. It would be especially galling where safe-place strategies were not being implemented adequately. Another feasible action is that, on learning of the poor state of health of the workers, management implements an appropriate strategy—but the question should then be asked why that strategy was not previously implemented.

The use of work people as sampling devices must, in principle, be an inferior strategy to the use of scientific instrumentation. Where adequate instrumentation does not exist 'biological monitoring' may have to be relied upon, but this fact should provide the stimulus for the development of appropriate instrumentation. Dinman (1974) has reviewed the role of routine medical examinations in relation to occupational carcinogenesis:

> The relatively poor record—even of the most technologically advanced medical capabilities—to affect cancer mortality rates

*For a full discussion of the development of workers' representation in safety see Williams (1960), and for a brief review of developments post-Williams see Atherley et al. (1975).

has one overriding implication: present diagnostic and therapeutic modalities, in men, for the early detection and cure of cancer are discouragingly inadequate. This immediately places an important limit upon the efficiency of medical control or periodic health examinations.

All this is not to say that medical procedures should be rejected but, plainly, undue reliance should not be placed upon them. They are, in short, adjuncts and not essentials in the control of danger. However, not all authorities see medical and biological strategies in this light. The American Conference of Governmental Industrial Hygienists stated in 1976:

> Other means exist and may be necessary for monitoring worker exposure other than reliance on the Threshold Limit Values for industrial air, namely, the Biologic Limit Values. These values represent limiting amounts of substances (or their effects) to which the worker may be exposed without hazard to health or well-being as determined in his tissues and fluids or in his exhaled breath. The biologic measurements on which the BLVs are based can furnish two kinds of information useful in the control of worker exposure: (1) measure of the individual worker's over-all exposure; (2) measure of the worker's individual and characteristic response. Measurements of response furnish a superior estimate of the physiologic status of the worker, and may be made of (a) changes in amount of some critical biochemical constituent, (b) changes in activity of a critical enzyme, (c) changes in some physiologic function. Measurement of exposure may be made by (1) determining in blood, urine, nails, in body tissues and fluids, the amount of substance to which the worker was exposed; (2) determination of the amount of the metabolite(s) of the substance in tissues and fluids; (3) determination of the amount of the substance in the exhaled breath. The biologic limits may be used as an adjunct to the TLVs for air, or in place of them. The BTLs, and their associated procedures for determining compliance with them, should thus be regarded as an effective means of providing health surveillance of the worker.

From a strategy standpoint it could be argued that BLVs are a substitute for safe-place strategies, rather than an adjunct to them.

(4) Collateral Strategies

Collateral strategies are those which exist in parallel to or alongside danger-directed strategies but which have no effect on the danger. Of these, compensation is by far the most important. The role of compensation in deterrence has been discussed fully by Atiyah (1975); he came to guarded conclusions about its value. But there can be no doubt that the Robens Committee was right in pointing out the way in which questions of compensation can sometimes dominate those of prevention. Chapter 18 of this book discusses the point specifically in relation to occupational cancers of the renal tract. As in so many strategy questions affecting health and safety the issue was raised in the last century and is still unresolved.

Jeremy Bugler (1977) in a discussion of airline safety in the wake of the Tenerife aircraft accident, in 1977, in which 562 people died, concluded with this comment:

> The overall moral drawn by Insight (my note: a *Sunday Times* team of journalists) is that American airlines are better because Americans make the planes, and when things go wrong, Americans *sue*.

This makes an interesting parallel with a statement by the Royal Commission on Railway Accidents in 1865 (Parliamentary Papers, 1867):

> Parliament has relied for the safe working of railways upon the efficiency of the common law and of Lord Campbell's Act which give persons injured and near relatives of persons killed a right to compensation. We consider that this course has been more conducive to the protection of the public than if the Board of Trade had been empowered to interfere in the detailed arrangements for the working of traffic.

REFERENCES

AMERICAN CONFERENCE OF GOVERNMENTAL INDUSTRIAL HYGIENISTS (1971). *Documentation of the Threshold Limit Values for Substances in Work Room Air*, 3rd Edition, Cincinatti: ACGIH.

AMERICAN CONFERENCE OF GOVERNMENTAL INDUSTRIAL HYGIENISTS (1976). *Threshold Limit Values for Chemical Substances and Physical Agents in the Work Room Environment with Intended Changes for 1976*, Cincinatti: ACGIH.

ARLIDGE, J. T. (1892). *The Hygiene, Disease and Mortality of Occupations*, London: Percival.

ATHERLEY, G. R. C. (1975). Strategies in health and safety at work, *The Production Engineer*, **54**, 50–5.

ATHERLEY, G. R. C., BOOTH, R. T. and KELLY, M. J. (1975). Workers' involvement in occupational health and safety in Britain, *International Labour Review*, **111**, 469–82.

ATIYAH, P. S. (1975). *Accidents, Compensation and the Law*, 2nd Edition, London: Weidenfeld and Nicolson.

BARR, T. (1886). Enquiry into the effects of loud sounds upon the hearing of boilermakers and others who work amid noisy surroundings, *Proceedings of the Glasgow Philosophical Society*, **17**, 223–39.

BRIDGE, J. (1933). A hundred years of industrial health, in *Annual Report HM Chief Inspector of Factories for the Year 1932*, London: HMSO.

BUGLER, J. (1977). Fear of flying, *New Statesman*, **93** (2402), 425.

COOKE-TAYLOR, W. (1844). *Factories and the Factory System*, London: Jeremiah How.

CROCKFORD, G. W. (1976). Personal protection—the last resort. *Annals of Occupational Hygiene*, **19**, 345–50.

DINMAN, B. D. (1973). Principles and use of standards of quality for the work environment, in *The Industrial Environment—Its Evaluation and Control* (anon) Washington: US Department of Health, Education and Welfare.

DINMAN, B. D. (1974). *The Nature of Occupational Cancer*, Springfield: Thomas.

FIFE, I. and MACHIN, E. A. (1976). *Redgrave's Health and Safety in Factories*, London: Butterworth.

HARRIES, T. G., MACKENZIE, F. A. F., SHEERS, G., KEMP, J. H., OLIVER, T. P. and WRIGHT, D. S. (1972). Radiological survey of men exposed to asbestos in naval dockyards, *British Journal of Industrial Medicine*, **29**, 274–9.

HATCH, T. F. (1972). Permissible levels of exposure to hazardous agents in industry, *Journal of Occupational Medicine*, **14**, 134–7.

HATCH, T. F. (1973). Criteria for hazardous exposure limits, *Archives of Environmental Health*, **27**, 231–5.

HEALTH AND SAFETY EXECUTIVE (1977). Threshold Limit Values for 1976, Guidance Note EH15/76, London: Health and Safety Executive.

HENRY, S. A. (1934). Preface to Sir Thomas Legge's *Industrial Maladies*, Oxford: University Press.

HOLT, E. E. (1882) Boilermakers deafness and hearing in a noise, *Transactions of the American Otological Society of Boston*, **3**, 34–44.

HOOVER, H. C. and HOOVER, L. H. (1912). Translations of Agricola, G. (1556) *De Re Metallica*, Basel; London: The Mining Magazine.

HUNTER, D. (1975). *The Diseases of Occupation*, London: English Universities Press.

LEGGE, T. (1934). *Industrial Maladies*, Oxford: University Press.

LEVINSON, C. (1975). *Work Hazard: Threshold Limit Values for Chemical Agents in the Workplace*, Geneva: ICF.

MARWICK, A. (1977). *Women at War 1914–1918*, London: Fontana and Imperial War Museum.

MEIKLEJOHN, A. (1956). Sixty Years of Industrial Medicine in Great Britain, *British Journal of Industrial Medicine*, **13**, 155–62.

MEIKLEJOHN, A. (1963). The successful prevention of lead poisoning in the glazing of earthenware in the N. Staffordshire potteries, *British Journal of Industrial Medicine*, **20**, 169–80.

NATIONAL ACADEMY OF SCIENCES (1975). Decision making for regulating chemicals in the environment, A report by the Committee on Principles of Decision Making for Regulating Chemicals in the Environment, Washington: National Academy of Sciences.

NOBLE, D. (1843). *Facts and Observations Relative to the Influence of Manufactures upon Health and Life*, London: John Churchiil.

PARLIAMENTARY PAPERS (1867) XXXVIII 77.

POCHIN, E., Sir, MCLEAN, A. S. and RICHINGS, L. D. G. (1976). *Radiological Protection Standards in the United Kingdom*, Report NRPB-R46, Harwell: National Radiological Protection Board.

PRIVY COUNCIL (1861). *Third Report of the Medical Officer of the Privy Council, 1860*, London: HMSO.

REPORTS OF THE INSPECTORS OF FACTORIES for the half-year ending 31 October 1845 (1846), Report by Leonard Horner dated 26 November 1845, London: HMSO.

ROSCHIN, A. V. and TIMOFEEVSKAYA, L. A. (1975). Chemical substances in the work environment: some comparative aspects of USSR and US Hygienic Standards, *Ambio*, **4**, 30–3.

SCHILLING, R. S. F. (ed.) (1973). *Occupational Health Practice*, London: Butterworth.

STOKINGER, H. E. (1969). Current problems of setting Occupational Exposure Standards, *Archives of Environmental Health*, **19**, 277–614.

STOKINGER, H. E. (1970). Criteria and procedures for assessing the toxic responses to industrial chemicals, in *Permissible Levels of Toxic Substances in the Working Environment*, ILO Occupational Safety and Health Series No. 20, Geneva: ILO.

WILLIAMS, J. L. (1960). *Accidents and Ill-Health at Work*, London: Staples Press.

WRIGHT, W. C. (1940). Translation of Ramazzini, B. (1713) *De Morbis Artificum Diatriba*, Geneva; Chicago: University Press.

Index

Accumulation,140
Acoustical energy, 250–56
Acro-osteolysis, 346
ACTH, 78
Adipose tissue, 139–40
Adopted Value, 377
Adrenal cortex, 78
Aerodynamic diameter, 29, 34, 36
Aerosols, 278
  deposition, 29, 34–36
  measurement, 42–55
  retention in respiratory pathway, 54
  sampling, 43
  sites of contact and principal effects, 46
  size selection, 51–54
   and personal protection, 54
  use of term, 29
Air sampling, 42–44
Airborne particles, 29
Albumins, 139
Alkylating agents, 114, 205
Allergies, 16, 175
Alpha particles, 238
Aluminium, 281
Alveolar deposition, 49
Alveoli, transport out of, 36
Amines, 259
Amino acids, 188
4-Aminodiphenyl, 262, 263
2-Aminofluorine, 224
2-Amino-1-naphthol, 263
Ammonia, 167, 342
Anabolism, 98, 104
Anaplasia, 206
Aniline, 259, 261, 262, 265
Anoxia, 328, 331, 334
Anthrax, 166
Antibodies, 72, 73, 75, 176, 177
Antigens, 71, 73, 74, 75, 176, 177
Antimony, 278, 281
α-1-Antitrypsin, 199

Aplasia, 201
Aromatic amines, 259–73
  compensation, 267
  early history of occupational cancer problem, 260–62
  history of action against, 265–66
  industrial and chemical background, 259
  recent history of occupational cancer problem, 263
  site of tumours, 262
  statutory control, 268
Aromatic compounds, 259
Arsenic, 276, 281
Arsine, 169, 281
Arthropods, 21
Asbestos, 46, 48, 156–58, 194, 290–302
  airborne particles, 291
  fibre dimension, 291
  history of action on, 295–300
  hygiene standard, 298
  regulations, 296, 298–300
  types of, 290
Asbestosis, 156, 194, 292, 295
Asthma, 180–82
Atheroma, 169
Atmospheric measurements, 44
Atmospheric pollution, 167
ATP, 120
Atrophy, 201
Auramine, 263, 268

Bacteria, 12
Bagassosis, 183
Barium, 281
Basophils, 60, 62
Benzene, 259, 333
Benzidine, 262–66, 270
3,4-Benzpyrene, 222
Beryllium, 168, 279, 281
Biochemical reactions, 100

Bio-dumping, 5, 17–19, 141–44
Biological half-time, 140, 141, 244
Biological processes, 15
  changes in, 5
Bladder, 101
  cancer, 261, 264, 268
  epithelium of, 212
  inflammation, 170
  papilloma, 271
  tumours, 261–70
'Blind staggers', 288
Blood, 168, 184–85, 212
  cells, 66, 139, 245
  concentration, 153, 277
  flow, 92
  -forming organs, 168, 184–85
  functions, 138–39
  supply, 207
  vessels, 207
  volume, 87
Body temperature, 11
Bone
  inflammation, 170
  marrow, 65–66, 152, 245
Boundaries
  concept of, 132
  crossing, 132
Brain, 82
Bright's disease, 169
Bronchi, 31
Brownian movement, 135
Brucellosis, 168
Byssinosis, 183–84, 309

Cable-making industry, 263, 267
Cadmium, 199, 276, 278, 280, 284
Cancer, 97, 200, 204, 205, 217
  cells, 114, 133
  occupational, 73, 101, 110, 219
  therapy, 114
  see also Carcinogenesis; Carcinogens;
    Necrosis; Neoplasia; Neo-
    plasms; Tumours; and under
    specific organs
Caplan's syndrome, 190, 305
Carbon dioxide, 330
Carbon monoxide, 169, 330, 339–41
Carbon tetrachloride, 170

Carcinogenesis, 155, 205, 217, 219, 222
  irritation theory of, 223
  occupational, 224
  somatic mutation theory, 224
  time dependency for, 155–56
Carcinogenic Substances (Prohibi-
    tion of Importation) Order
    1967, 268
Carcinogenic Substances Regula-
    tions, 268
Carcinogens, 155, 218, 222, 224
Catabolism, 78, 98, 104
Cataract induction by microwaves,
    250
Cavitation effects, 251
Cell
  division, 110, 112–17, 121, 123
  functions, 14, 118
  genetics, 15, 118–29
  -mediated response, 72
  membranes, 133–35
  necrosis, 162
  structure, 118–21
  transport mechanisms, 135–38
Cellular biochemistry, 85
Cellular life cycle, 112, 115
Cellular theory, 96
Cerebral cortex, 92
Cheloids, 191
Chemical groupings, 100
Chemical reactions, 100
Chemical substances, 10, 97, 147, 222
Chemotaxis, 58, 74
Chest, 27
Chlor-acne, 166
Chlorine, 166, 167, 337–39
Cholesterol, 169
Chromatin, 114, 115
Chromium, 275, 279, 284
Chromosome aberrations, 127–28
Chromosomes, 109, 114, 115, 206
Chronic bronchitis, 167, 168, 184,
    308, 309, 315, 316
Cigarette smoking, 184, 194, 309
Cilia, 32
Circulatory failure, 87
Circulatory system, 37, 138, 168, 184–
    85
Cirrhosis, 191

Climates, 11
Coal
  dust in lung, 307
  industry, respiratory disease in, 167, 303–27
  mines, dust levels in, 316
  -mining, 385
Coal Mines (Respirable Dust) Regulations 1975, 319–25
Cobalt, 275, 284
Cold
  altered sensitivity to, 185
  effects, 88, 92
Collagen, 62, 188, 195, 292, 306
  -centred repair processes, 188
  disease, 190
Collagenous pneumoconiosis, 193, 196–97
  classification of severity, 197–98
Commensals, 20
Compensation, 267, 304, 305, 309–13, 316
Complement, 75, 114, 177
Congenital abnormalities, 203
Connective tissues, 211
  tumours in, 215
Control strategies, 381–95
  adjunctive strategies, 393–94
  classification, 390–95
  collateral strategies, 394
  essential strategies, 390–92
  hierarchical classification of, 6
  historical development, 382–90
  prerequisite strategies, 392
Copper, 275, 280, 285
Cortisol, 79, 80, 83, 117
Cotton dust, 183–84
Cramp, 86
Critical concentration, 152
Critical effect, 151
Critical organ, 151, 152
  concentration, 151, 152
Croton oil, 222
Cyanosis, 260, 264
Cyclone, 51
Cytoplasm, 62
Cytotoxic substances, 195

Deafness, 155, 173

Death, 17
Defence processes, 13–14, 23, 97, 147–48
  respiratory, 26–42
Defensive cells
  and their functions, 56–67
  in tissues, 62–65
  origin of, 65–67
Degeneration, 162
Deposition dynamics, 29
Dermatitis, 163–64, 179, 180, 237, 279, 280, 284, 285, 287
Detoxication, 100–1, 104
Dianisidine, 268
o-Dianisidine, 263
Diatomite, 194
Dibenzanthracene, 224
Dichapetalum cymosum, 100
Dichapetalum toxicarium, 100
Dichlorbenzidine, 263, 268
Diffusion, 29, 133–37
Disablement, 17
Disease, 16
  manifestation, 5
  processes, 96
DNA, 121–28, 205, 221, 222, 242, 243
Dose
  definition, 153
  –effect relation, 45, 46, 148–51, 159, 173, 174
  –output relation, 151
  –rate dependency, 158–59
  relations, time related law for, 156
  –response relation, 148, 159, 224, 225
  specification for carcinogens, 155
Dust, 36, 192, 278
  coal-mining, 316, 317
  fibrogenic, 46, 193, 196
  inert, 194, 197
  measurement, 48, 52–54, 319
  mixed dust fibrosis, 194
  reactions, 183
  respirable, 52–54, 320
  sampling, 52–54, 318, 324, 325
Dyes, 260
Dyestuffs industry, 269
Dysplasia, 202–3

Effective decay, 244
Effective half-life, 244
Effectors, 82, 91
Effects, 279
  use of term, 148, 151
Electrical energy, 246
Electrical fields, 248
Elimination, 141
  constant, 140
Elutriator, 51
Emotional factors, 179
Emphysema, 198–99, 308, 309
Endocrine system, 82
Endothelia, 211
Energies and energy transformations,
  11
Environment, 11
Environmental factors, 3
Enzymes, 59, 85, 101, 121, 125, 195,
  276
Eosinophils, 60
Epidemiological research, 387
Epigenetic change, 222
Epithelia, 211
Epithelium, 211
  neoplasms in, 214
  tumours in, 214
Equivalent unit density sphere
  (EUDS); 29, 34, 291
Esotropy, 136
Ethyl methanesulphanate, 205
Evaporation, 91
Exhalation, 44
Exotropy, 136
External allergic alveolitis, 182–83
Eye
  inflammation, 171
  sensitivity alteration, 185

Facilitated transportation, 137–38
Farmers' lung, 182, 183
Fatigue, 17
Fatty acids synthesis, 101
Fibres, 43
Fibroblasts, 62, 188, 189
Fibrosis, 188–93, 198, 256, 292, 305
Fick's law, 135

Finger clubbing, 295
Folliculitis, 166
Foundrymen, respiratory disease in,
  168
Frost-bite, 88
Fume fevers, 332
Fumes, 278

Gamma radiation, 239
Gangrene, 69
Gas exchange, 26
Gassing accidents, 167, 328–43
  ammonia, 342
  anoxia, 328, 331, 334
  carbon monoxide, 339–41
  case studies, 336–42
  central nervous system equilibrium
    effects, 332
  chlorine, 337–39
  classification according to nature
    of accident, 329
  classification according to physical
    effects, 330
  combined toxic effects, 333
  first-aid, 336
  fume fevers, 332
  hydrogen sulphide, 340
  irritation of respiratory tract, 331
  metal fume fever, 338
  nitrous fumes, 333–35, 337
  oxy-acetylene torches, 342
  polytetrafluoroethylene (PTFE),
    340
  prevention, 336
  sodium sulphide solution, 340
Gastritis, 171
Gastro-enteritis, 171
Generative mechanism, 9
Genes, 115
Genetic change, 222
Genetic information, 123, 125
Giant cells, 70, 189
Glanders, 166
Globulins, 139
Glomerulonephritis, 185
Gluconeogenesis, 79
Glycine, 188

Gold, 285
Granulation tissue, 189
Granulocytes, 62
Graphite pneumoconiosis, 197
Growth
    disorders, 16, 200–27
    of living tissues, 15, 109–17
Gut, inflammation, 171

Haematological depression, 240
Haemoglobin, 264–65
Haemoglobinuria, 169
Haemolysis, 169
Haemorrhage, 87
Haempoiesis, 245
'Hard metal' disease, 284, 289
Health and Safety at Work Act 1974,
    268
Health and Safety Executive, 324
Hearing loss, 148–49, 159–61, 252–55
Heart disease, 77
Heat
    cramps, 86–87
    effects, 85
    exhaustion, 87
    exposure, 94
    loss, 89–93
    production, 89–90, 92
    strain, 93
    stress, 93, 94
    stroke, 87
Helminths, 21
Hemangiosarcoma, 345, 348
Homeostasis, 14, 80–82, 91, 141, 142,
    169
Homeotherms, 89
Horizontal sampler, 51
Hormones, 78, 122
Human safety, 7
Hyaline cartilage, 31
Hydrogen sulphide, 340
Hydrolysis, 104
Hydroxyproline, 188
Hygiene limits, 371–81
    general comments, 372
    problems associated with, 375–80
    qualifications needed with, 379

Hygiene limits (contd.)
    restricted applicability outside
        country of origin, 376
    setting of, 373–75
    socio–economic decision making,
        373–75
    terminology, 376–79
    uncertainty concerning, 379
Hygiene Standard, 156
Hyperplasia, 202
Hypersensitivity, 173
    delayed, 177
Hypertrophy, 191, 202
Hypoplasia, 201
Hypothalamic thermostat, 89
Hypothalamus, 82, 89, 91, 92

Immune response, 14, 71–73, 175–78,
        184, 185, 222–23
    disorders of, 72
    steps in, 73–75
    types of, 72
Immunoglobulins, 73, 74
Immunopathological   mechanisms,
        175–78
Immunosurveillance, 125
Impaction, 29
Implantation, 12
Infestation, 21
Inflammation, 68, 72–73
    acute, 68, 69, 162
    chronic, 68, 70, 162, 167
    harmful features of, 162–72
Inflammatory response, 14, 68–70
Information input, 13
Infrasound, 255–56
Ingestion, 12
Inhalation, 12, 26–42, 44
Inheritance, 125
Injury, 17
Input–output model, 4–5, 7–21
    classes of inputs, 10
    defence processes, 13–14
    disease condition, 16
    elements of, 9–19
    input measurement, 18
    intervention, 19–20
    modes of action, 15

Input–output model (*contd.*)
  modes of input, 12
  output measurement, 18
  processes, 15–16
Insects, 12, 21
Insulin, 78
Integrating mechanisms, 82, 91
Interception, 29
Interstitial spaces, 41, 42
Intoxication, 100–1, 105
Intra-cellular transport, 133–38
Ionizing radiations, *see* Radiation exposure
Iron, 276, 285
Irradiation, *see* Radiation exposure
Isokinetics, 43

Kidney, 185, 287
  function, 141–44
  inflammation, 169–70
Kuppffer cells, 170

Lasers, 250
Lead, 101, 141, 276–79, 285, 388, 390
Lethal concentration, 152
Lethal synthesis, 101
Leucocytes, 60
Lipids, 140
Lithium, 286
Liver
  cirrhosis, 191
  damage, 262
  inflammation, 170
  necrosis, 190
Luminous paint, 233, 235
Lung
  cancer, 156–58, 194, 223–24, 233, 234, 243, 264, 293
  diseases, 56, 199
  fibrosis, 193, 198
Lungs, 27, 28, 36
  coal dust in, 307
  inflammation, 166
  lymphatic circulation, 39–42
Lymph, 39
  glands, 42

Lymph (*contd.*)
  nodes, 39, 72, 74, 210
Lymphatic circulation, 37–39
  in lung, 39–42
Lymphatic drainage, 138
Lymphatic system, 207, 210
Lymphocytes, 62
Lympho-reticular system, neoplasms of, 216
Lysosomes, 59, 195

Macrophages, 36–37, 46, 56, 59, 168, 170, 188, 195
Magenta, 263, 268
Magnesium, 280, 286
Malnutrition, 21
Manganese, 101, 275, 281, 286
Mass median diameter (MMD), 54
Maximum Allowable Concentration, 376
Median lethal dose, 105
Meiosis, 117
Membranes, 132–36
Mercuric chloride, 144
Mercury, 276, 278, 286
Mesothelia, 211
Mesothelioma of pleura and peritoneum, 294
Metabolic conjugation, 96–106
Metabolic transformation, 14, 96–106
Metabolism, 105
Metal fume fever, 184, 280, 289, 332, 338
Metals, 274–89
  target organs and effects, 280
  see also under specific metals
Metaplasia, 202
Metastasis, 207, 216
Methylcholanthrene, 224
Methyl mercury, 286
Micronutrients, 275, 276
Microphages, 60
Micropredators, 11, 20
Microvilli, 133
Microwaves, 247–50
Minamata disease, 287
Miners, 86
Mitochondria, 120

Mitosis, 110, 112–17, 121
Mitotic figures, 117, 206
Mitotic rate, 117
Modes of action, 279
Moisture effects, 43
Molybdenum, 275, 287
Monocytes, 60, 62
Mucous membrane, 31–32
Multiple hyperkeratosis, 166
Muscle tissue, 212
  tumours of, 216
Mutagenesis, 110, 203
Mutation, 125–26, 222

Naphthylamine, 261, 267
1-Naphthylamine, 262–70
2-Naphthylamine, 262–70
National Coal Board, 312, 316
National Health Service, 304
National Insurance (Industrial In-
        juries) Act 1946, 304
Necrosis, 190, 207
Neoplasia, 204, 205
  classification of, 210–17
  occupation-linked, 217–27
Neoplasms, 205
  benign and malignant, 205–7, 212–
        17
  in epithelium, 214
  in lympho-reticular system, 216
  in nervous system, 215
  secondary, 207, 210
Neoplastic cells, 205, 206
Nephrotoxic agents, 170
Nerve tissue, 212
Nervous system, 82, 246
  neoplasms in, 215
Neutrophil leucocytes, 60
Neutrophils, 62
Nickel, 276, 279, 287
Nickel carbonyl, 333
Nitric acid, 333
4-Nitrodiphenyl, 263
Nitrogen oxides, 167
Nitrous fumes, 333–35, 337
Noise, 76–78, 148, 149, 159–61, 173,
        252–55

Non-nutrients, 10, 98, 100, 101, 275,
        279
Nonox S, 267
Nucleus, 62
Nutrients, 10, 98, 101, 120, 275, 279
Nystagmus, 304

Occupational health and safety, 4
  selection and implementation of
        action in, 5
Organometallic compounds, 279
Osmosis, 135
Oxidation, 105
Oxy-acetylene torches, 342

Paints
  lead, 390
  luminous, 233, 235
Parasitic infestation, 21
Particle mass, 43
Pathobiology, 96
Pathogenic substances, 148
Pathogenicity, 97–98, 101
Pathogens, 20
Pathological processes, 15
Penicillin, 240
Permeation by solution, 135
Personal samplers, 43, 51
Pervasion, 12
Phagocytosis, 56–59, 72, 74, 134, 188,
        195
Phagosomes, 195
Phosphine, 288, 332
Phospholipids, 169
Phosphorus, 287
Phosphorylation, 101
Phossy jaw, 287–88
Photopthalmia, 171
Pinocytosis, 57
Plasma, 39, 139, 168, 184–85
Platinum, 276, 279, 288
Pleomorphism, 206
Pleura, 42
Pleural plaques, 294
Plutonium, 236, 243

Pneumoconiosis, 52, 191, 192, 194, 292
  coal-workers, 305–16
    categories, 313–16
    compensation, 310–13, 316
    progression of, 323
    social consequences of, 309
  collagenous, 193, 196–98
  definition of, 309
  radiological, 319
Pneumoconiosis Medical Panels, 315
Poikilotherms, 89
Polarity, 100, 105
Polymer fume fever, 332, 341
Polymorphonuclear granulocytes, 62
Polymorphs, 62
Polyploidy, 206
Polytetrafluoroethylene (PTFE), 340
Polyvinyl chloride, 345
Preselectors, 49–50
Progressive massive fibrosis, 196, 306, 307
Proline, 188
Protein catabolism, 78
Protein synthesis, 121, 122
Proteins, 85
  as enzymes, 121
  as secretions, 122
  as structural components, 121
  roles of, 121–22
Pruritis, 179
Psycho–social climate, 11
Pulmonary fibrosis, 183, 193
Pyrogens, 89

Quartz, 183, 306, 324

Radiation
  damage to white blood cells, 66
  dose, 245
  exposure, 13, 56, 114–15, 127, 158–59, 169–70, 231–58
    acoustical energy, 250–56
    case histories, 231–40
    cellular response to non-ionizing energies, 246–47
    free radical reactions, 242
    implantation mode of input, 236

Radiation (contd.)
  exposure (contd.)
    ingestion mode of input, 234
    inhalation input mode, 232
    ionic solution, 241
    irradiation, 238–40
    lasers, 250
    microwaves, 247–50
    modes of action
      cellular level, 242
      external emitters, 245
      in relation to transport and storage, 243–44
      internal emitters, 243
      non-ionizing radiations and other forms of energy, 247–56
      of ionizing radiations, 241–45
    part-body, 245
    regulations, 232
    skin pervasion, 235
    somatic effects, 245
    surface penetration mode of input, 236
    whole-body, 245
  sickness, 245
  toxicology, 243–44
Radioactive half-life, 244
Radioactive particles, 54
Radioactivity, 233
  decay, 244
Radiomimetic agents, 114
Radionuclides, 243
Radium, 233–35, 243
Radon, 233, 243
Red blood cells, 169, 265
Reduction, 105
Renal system, 141–43
Renal tract
  cancer, 259–73
  inflammation, 170
Repair disorders, 187–99
  respiratory tract, 190, 191
  skin, 190, 191
  types of, 189
Repair processes, 187–89
  collagen-centred, 188
Respirable dust index, 320, 325

Respirable fraction, 53
Respiratory clearance, 36
Respiratory defensive processes, 26–42
Respiratory deposition and clearance, 37–39
Respiratory disease
  in coal industry, 167, 303–27
  in foundrymen, 168
Respiratory filtration, 14, 27–28, 44
Respiratory tract, 30
  inflammation, 166
  irritation, 331, 333
  repair disorders, 190–91
  sensitivity alteration, 180–84
Retention, 140
Reticulo-endothelial system, 59
Rheumatoid pneumoconiosis, 305–6
RNA, 123, 124, 126, 128, 242
Rubber industry, 263, 267, 269

Salt loss, 86, 87
Sampling instruments, 43, 48
Scar tissue, 188–91, 193
Scarring, 70–71
Schlerotic syndrome, 345
Sedimentation, 29
Selection, 125
Selenium, 276, 288
Self-disposal, 72
Self/non-self recognition, 75, 175
Sensitivity, 173, 184
  alteration in, 173–86
Sensitization, 173, 178
Sensors, 82, 91
Shaver's disease, 281
Shell filling, 389
Silica, 46, 194, 306
Silicic acid, 196
Silicosis, 183, 194–96, 310
  theories, 196
Silver, 288
Skin, 132
  cancer, 237
  disease, 179
  inflammation, 163–66
  pervasion, radiation exposure, 235

Skin (contd.)
  repair disorders, 190–91
  sensitivity increase, 179–80
  structures, 163
Social policy, 4
Sodium sulphide solution, 340
Solar energy, 94
Spleen, 72
Sri Lankan graphite industry, 196–97
Stem cells, 65
Stokes diameter, 29
Strategies, see Control strategies
Stress
  and stress responses, 76–84, 179
  resistance, 14
Strontium, 243
Sub-critical concentrations, 153
Sub-critical effects, 153
Sulphur dioxide, 167
Sulphydryl groups, 275
Summation law, 155
Surface penetration, 13
Surfactant, 36
Surveillance, 71
Susceptibility, 173
Sweat, 86, 91
Symbionts, 20

Target organs, 153, 154, 163, 211, 277, 279, 280
Temperature effects, 85, 87
Teratogenesis, 110, 203
Textiles, 386
Thallium, 288
Thermal comfort/discomfort, 94
Thermal environmental standard, 92
Thermal stress, 93
Thermoregulation, 14, 85–95
  disorders of, 86
  processes, 89
Threshold Limit Values, 94, 173, 181, 376
  categories of, 378–79
  difficulties with enforcement, 380
  documentation of, 380
Thymus, 72
Time-weighted average, 159, 376, 377, 380

Tin, 288
Tissue
  damage, 73, 187–89
  fluid, 39
  gaps, 189
  growth, 110–12
  injury, repair response to, 16
  regeneration, 189
  spaces, 41
Tissues, classification of, 210–17
o-Tolidine, 263
Toxicity, 97, 101
  use of term, 10
Tracheo-bronchial deposition, 35
Transamination, 111
Transformation, 105
Transmembranosis, 134
Transport
  across and out of cells, 135–38
  systems, 15, 130–44
Trench-foot, 88
2,4,6-Trinitrotoluene, 389
Tuberculosis, 193, 194, 307–8
Tumour cells, 204
Tumours
  in connective tissues, 215
  in epithelium, 214
  in muscle tissue, 216
  mixed, 216
Tungsten, 288
Type I reactions, 176, 180, 182
Type II reactions, 176
Type III reactions, 177, 182
Type IV reactions, 177

Ultra-filtration, 135
Ultrasound, 250–52
Ultra-violet radiation, 232
Undulant fever, 168
Uranium, 233, 243
Urinary tract, 185
  inflammation, 169–70

Vanadium, 278, 289
Vibration, 255, 256
  -induced 'white finger', 185, 256
Vinyl chloride, 287
  carcinogenicity of, 348
  case history, 344–67
  Code of Practice, 353, 364
  disease, 345, 346, 348
  education and training, signs and
      information, 360
  emergency plan, 359
  employment restrictions, 365
  environmental monitoring, 354
  general requirements, 363
  health precautions, 349–53
  joint consultation, 364
  limits, 348, 350, 353
  medical surveillance, 361
  narcosis, 347
  pathogenicity, 333
  properties of, 345
  regulated areas/controlled areas,
      355, 357
  reports and record keeping, 361
  respiratory protection, 356
  responsibilities, 362
Viruses, 12, 223
Vitamin
  B12, 20
  deficiency disease, 20

Wet bulb–globe temperature index
    (WBGT), 94
White blood cells, 60, 168
  radiation damage, 66
Worms, 21

X-ray films, 197–98, 237, 239, 242,
    312–14, 323, 325

Zinc, 275, 280, 289, 338